The Challenge of Highly Pathogenic Microorganisms

AF182684

Avigdor Shafferman
Arie Ordentlich
Baruch Velan
Editors

The Challenge of Highly Pathogenic Microorganisms

Mechanisms of Virulence and Novel Medical Countermeasures

Springer

Editors
Avigdor Shafferman, Ph.D.
Israel Institute for Biological Research
74100 Ness-Ziona
Israel
avigdors@iibr.gov.il

Arie Ordentlich, Ph.D.
Israel Institute for Biological Research
P.O. Box 19
74100 Ness-Ziona
Israel
arieo@iibr.gov.il

Baruch Velan, Ph.D.
Israel Institute for Biological Research
P.O. Box 19
74100 Ness-Ziona
Israel
baruchv@iibr.gov.il

ISBN 978-94-007-9911-0 ISBN 978-90-481-9054-6 (eBook)
DOI 10.1007/978-90-481-9054-6
Springer Dordrecht Heidelberg London New York

© Springer Science+Business Media B.V. 2010
Softcover re-print of the Hardcover 1st edition 2010
No part of this work may be reproduced, stored in a retrieval system, or transmitted in any form or by any means, electronic, mechanical, photocopying, microfilming, recording or otherwise, without written permission from the Publisher, with the exception of any material supplied specifically for the purpose of being entered and executed on a computer system, for exclusive use by the purchaser of the work.

Printed on acid-free paper

Springer is part of Springer Science+Business Media (www.springer.com)

Preface

The OHOLO conferences are sponsored by the Israel Institute for Biological Research and take their name from the site of the first meeting on the shores of Lake Kinnereth. The purpose of these meetings is, as it was at their inception over 50 years ago, "to foster interdisciplinary communication between scientists in Israel, and to provide added stimulus by the participation of invited scientists from abroad".

The core of the organizers of the OHOLO conferences are scientists from the Israel Institute for Biological Research. From time to time a particular OHOLO conference cooperates with an international scientific organization. The present 46th OHOLO Conference marks the resumption of the OHOLO tradition after 8 years of interruption caused by events beyond our control. It is my belief that our uncompromising commitment to excellence in research and development in the various areas of science in Israel is essential to our survival in this troubled region. The OHOLO conference tradition is a reflection of this conviction.

The present 46th OHOLO Conference entitled: *The Challenge of Highly Pathogenic Microorganisms – Mechanisms of Virulence and Novel Medical Countermeasures* intends to address the unique virulence features and host-pathogen interactions of microorganisms constituting emerging biothreat with emphasis on *Y. pestis, B. anthracis, F. tularensis* and Orthopox viruses. Accordingly we selected classical microbiological as well as genomic, proteomic & transcriptomic approaches towards developments of novel prophylactic and post-exposure treatment, as well as updated strategies of diagnostics and bioforensics.

I wish to thank the members of the international Scientific Advisory committee: Elisabeth Carniel, Arthur Friedlander, Paul Keim, Johannes Löwer, Michèle Mock and Anders Sjöstedt who helped us to formulate scientific content of this meeting as well as the local organizers Arie Ordentlich, Baruch Velan and Sara Cohen from IIBR.

Ness-Ziona, Israel Avigdor Shafferman

Contents

Contributors

Moshe Aftalion Department of Biochemistry and Molecular Genetics, Israel Institute for Biological Research, P.O. Box 19, Ness Ziona 74100, Israel, mosheav@iibr.gov.il

S.A. Ageev State Research Center for Applied Microbiology and Biotechnology, Obolensk, 142279, Moscow Region, Russia

Ronit Aloni-Grinstein Department of Biochemistry and Molecular Genetics, Israel Institute for Biological Research, P.O. Box 19, Ness-Ziona 74100, Israel, ronita@iibr.gov.il

Zeev Altboum Departments of Infectious Diseases, Israel Institute for Biological Research, Ness-Ziona 74100, Israel, zeeva@iibr.gov.il

Andrey P. Anisimov State Research Center for Applied Microbiology and Biotechnology, Obolensk, 142279, Moscow Region, Russia, a-p-anisimov@yandex.ru

Naomi Ariel Departments of Biochemistry & Molecular Genetics, Israel Institute for Biological Research, P.O. Box 19, Ness-Ziona 74100, Israel, naomia@iibr.gov.il

Patrick Avé Unité de Recherche et d'Expertise d'Histotechnologie et Pathologie, Institut Pasteur, 28 rue du Docteur Roux, 75724 Paris Cedex 15, France, patrick.ave@pasteur.fr

Menachem Banai Department of Bacteriology, Kimron Veterinary Institute, P.O. Box 12, Bet Dagan 50250, Israel, menachemba@moag.gov.il

Svetlana Bardenstein Department of Bacteriology, Kimron Veterinary Institute, P.O. Box 12, Bet Dagan 50250, Israel, svetab@moag.gov.il

Erez Bar-Haim Department of Biochemistry and Molecular Genetics, Israel Institute for Biological Research, P.O. Box 19, Ness Ziona 74100, Israel, erezb@iibr.gov.il

Ada Barnea Departments of Biotechnology, Israel Institute for Biological Research, Ness-Ziona 74100, Israel, adab@iibr.gov.il

Einat Ben-Arie Departments of Biochemistry and Molecular Genetics, Israel Institute for Biological Research, Ness-Ziona 74100, Israel, einatb@iibr.gov.il

Zain Bengali Laboratory of Viral Diseases, National Institute of Allergy and Infectious Diseases, National Institutes of Health, Bethesda, MD 20892, USA

David Ben-Nathan Department of Infectious Diseases Israel Institute for Biological Research, P.O. Box 19 Ness Ziona, 74100, Israel

Raphael Ber Department of Biochemistry and Molecular Genetics, Israel Institute for Biological Research, P.O. Box 19, Ness Ziona 74100, Israel, raphaelb@iibr.gov.il

Adi Beth-Din Departments of Biochemistry & Molecular Genetics, Israel Institute for Biological Research, P.O. Box 19, Ness-Ziona 74100, Israel, adib@iibr.gov.il

Himani Bisht Laboratory of Viral Diseases, National Institute of Allergy and Infectious Diseases, National Institutes of Health, Bethesda, MD 20892, USA

James B. Bliska Center for Infectious Diseases and Department of Molecular Genetics and Microbiology, Stony Brook University, Stony Brook, NY 11794, USA, jbliska@ms.cc.sunysb.edu

Jeanette E. Bröms Department of Clinical Microbiology, Clinical Bacteriology, and Laboratory for Molecular Infection Medicine Sweden (MIMS), Umeå University, SE-901 85 Umeå, Sweden

Elisabeth Carniel Unité des Yersinia, Institut Pasteur, 28 rue du Docteur Roux, 75724 Paris Cedex 15, France, carniel2@pasteur.fr

Theodor Chitlaru Department of Biochemistry and Molecular Genetics, Israel Institute for Biological Research, P.O. Box 19, Ness-Ziona 74100, Israel, theodorc@iibr.gov.il

Sara Cohen Department of Biochemistry and Molecular Genetics, Israel Institute for Biological Research, P.O. Box 19, Ness-Ziona, 74100, Israel, sarac@iibr.gov.il

Ofer Cohen Department of Biochemistry and Molecular Genetics, Israel Institute for Biological Research, P.O. Box 19, Ness-Ziona 74100, Israel, oferc@iibr.gov.il

Hila Cohen Department of Biochemistry and Molecular Genetics, Israel Institute for Biological Research, P.O. Box 19, Ness Ziona 74100, Israel, hilac@iibr.gov.il

Wayne Conlan National Research Council of Canada, Institute for Biological Sciences, Ottawa, ON K1C 2M7, Canada, wayne.conlan@nrc-cnrc.gc.ca

Svetlana V. Dentovskaya State Research Center for Applied Microbiology and Biotechnology, Obolensk, 142279, Moscow Region, Russia

Sharon Ehrlich Department of Biochemistry and Molecular Genetics, Israel Institute for Biological Research, P.O. Box 19, Ness-Ziona 74100, Israel, sharone@iibr.gov.il

Eitan Elhanany Department of Biochemistry and Molecular Genetics, Israel Institute for Biological Research, P.O. Box 19, Ness-Ziona 74100, Israel, eitane@iibr.gov.il

Noam Erez Department of Infectious Diseases, Israel Institute for Biological Research, P.O. Box 19, Ness-Ziona 74100, Israel, noame@iibr.gov.il

VA Feodorova Scientific and Research Institute for Medical and Veterinary Biotechnology, Branch in Saratov, 9, Proviantskaya Street, Box 1580, Saratov, 410028, Russia; Russian Research Anti-Plague Institute "Microbe", Saratov, Russia, feodorovav@mail.ru

Morly Fisher Department of Infectious Diseases, Israel Institute for Biological Research, P.O. Box 19, Ness-Ziona 74100, Israel, morlyf@iibr.gov.il

Yehuda Flashner Department of Biochemistry and Molecular Genetics, Israel Institute for Biological Research, P.O. Box 19, Ness-Ziona, 74100, Israel, yehudaf@iibr.gov.il

Arthur M. Friedlander US Army Medical Research Institute of Infectious Diseases, 1425 Porter Street, Frederick, MD 21702, USA; Uniformed Services University of the Health Sciences, Bethesda, MD 20814, USA, arthur.friedlander@amedd.army.mil

Gila Friedman Department of Biochemistry and Molecular Genetics, Israel Institute for Biological Research, P.O. Box 19, Ness-Ziona 74100, Israel, gilaf@iibr.gov.il

Orit Gat Department of Biochemistry and Molecular Genetics, Israel Institute for Biological Research, P.O. Box 19, Ness-Ziona 74100, Israel, oritg@iibr.gov.il

Ian J. Glomski Department of Microbiology, University of Virginia, Charlottesville, VA 22908, USA, iglomski@virginia.edu

Pierre L. Goossens Unité des Toxines et Pathogénie Bactérienne, URA 2172 CNRS, Institut Pasteur, 75015 Paris, France, pierre.goossens@pasteur.fr

Haim Grosfeld Department of Biochemistry and Molecular Genetics, Israel Institute for Biological Research, P.O. Box 19, Ness-Ziona 74100, Israel, haimg@iibr.gov.il

Roland Grunow Director and Professor, Head of Division Highly Pathogenic Microorganisms (ZBS 2), Centre for Biological Security, Robert Koch-Institute, Berlin, Germany, GrunowR@rki.de

Françoise Guinet Unité des *Yersinia*, Institut Pasteur, 28 rue du Docteur Roux, 75724 Paris Cedex 15, France, francoise.guinet@pasteur.fr

David Gur Department of Biochemistry and Molecular Genetics, Israel Institute for Biological Research, P.O. Box 19, Ness Ziona 74100, Israel, gurd@iibr.gov.il

Gideon Halperin Department of Biotechnology, Israel Institute for Biological Research, Ness-Ziona, 74100, Israel, gideonh@iibr.gov.il

O Holst Research Center Borstel, D-23845, Borstel, Germany

Michel Huerre Unité de Recherche et d'Expertise d'Histotechnologie et Pathologie, Institut Pasteur, 28 rue du Docteur Roux, 75724 Paris Cedex 15, France, michel.huerre@pasteur.fr

Izhak Inbar Department of Biochemistry and Molecular Genetics, Israel Institute for Biological Research, P.O. Box 19, Ness-Ziona 74100, Israel, itzacki@iibr.gov.il

Ofir Israeli Departments of Biochemistry & Molecular Genetics, Israel Institute for Biological Research, P.O. Box 19, Ness-Ziona 74100, Israel, ofiri@iibr.gov.il

Tomer Israely Department of Infectious Diseases, Israel Institute for Biological Research, P.O. Box 19, Ness-Ziona 74100, Israel, tomeri@iibr.gov.il

Daniela Jacob Division "Highly Pathogenic Microorganisms" (ZBS 2), Centre for Biological Security, Robert Koch-Institut, Berlin, Germany

Louis Jones Groupe Logiciels et Banques de Données, Institut Pasteur, 28 rue du Docteur Roux, 75724 Paris Cedex 15, France, louis.jones@pasteur.fr

Paul Keim Microbial Genetics and Genomics Center, Northern Arizona University, Flagstaff, AZ 86011-4073, USA; Pathogen Genomics Division, The Translational Genomics Research Institute, Phoenix, AZ, 85004, USA, Paul.Keim@nau.edu

YA Knirel N. D. Zelinsky Institute of Organic Chemistry, Russian Academy of Sciences, 47, Leninskii Prospect, Moscow, 119991, Russia

David Kobiler Department of Infectious Diseases, Israel Institute for Biological Research, P.O. Box 19, Ness-Ziona 74100, Israel, davidk@iibr.gov.il

Nina A. Kocharova N. D. Zelinsky Institute of Organic Chemistry, Russian Academy of Sciences, 119991 Moscow, Russia

AN Kondakova N. D. Zelinsky Institute of Organic Chemistry, Russian Academy of Sciences, 47, Leninskii Prospect, Moscow, 119991, Russia

NP Konnov Russian Research Anti-Plague Institute "Microbe", 46, Universitetskaya Street, Saratov, 410005, Russia

Melanie Kremer Institut für Infektionsmedizin und Zoonosen, Ludwig-Maximilians-Universität, Veterinärstr. 13, 80539 München, Germany

Andreas Kurth Centre for Biological Safety 1, German Consultant Laboratory for Poxviruses, Robert Koch Institute, Nordufer 20, 13353 Berlin, Germany, kurtha@rki.de

OS Kuznetsov Russian Research Anti-Plague Institute "Microbe", 46, Universitetskaya Street, Saratov, 410005, Russia

Andrea Kühn Division "Highly Pathogenic Microorganisms" (ZBS 2), Centre for Biological Security, Robert Koch-Institut, Berlin, Germany

Batel Lachmi Department of Infectious Diseases Israel Institute for Biological Research, P.O. Box 19, Ness Ziona, 74100, Israel, bat-ell@iibr.gov.il

Jason P. Laliberte Laboratory of Viral Diseases, National Institute of Allergy and Infectious Diseases, National Institutes of Health, Bethesda, MD 20892, USA

Moa Lavander Department of Clinical Microbiology, Clinical Bacteriology, and Laboratory for Molecular Infection Medicine Sweden (MIMS), Umeå University, SE-901 85 Umeå, Sweden

Shirley Lazar Department of Biochemistry and Molecular Genetics, Israel Institute for Biological Research, P.O. Box 19, Ness Ziona 74100, Israel, shirleyl@iibr.gov.il

Michael H. Lehmann Institut für Infektionsmedizin und Zoonosen, Ludwig-Maximilians-Universität, Veterinärstr. 13, 80539 München, Germany

Moshe Leitner Department of Biochemistry and Molecular Genetics, Israel Institute for Biological Research, P.O. Box 19, Ness-Ziona 74100, Israel, moshel@iibr.gov.il

Reuven Levin Department of Infectious Diseases Israel Institute for Biological Research, P.O. Box 19 Ness Ziona, 74100, Israel, reuvenl@iibr.gov.il

Haim Levy Department of Infectious Diseases, Israel Institute for Biological Research, P.O. Box 19, Ness-Ziona 74100, Israel, haiml@iibr.gov.il

Yinon Levy Department of Biochemistry and Molecular Genetics, Israel Institute for Biological Research, P.O. Box 19, Ness Ziona 74100, Israel, levy-ynon@iibr.gov.il; yinonl@iibr.gov.il

Buko Lindner Research Center Borstel, D-23845, Borstel, Germany, blindner@fz-borstel.de

Shlomo Lustig Department of Infectious Diseases, Israel Institute for Biological Research, P.O. Box 19, Ness-Ziona 74100, Israel, shlomol@iibr.gov.il

Emanuelle Mamroud Department of Biochemistry and Molecular Genetics, Israel Institute for Biological Research, Ness-Ziona 74100, Israel, emmym@iibr.gov.il

Hadar Marcus Departments of Biotechnology, Israel Institute for Biological Research, Ness-Ziona 74100, Israel, hadarm@iibr.gov.il

Gal Markel Department of Biochemistry and Molecular Genetics, Israel Institute for Biological Research, P.O. Box 19, Ness Ziona 74100, Israel, galm@iibr.gov.il

Ohad Mazor Departments of Biochemistry and Molecular Genetics, Israel Institute for Biological Research, Ness-Ziona 74100, Israel, ohadm@iibr.gov.il

Adva Mechaly Departments of Infectious Diseases, Israel Institute for Biological Research, Ness-Ziona 74100, Israel, advam@iibr.gov.il

Sharon Melamed Department of Infectious Diseases, Israel Institute for Biological Research, P.O. Box 19, Ness-Ziona 74100, Israel, sharonm@iibr.gov.il

Itai Mendelson Department of Biochemistry and Molecular Genetics, Israel Institute for Biological Research, P.O. Box 19, Ness-Ziona 74100, Israel, itaim@iibr.gov.il

Michèle Mock Unité des Toxines et Pathogénie Bactérienne, URA 2172 CNRS, Institut Pasteur, 75015 Paris, France, michele.mock@pasteur.fr

Bernard Moss Laboratory of Viral Diseases, National Institute of Allergy and Infectious Diseases, National Institutes of Health, Bethesda, MD 20892, USA, bmoss@nih.gov

VL Motin University of Texas Medical Branch, Galveston, TX 77555, USA

Herbert Nattermann Division “Highly Pathogenic Microorganisms” (ZBS 2), Centre for Biological Security, Robert Koch-Institut, Berlin, Germany

Andreas Nitsche Centre for Biological Safety 1, German Consultant Laboratory for Poxviruses, Robert Koch Institute, Nordufer 20, 13353 Berlin, Germany, nitschea@rki.de

Udy Olshevsky Department of Infectious Diseases Israel Institute for Biological Research, P.O. Box 19, Ness Ziona, 74100, Israel

Arie Ordentlich Department of Biochemistry and Molecular Genetics, Israel Institute for Biological Research, P.O. Box 19, Ness-Ziona 74100, Israel, arieo@iibr.gov.il

LN Pan’kina Russian Research Anti-Plague Institute “Microbe”, 46, Universitetskaya Street, Saratov, 410005, Russia

Nir Paran Department of Infectious Diseases, Israel Institute for Biological Research, P.O. Box 19, Ness-Ziona, 74100, Israel, nirp@iibr.gov.il

Avi Pass Departments of Biotechnology, Israel Institute for Biological Research, Ness-Ziona 74100, Israel, avip@iibr.gov.il

Talima Pearson Pathogen Genomics Division, The Translational Genomics Research Institute

GB Pier Channing Laboratory, Department of Medicine, Brigham and Women’s Hospital, Harvard Medical School, Boston, MA, USA, gpier@rics.bwh.harvard.edu

Boaz Politi Department of Infectious Diseases, Israel Institute for Biological Research, P.O. Box 19, Ness-Ziona, 74100, Israel, boazp@iibr.gov.il

Shaul Reuveny Department of Biotechnology, Israel Institute for Biological Research, P.O. Box 19, Ness-Ziona 74100, Israel, shaulr@iibr.gov.il

Ronit Rosenfeld Departments of Biochemistry and Molecular Genetics, Israel Institute for Biological Research, Ness-Ziona 74100, Israel, ronitr@iibr.gov.il

Nili Rothschild Departments of Infectious Diseases, Israel Institute for Biological Research, Ness-Ziona 74100, Israel, nilir@iibr.gov.il

P.S. Satheskumar Laboratory of Viral Diseases, National Institute of Allergy and Infectious Diseases, National Institutes of Health, Bethesda, MD 20892, USA

EP Savostina Russian Research Anti-Plague Institute "Microbe", 46, Universitetskaya Street, Saratov, 410005, Russia

LV Sayapina Tarasevich State Institute for Standartisation and Control of Biomedical Preparations, 41, Sivtsev Vrazhek, Moscow 121002, Russia

Josef Schlomovitz Departments of Infectious Diseases, Israel Institute for Biological Research, Ness-Ziona 74100, Israel

Paula Schneider Department of Infectious Diseases Israel Institute for Biological Research, P.O. Box 19 Ness Ziona, 74100, Israel, paulas@iibr.gov.il

Avigdor Shafferman Department of Biochemistry and Molecular Genetics, Israel Institute for Biological Research, P.O. Box 19, Ness-Ziona 74100, Israel, avigdors@iibr.gov.il

Rima Z. Shaikhutdinova State Research Center for Applied Microbiology and Biotechnology, Obolensk, 142279, Moscow Region, Russia

Ohad Shifman Departments of Biochemistry & Molecular Genetics, Israel Institute for Biological Research, P.O. Box 19, Ness-Ziona 74100, Israel, ohads@iibr.gov.il

Anders Sjöstedt Department of Clinical Microbiology, Clinical Bacteriology, and Laboratory for Molecular Infection Medicine Sweden (MIMS), Umeå University, SE-901 85 Umeå, Sweden, anders.sjostedt@climi.umu.se

Geoffrey L. Smith Department of Virology, Faculty of Medicine, Imperial College London, St. Mary's Campus, Norfolk Place, London W2 1PG, England, geoffrey.l.smith@imperial.ac.uk

Dana Stein Departments of Biochemistry & Molecular Genetics, Israel Institute for Biological Research, P.O. Box 19, Ness-Ziona 74100, Israel, danas@iibr.gov.il

Yasemin Suezer Paul-Ehrlich-Institut, 63225 Langen, Germany

Gerd Sutter Institut für Infektionsmedizin und Zoonosen, Ludwig-Maximilians-Universität, Veterinärstr. 13, 80539 München, Germany, gerd.sutter@lmu.de

Arnon Tal Departments of Biotechnology, Israel Institute for Biological Research, Ness-Ziona 74100, Israel, arnont@iibr.gov.il

Avital Tidhar Department of Biochemistry and Molecular Genetics, Israel Institute for Biological Research, P.O. Box 19, Ness-Ziona, 74100, Israel, avitalt@iibr.gov.il

Lhousseine Touqui Unité de Défense innée et Inflammation, Unité INSERM U874, Institut Pasteur, 75015 Paris, France, lhousseine.touqui@pasteur.fr

Shay Weiss Department of Infectious Diseases, Israel Institute for Biological Research, P.O. Box 19, Ness-Ziona 74100, Israel, shayw@iibr.gov.il

Baruch Velan Department of Biochemistry and Molecular Genetics, Israel Institute for Biological Research, P.O. Box 19, Ness-Ziona 74100, Israel, baruchv@iibr.gov.il

Galia Zaide Department of Biochemistry and Molecular Genetics, Israel Institute for Biological Research, P.O. Box 19, Ness-Ziona 74100, Israel, galiaz@iibr.gov.il

Ayelet Zauberman Department of Biochemistry and Molecular Genetics, Israel Institute for Biological Research, Ness-Ziona, 74100, Israel, ayeletz@iibr.gov.il

Anat Zvi Department of Biochemistry and Molecular Genetics, Israel Institute for Biological Research, P.O. Box 19, Ness-Ziona 74100, Israel, anatz@iibr.gov.il

Chapter 1
The Anthrax Capsule: Role in Pathogenesis and Target for Vaccines and Therapeutics

Arthur M. Friedlander

Abstract The polyglutamic acid capsule of *Bacillus anthracis* is a well-established virulence factor, conferring antiphagocytic properties on the bacillus. We have shown that the capsule also confers partial resistance to killing by human defensins. In our research we targeted the anthrax capsule for developing medical countermeasures, first using the capsule as a vaccine, similar to successful efforts with other bacteria, and secondly, by developing a novel therapeutic against the capsule. Our experiments showed that a capsule vaccine is protective in the mouse model and its efficacy could be enhanced by conjugation to a protein carrier. In initial experiments using high challenge doses, a capsule conjugate vaccine was not protective in rabbits but did show some protection in nonhuman primates. This suggests it may be useful as an addition to a protective antigen-based vaccine. We are also developing the use of the *B. anthracis* capsule-depolymerizing enzyme, CapD, as a therapeutic. We demonstrated that in vitro treatment of the encapsulated anthrax bacillus with CapD enzymatically removed the capsule from the bacterial surface making it susceptible to phagocytic killing. Initial experiments in vivo showed that CapD could be used successfully to treat experimental anthrax infections. Such a novel approach to target the capsule virulence factor might be of value in the treatment of infections due to antibiotic-resistant strains.

Keywords Anthrax · Capsule · Vaccine · Therapy · Capsule depolymerase

The recent renewed interest in anthrax after the cases of anthrax that developed in 2001 after the mailing of letters containing anthrax spores has stimulated research to develop new medical countermeasures.

In this review I will give an overview of the anthrax capsule and its role in pathogenesis and describe our efforts to develop a capsule-based vaccine and therapeutic.

A.M. Friedlander (✉)
US Army Medical Research Institute of Infectious Diseases, 1425 Porter Street, Frederick, MD 21702, USA; Uniformed Services University of the Health Sciences, Bethesda, MD 20814, USA
e-mail: arthur.friedlander@amedd.army.mil

A. Shafferman et al. (eds.), *The Challenge of Highly Pathogenic Microorganisms*,
DOI 10.1007/978-90-481-9054-6_1, © Springer Science+Business Media B.V. 2010

1.1 Overview of Anthrax Pathogenesis

The infectious form of anthrax, the spore, germinates within the infected host to the vegetative bacillus, and in this regard it is similar to the pathogenic dimorphic fungi. *Bacillus anthracis* possesses three major recognized virulence factors: the two exotoxins, lethal and edema toxins and the antiphagocytic polyglutamic acid capsule. In recent years other virulence factors have been identified including several whose deletion results in 2 or more logs of attenuation. These include a capsule depolymerase (CapD) responsible for attachment of the nascent capsule filament to the cell wall peptidoglycan (Candela and Fouet, 2005), a manganese ATP-binding cassette transporter (Gat et al., 2005), anthrachelin siderophore biosynthesis genes (Cendrowski et al., 2004), nitric oxide synthase (Shatalin et al., 2008) and the caseinolytic protease component ClpX (McGillivray et al., 2009). Several other genes have also been identified that contribute to virulence but whose deletion results in less attenuation (<2 logs).

The spore enters the skin, gastrointestinal tract, or lung. Germination occurs locally extracellularly or in phagocytic cells during or after transport to regional lymph nodes. Some organisms are killed while others remain viable and proliferate. Germination is not synchronous and toxins and capsule are synthesized early after germination. Bacilli located extracellularly or after escape from a phagocyte become encapsulated and resistant to subsequent phagocytosis. Local production of toxins leads to the pathological effects of edema and necrosis. The lethal and edema toxins likely act early in the infectious process, intracellularly and extracellularly, to subvert host innate immune mechanisms by inhibiting phagocytic cells and other cell types (Tournier et al., 2009). The organism spreads from the lymph node, resulting in bacteremia and subsequent toxemia. Death after inhalational anthrax is likely due to lymphatic/vascular obstruction, pulmonary hemorrhage/edema, pleural effusions and toxicity.

1.2 Role of Capsule in Virulence

Since the early 1900s (Preisz, 1909), it has been known that capsule expression is associated with virulence in *B. anthracis* and that strains lacking the capsule are attenuated. When grown with bicarbonate and carbon dioxide *B. anthracis* produces an antiphagocytic capsule. This results in smooth glistening colonies, while capsule-negative mutants, such as Sterne-like strains lacking the capsule encoding pX02 plasmid, appear rough and dry. Capsule can bevisualized by negative staining with India ink. As indicated in Fig. 1.1, the wild-type Ames strain has a typical crinkled appearance while an isogenic, unencapsulated strain is difficult to see.

The five genes necessary for capsule synthesis and formation are encoded in an operon on the pX02 plasmid. CapB and CapC are thought to be responsible for synthesis of the polyglutamate filament, CapA and CapE for translocation across the cell membrane, and CapD for covalent attachment to meso-diaminopimelic acid of the peptidoglycan (Candela and Fouet, 2006; Richter et al., 2009). The capsule is

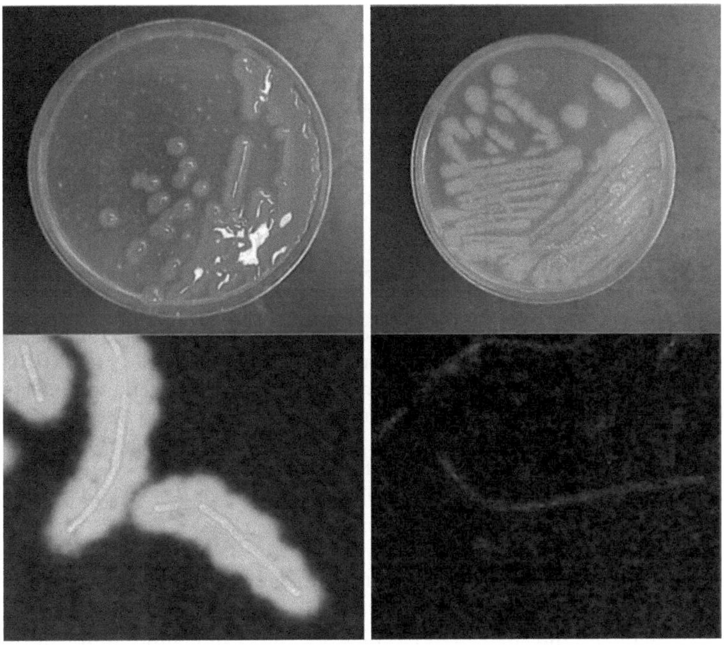

Fig. 1.1 *B. anthracis* grown on NBY/bicarbonate agar with 20% CO_2 at 37°C. The colonial morphology (*top panels*) and India ink preparation (1000×) of bacilli (*bottom panels*) of wild-type encapsulated *B. anthracis* (*left panels*) and an unencapsulated isogenic strain (*right panels*) are shown

located external to the S-layer of the bacillus. As with other extracellular bacterial pathogens, the anthrax capsule has long been known to be antiphagocytic. It also blocks phage attachment (McCloy, 1951) and access of antibodies to underlying bacillus antigens (Mesnage et al., 1998). Synthesis of the capsule is regulated by factors on both pX01 and pX02 and is induced by carbon dioxide and serum. The capsule is a homopolymer of gamma-linked polyglutamic acid composed entirely of the D enantiomer. As such it is resistant to proteases and a poorly immunogenic T-independent antigen. Its mass varies from 100,000 to 1,000,000 daltons.

The antiphagocytic nature of the capsule is evidenced by the fact that there is minimal binding of encapsulated bacilli to macrophages without addition of anti-capsule antibodies. There is also some evidence that low molecular weight capsule fragments are released from the bacillus and may contribute to virulence (Makino et al., 2002). Evidence from our laboratory suggests that the capsule provides some protection against the bactericidal effects of some of the antimicrobial peptides, including human beta defensins. The defensins are cationic peptides with intramolecular disulfide bonds and are an important part of the innate immune system. Beta defensins occur in the epithelium of the skin, lung, gastrointestinal, and genitourinary tracts.

We found that the beta defensins 1, 2, and 3 have some antimicrobial activity against the encapsulated Ames strain, but that they are significantly more active

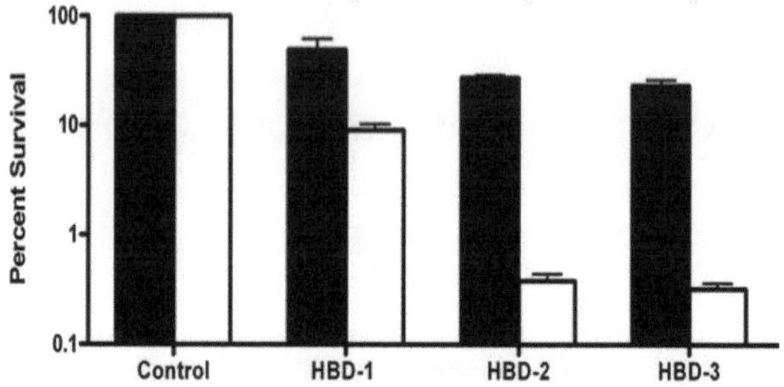

Fig. 1.2 Antimicrobial activity of human beta defensins against encapsulated and unencapsulated *B. anthracis* strains. Bacilli were incubated with human beta defensins-1, 2, or 3 (HBD-1,2,3) at a final concentration of 20 μg/ml at 37°C in 5% CO_2 for 2 h and bacteria were then plated for cfu. Survival percentages were calculated by comparing the cfu at 2 h to a control containing no defensins. Results are expressed as the mean + SEM of triplicate samples. Black bars represent wild-type and white bars represent unencapsulated bacilli. The differences between encapsulated and unencapsulated strains were significant for all defensins tested ($P < 0.0001$)

against an unencapsulated isogenic strain (Fig. 1.2) (O'Brien et al. manuscript in preparation). The mechanism of this inhibition is under study but preliminary data suggest the capsule on the surface of the bacillus may bind some of the antimicrobial peptide and thus prevent it from binding to its presumed target on the cytoplasmic membrane.

1.3 Capsule as a Vaccine Target: Development of a Capsule Vaccine

Essentially all licensed vaccines against extracellular bacteria are based upon the bacterium's capsule. However no work had been done with the anthrax capsule until research in our laboratory demonstrated for the first time that the capsule is effective as a vaccine against challenge with an encapsulated nontoxinogenic *B. anthracis* strain that is virulent for the mouse (Chabot et al., 2004). Mice vaccinated with two doses of capsule were protected against subcutaneous challenge with approximately 100 LD_{50} with 7/12 animals surviving compared to 0/12 controls (Table 1.1, $P = 0.014$). The capsule vaccine was unable to protect against challenge with the fully virulent encapsulated and toxinogenic Ames strain as was protective antigen by itself, as has been reported previously. However, the combination of capsule and protective antigen was protective (9/11 surviving) in mice suggesting that capsule and protective antigen might be synergistic. In an attempt to convert the capsule from a T-independent to a T-dependent antigen resulting in a more mature IgG response and immunological memory, as has been done with polysaccharide homopolymer antigens, we conjugated the capsule to an immunogenic protein carrier (bovine serum albumin). This resulted in the expected increased IgG response but did not protect mice, likely because the carbodiimide conjugation procedure

Table 1.1 Efficacy of capsule vaccine against anthrax infection in the mouse[a]

Experiment	Vaccine	No. survivors/no. challenged
1. Challenge with *B. anthracis* delta Ames (cap+, tox-)	Capsule	7/12
	PBS	0/12
2. Challenge with *B. anthracis* Ames (cap+, tox+)	Protective antigen	1/12
	Capsule	0/12
	Protective antigen+capsule	9/11
	PBS	0/12
3. Challenge with *B. anthracis* Ames (cap+, tox+)	Capsule conjugate	9/9
	Alum control	1/10

[a]Mice were vaccinated with two doses of vaccine and challenged subcutaneously 4–8 weeks later with 116 LD_{50}, 78 LD_{50} or 20,000 spores in Experiments 1, 2, or 3, respectively. Data and details for experiments 1 and 2 were from Chabot et al. 2004 and for experiment 3 from Joyce et al. 2006. The differences in survival between capsule and PBS in experiment 1, protective antigen+capsule and PBS in experiment 2, and capsule conjugate and alum control in experiment 3 were all significant ($P = 0.014$, $P < 0.001$, and $P < 0.001$, respectively, Fisher's exact test).

reduced the size of the capsule and destroyed antigenic epitopes. However, using a novel controlled conjugation procedure to couple the capsule to the outer membrane protein complex of *Neisseria meningitidis* serotype B as the carrier, the high molecular mass and antigenic epitopes on the capsule were preserved (Joyce et al., 2006). This capsule conjugate was now able to protect mice against challenge with the fully virulent encapsulated toxinogenic strain (Table 1.1). Note that this degree of protection in mice was greater than what is observed with the protective antigen.

To further investigate the effectiveness of the capsule conjugate, we tested the vaccine against an aerosol challenge with the encapsulated toxinogenic Ames strain in the rabbit and nonhuman primate models. The vaccine was highly immunogenic in both species and induced opsonic anticapsule antibodies. Rabbits vaccinated with two doses of the conjugate vaccine and challenged with 680 LD_{50} were not protected, with none of 10 animals surviving. However, significant protection was seen in the nonhuman primate challenged with 123 LD_{50} with 3/5 and 2/5 animals surviving after two or one dose of the vaccine compared to 0/5 controls ($P = 0.02$ and $P = 0.04$, respectively) (Chabot et al., 2009). The results suggest that including the capsule might increase the potency of a protective antigen-based vaccine. Such a multicomponent vaccine might be of value against strains resistant to vaccination with one antigen and as the mouse experiments suggest, there may be synergy between the capsule and protective antigen components targeting different virulence factors.

1.4 Capsule as a Therapeutic Target

Vaccines against bacterial capsules function by inducing antibodies that are opsonic and enable phagocytic killing as described above for anthrax. Antibodies against the capsule have also been developed as a therapeutic and have shown some efficacy in experimental animals (Kozel et al., 2004; Joyce et al., 2006). A novel alternative approach we have taken is to convert the encapsulated bacillus to a form susceptible to phagocytic killing by removing the capsule from the bacterial surface using a capsule-degrading enzyme. This builds on the concept of using microbial enzymes as antibacterials developed many years ago, and subsequently demonstrated with the use of capsule degrading enzymes to treat with varying success, pneumococcal (Avery and Dubos, 1931), *E. coli* (Mushtaq et al., 2005) and cryptococcal infections (Gadebusch, 1960). Another approach recently explored is the development of drugs that inhibit capsule attachment (Richter et al., 2009).

As indicated above, *B. anthracis* expresses CapD, which is required for covalently attaching the nascent polyglutamic acid filament to the peptidoglycan and CapD null mutants are attenuated (Uchida et al., 1993; Candela and Fouet, 2005). However, this enzyme, a gamma glutamyltranspeptidase, when added to purified capsule, can carry out a hydrolytic reaction and degrade the high molecular weight polyglutamate to low molecular species using water or amino acids as an acceptor as shown in Fig. 1.3.

We used this property to show that purified CapD added externally to encapsulated *B. anthracis* is capable of removing the capsule from the surface (Fig. 1.4), making the bacillus susceptible to phagocytosis and neutrophil killing. Pretreating encapsulated bacilli with CapD before incubation greatly enhanced their adherence

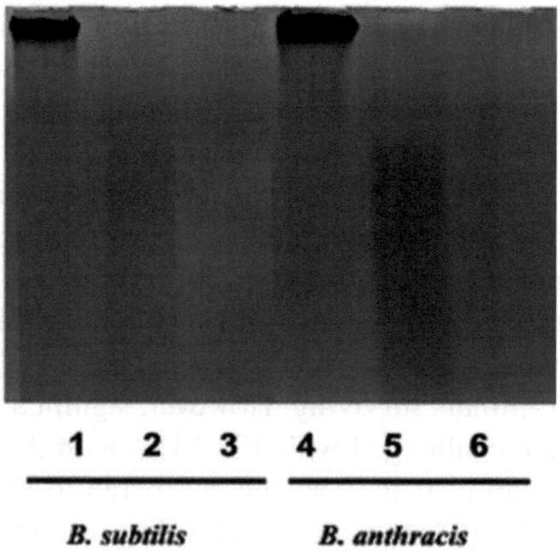

Fig. 1.3 Degradation of capsules from *B. anthracis* and *B. subtilis* by CapD. Capsule purified from *B. subtilis* (lanes 1–3) or *B. anthracis* (lanes 4–6) were incubated with CapD and examined by SDS-polyacrylamide gel electrophoresis. CapD was present at the following concentrations: lanes 1 and 4, 0.35 ug/ml; lanes 2 and 5, 3.5 μg/ml; and lanes 3 and 6, 35 μg/ml together with capsule at 200 μg/ml. Data adapted from Scorpio et al., 2007

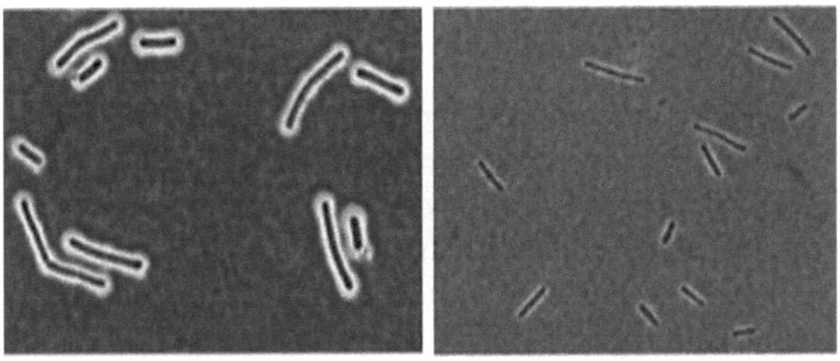

Fig. 1.4 Removal of capsule from encapsulated *B. anthracis* by CapD. Encapsulated bacilli were treated for 20 min at 37°C with PBS or CapD (50 μg/ml) and examined by India ink and phase-contrast microscopy (1000×). Adapted from Scorpio et al., 2007

to macrophages and resulted in >2 logs killing by human neutrophils compared with minimal to no killing in untreated controls (Scorpio et al., 2007).

We then extended these observations to determine whether CapD could be used to treat experimental anthrax infections (Scorpio et al., 2008). We used two different models of anthrax in the mouse model. In the first series of experiments, mice were infected with the encapsulated but nontoxinogenic delta Ames strain that is almost as virulent as the wild-type encapsulated toxinogenic parental Ames strain. In this model, to optimize the chance of success, mice were injected intraperitoneally (ip) 6 h before challenge with a 2% starch solution to induce neutrophil infiltration. They were then challenged ip with 1000 colony forming units (cfu) of delta Ames bacilli together with either 4 μg of CapD or PBS. This single dose of CapD given at the time of challenge resulted in a statistically significant increase in survival. All seven animals given CapD survived compared to three of seven given PBS ($P = 0.035$, Fisher's exact test). In a similar experiment using the encapsulated toxinogenic Ames strain for the challenge (500 cfu), CapD injected together with the bacteria ip also was protective. All eight mice given CapD survived compared to none of eight in the control group, consisting of heat-inactivated CapD in this experiment ($P = 0.0005$, Fisher's exact test). Similar protection was afforded when CapD was given 10 min after bacterial challenge. When higher challenge doses were used, there was no protection with the same treatment regimen.

In another experimental design, we determined if CapD could treat an established infection using a spore challenge. Mice were challenged with 6,600 delta Ames spores ip and treated 30 h later with 400 μg of CapD given both ip and intravenously. Significant protection was observed with 10 of 10 survivors in the CapD group compared to five of nine controls ($P = 0.033$, Fisher's exact test). No protection was observed using an Ames spore challenge. These initial experiments showing some success in treating anthrax were accomplished using treatment with only one or two doses, which is very encouraging. Improvements in the pharmacokinetic properties of CapD may allow for greater efficacy in treating anthrax.

1.5 Summary and Conclusions

The antiphagocytic polyglutamic acid capsule of *B. anthracis* is a well-established virulence factor. We have focused on the use of the capsule both as a vaccine and a therapeutic target. A capsule vaccine was effective in the mouse model and conjugation to a protein carrier enhanced its efficacy. Initial experiments using high aerosol spore challenges revealed no protection in rabbits but significant protection in nonhuman primates, suggesting that adding the capsule to protective antigen-based vaccines may be advantageous. We also showed that enzymatic removal of the capsule from the bacterial surface by treatment with CapD was effective in allowing neutrophils to phagocytize and kill the bacillus. Initial experiments in mice showed that CapD was effective in treating anthrax infections, suggesting this approach might be of value in the treatment of infections caused by antibiotic-resistant strains.

Acknowledgments I thank the Defense Threat Reduction Agency for funding our research and Donald Chabot for help with several of the figures. Opinions, interpretations, conclusions, and recommendations are those of the author and are not necessarily endorsed by the U.S. Army.

References

Avery, O. T. & Dubos, R. (1931) The protective action of a specific enzyme against type III pneumococcus infection in mice. *J Exp Med*, 54, 73–89.

Candela, T. & Fouet, A. (2005) *Bacillus anthracis* CapD, belonging to the γ-glutamyltranspeptidase family, is required for the covalent anchoring of capsule to peptidoglycan. *Mol Microbiol,* 57, 717–26.

Candela, T. & Fouet, A. (2006) Poly-gamma-glutamate in bacteria. *Mol Microbiol,* 60, 1091–8.

Cendrowski, S., Macarthur, W. & Hanna, P. (2004) *Bacillus anthracis* requires siderophore biosynthesis for growth in macrophages and mouse virulence. *Mol Microbiol,* 51, 407–17.

Chabot, D., Joyce, J., Caulfield, M., Wang, S., Vietri, N., Leffel, E., Pitt, L., Cook, J., Hepler, R., Ruthel, G. & Friedlander, A. (2009) Efficacy of a poly-γ-D-glutamic acid capsule conjugate vaccine against inhalational anthrax in rabbits and non-human primates. Bacillus-ACT (2009): The international *Bacillus anthracis, B. cereus, B. thuringiensis* Conference. Santa Fe, NM. Abstract S5:4.

Chabot, D. J., Scorpio, A., Tobery, S. A., Little, S. F., Norris, S. L. & Friedlander, A. M. (2004) Anthrax capsule vaccine protects against experimental infection. *Vaccine,* 23, 43–7.

Gadebusch, H. H. (1960) Specific degradation of *Cryptococcus neoformans* 3723 capsular polysaccharide by a microbial enzyme. II. Biological activity of the enzyme. *J Infect Dis,* 107, 402–5.

Gat, O., Mendelson, I., Chitlaru, T., Ariel, N., Altboum, Z., Levy, H., Weiss, S., Grosfeld, H., Cohen, S. & Shafferman, A. (2005) The solute-binding component of a putative Mn(II) ABC transporter (MntA) is a novel *Bacillus anthracis* virulence determinant. *Mol Microbiol,* 58, 533-51.

Joyce, J., Cook, J., Chabot, D., Hepler, R., Shoop, W., Xu, Q., Stambaugh, T., Aste-Amezaga, M., Wang, S., Indrawati, L., Bruner, M., Friedlander, A., Keller, P. & Caulfield, M. (2006) Immunogenicity and protective efficacy of *Bacillus anthracis* poly-gamma-D-glutamic acid capsule covalently coupled to a protein carrier using a novel triazine-based conjugation strategy. *J Biol Chem*, 281, 4831–43.

Kozel, T. R., Murphy, W. J., Brandt, S., Blazar, B. R., Lovchik, J. A., Thorkildson, P., Percival, A., Lyons, C. R. (2004) mAbs to *Bacillus anthracis* capsular antigen for immunoprotection in anthrax and detection of antigenemia. *Proc Nat Acad Sci USA,* 101, 5042–7.

Makino, S., Watarai, M., Cheun, H. I., Shirahata, T. & Uchida, I. (2002) Effect of the lower molecular capsule released from the cell surface of *Bacillus anthracis* on the pathogenesis of anthrax. *J Infect Dis,* 186, 227–33.

McGillivray, S. M., Ebrahimi, C. M., Fisher, N., Sabet, M., Zhang, D. X., Chen, Y., Haste, N. M., Aroian, R. V., Gallo, R. L., Guiney, D. G., Friedlander, A. M., Koehler, T. M. & Nizet, V. (2009) ClpX Protease contributes to innate defense Peptide resistance and virulence phenotypes of *Bacillus anthracis. J Innate Immun,* 1, 494–506.

McCloy, E. W. (1951) Studies on a lysogenic *Bacillus* strain. I. A bacteriophage specific for *Bacillus anthracis. J Hyg* 49, 114–25.

Mesnage, S., Tosi-Couture, E., Gounon, P., Mock, M. & Fouet, A. (1998) The capsule and S-layer: two independent and yet compatible macromolecular structures in *Bacillus anthracis. J Bacteriol,* 180, 52–8.

Mushtaq, N., Redpath, M. B., Luzio, J. P. & Taylor, P. W. (2005) Treatment of experimental *Escherichia coli* infection with recombinant bacteriophage derived capsule depolymerase. *J Antimicrob Chemother,* 24, 160–5.

Preisz, H. (1909) Experimentelle Studien Åber Virulenz, Empfänglichkeit und Immunität beim Milzbrand. Zeitschr. Immunität Forsch, 5, 341–452.

Richter, S., Anderson, V. J., Garufi, G., Lu, L., Budzik, J. M., Joachimiak, A., He, C., Schneewind, O. & Missiakas, D. (2009) Capsule anchoring in *Bacillus anthracis* occurs by a transpeptidation reaction that is inhibited by capsidin. *Mol Microbiol,* 71, 404–20.

Scorpio, A., Chabot, D. J., Day, W. A., O'Brien, D. K., Vietri, N. J., Itoh, Y., Mohamadzadeh, M. & Friedlander, A. M. (2007) Poly-{gamma}-glutamate capsule-degrading enzyme treatment enhances phagocytosis and killing of encapsulated *Bacillus anthracis. Antimicrob Agents Chemother,* 51, 15–22.

Scorpio, A., Tobery, S. A., Ribot, W. J. & Friedlander, A. M. (2008) Treatment of experimental anthrax with recombinant capsule depolymerase. *Antimicrob Agents Chemother,* 52, 1014–20.

Shatalin, K., Gusarov, I., Avetissova, E., Shatalina, Y., McQuade, L. E., Lippard S. J. & Nudler, E. (2008) *Bacillus anthracis*-derived nitric oxide is essential for pathogen virulence and survival in macrophages. *Proc Natl Acad Sci USA,* 105, 1009–13.

Tournier, J. -N., Paccani, S. R., Quesnel-Hellmann, A. & Baldari, C. T. (2009) Anthax toxins: a weapon to systematically dismantle the host immune defenses. *Mol Aspects Med* 2009, 30, 456–66.

Uchida, I., Makino, S., Sasakawa, C., Yoshikawa, M., Sugimoto, C. & Terakado, N. (1993) Identification of a novel gene, *dep*, associated with depolymerization of the capsular polymer in *Bacillus anthracis. Mol Microbiol,* 9, 487–96.

Chapter 2
Proteomic Studies of *Bacillus anthracis* Reveal In Vitro CO_2-Modulation and Expression During Infection of Extracellular Proteases

Theodor Chitlaru, Orit Gat, Galia Zaide, Haim Grosfeld, Izhak Inbar, Sharon Ehrlich, Moshe Leitner, Ofer Cohen, and Avigdor Shafferman

Abstract A comparative proteomic study of secretomes of virulent and avirulent *Bacillus anthracis* strains in various culturing conditions, including those encountered in the host (high CO_2/bicarbonate), enabled identification of approximately 70 proteins representing collectively more than 99% of the secretome. In-vivo expression of 50 proteins was established by 2-dimension Western-analysis using anti *B. anthracis* immune sera. Many of the abundant proteins harbor features characteristic of virulence determinants and exhibit different patterns of expression. In minimal medium, virulent and avirulent *B. anthracis* strains manifest similar protein signatures and the metalloprotease NprA, (previously suggested to act in the context of a starvation-induced mechanism), represents 90% of the total secretome. Under high CO_2/bicarbonate, NprA is repressed (possibly by a mechanism which preserves toxin integrity), while other proteins, including the bacterial toxins, are induced. One of the immunogens observed to be induced under high CO_2-tension, was HtrA. We investigated the phenotype associated with disruption of HtrA by biochemical and proteomic approaches. The HtrA$^-$ bacteria are severely affected in their ability to respond to stress and fail to secrete the most abundant extracellular protease NprA. Most surprisingly, HtrA$^-$ cells do not possess the characteristic S-layer. This unique phenotype may have important implications for the role of HtrA in manifestation of *B. anthracis* virulence. Furthermore, the data show that distinct CO_2/bicarbonate responsive chromosome-and plasmid-encoded regulatory factors modulate the secretion of potential novel virulence factors, most of which are associated with extracellular proteolytic activities.

Keywords *Bacillus anthracis* · Patterns of expression · Proteases · Proteomic/serological analysis · Virulence

---

T. Chitlaru (✉)
Department of Biochemistry and Molecular Genetics, Israel Institute for Biological Research, P.O. Box 19, Ness-Ziona 74100, Israel
e-mail: theodorc@iibr.gov.il

A. Shafferman et al. (eds.), *The Challenge of Highly Pathogenic Microorganisms*,
DOI 10.1007/978-90-481-9054-6_2, © Springer Science+Business Media B.V. 2010

2.1 Introduction

The bacterium *Bacillus anthracis* is a Gram positive spore-forming obligatory-pathogen. It is the etiologic agent of anthrax, a severe disease, characterized by toxemia and septicemia initiated by spore entry into the host followed by their germination into rapidly multiplying toxin-secreting bacilli (Leppla, 1995; Mock and Fouet, 2001). Anthrax mainly affects mammalian herbivores, yet in humans, depending on the site of entry of the spores, three manifestations of the disease are known-coutaneous, gastroenteric and inhalatory; the latter is the most severe form, and potentially associated with bio-terror oriented dissemination of spores. The disease is toxinogenic in the sense that the bacterial binary exo-toxin is necessary for the onset of the disease and its symptoms can be recapitulated by the administration of pure preparations of toxin (Lacy and Collier, 2002), yet other factors may be required for the colonization and expansion of the bacteria in the host (Fellows et al., 2001; Brossier et al., 2002; Cendrowski et al., 2004; Gat et al., 2005, 2008; Fisher et al., 2006). The toxin is composed of three proteins: protective antigen (PA, which mediates binding to the receptor on target cells and internalization of the toxin components, Bradley et al., 2001; Collier, 2009), lethal factor (LF, a zinc protease targeting several MAP kinases, Klimpel et al., 1994) and edema factor (EF, a calmodulin dependent adenylate cyclase; Leppla, 1982; Lacy and Collier, 2002). The genes encoding for the three exotoxin components are located on the native virulence plasmid, pXO1. Genes encoding for functions involved in the synthesis of the second major *B. anthracis* virulence determinant, a polyglutamyl immunologically-inert capsule that protects bacteria from phagocytosis, are located on a second native virulence plasmid, pXO2 (Leppla, 1995). The non-toxic subunit of the bacterial toxin, PA is a strong immunogen able to elicit a protective humoral immune response, therefore anthrax vaccines are based on the administration of various preparations of PA (Friedlander et al., 2002).

In addition to the capsule, *B. anthracis* exhibits another well documented surface entity- the chromosomally encoded S-layer (Mesnage et al., 1997; Fouet, 2009), which is composed of two proteins, Sap and EA1, sequentially expressed during the bacterial logarithmic and stationary growth phases respectively. The S-layer proteins represent the prevalent components of the proteinous make-up of the cells, and are highly abundant in the bacterial secretome as well (Chitlaru et al., 2004, 2006). Furthermore they are highly immunogenic indicating that they are abundantly expressed during infection in a form and compartment which is readily amenable to interaction with the host immune-recognition mechanisms. The physiological role of the *B. anthracis* S-layer is not understood nor is the significance of the coexistance of both surface entities -the S-layer and the capsule-in the bacterial expansion in the host. Yet, we and others have documented that S-layer association (mediated by S layer homology [SLH] domains) represents a prevalent anchorage modality of exposed proteins on the surface of *B. anthracis* (Chitlaru et al., 2004).

The virulence of the bacteria entails intricate regulatory cross-talk mechanisms which coordinate expression of plasmid and chromosome genes (See: Perego and Hoch, 2008; Fouet and Mock, 2006; Koehler, 2009, for a full coverage of virulence regulatory circuits in *B. anthracis*). The AtxA regulator, encoded by pXO1, is essential for expression of the toxin and capsule synthesis genes in vivo (a situation which can be mimicked by high bicarbonate-CO_2, conditions (Wilson et al., 2008)). Synthesis of AcpA and AcpB, the products of pXO2-located genes which affect capsule synthesis, is regulated by AtxA. AtxA was found to influence directly or via AcpA and AcpB, expression of the chromosomally located genes coding for S-layer proteins as well as chromosomal genes involved in sporulation. The chromosomally-encoded protein AbrB, negatively controls the toxin-gene promoters. Finally, some chromosomally-encoded extracellular proteases may be controlled by Cot43, a regulator encoded by pXO1 (Aronson et al., 2005). In addition to these cross-talk regulatory mechanisms, *B. anthracis* is distinguished by the evolutionary silencing of the pleiotropic factor PlcR which regulates secretion of virulence-essential effectors in the philogenetically related strains of the *B. cereus* group (Gohar et al., 2008). In *B. anthracis*, PlcR is inactive due to a mutation which abrogates its expression (Mignot et al., 2001). This evolutionary inactivation of the PlcR regulon is probably due to its incompatibility with the AtxA regulon preventing efficient toxin synthesis and sporulation. The dormant state of the PlcR-inducible genes imply that the PlcR target genes are dispensable for anthrax pathogenicity and strengthens the notion that *B. anthracis* has evolved its own set of virulence mechanisms.

We have documented in the past global surveys based on functional genomic and proteomic studies combined with serological study of *B. anthracis* (Ariel et al., 2002, 2003; Gat et al., 2006, 2007; Chitlaru et al., 2006, 2007; Chitlaru and Shafferman, 2009; see also Chapter 32 on Reverse Vaccinology, authored by Shafferman et al. in this volume), for identification of vaccine and diagnostic marker candidates among extracellular (secreted or membranal) proteins, as well as targets for therapeutic intervention. A bioinformatic survey of the *B. anthracis* chromosome (Ariel et al., 2002, 2003) was initially complemented by a direct proteomic approach which confirmed some of the bioinformatic predictions regarding membranal localization of proteins. In these studies, the protein signature of a *B. anthracis* subcellular fraction enriched in membranal proteins was determined by two-dimensional electrophoretic separation, and identification of proteins by MALDI-TOF-MS (matrix-assisted-laser-desorption-ionization time-of-flight mass spectrometry). This analysis was followed by probing the proteins with immune sera on Western blots. We document that the vast majority of the proteins exhibited on the surface of *B. anthracis* belong to the category of proteins which exploit SLH (S-layer homology) domains for their membranal anchorage and many of them reacted with sera derived from immune animals demonstrating that these proteins arc expressed in vivo and are able to elicit a humoral immune response. These initial studies were expanded in a comprehensive proteomic study of secreted proteins. Based on their immunogenicity, abundance under various culture conditions, and functional relatedness to infection, these studies generated a list of proteins that can serve as candidates for further evaluation as vaccines and diagnostic tools, and on the other hand, for a better understanding of the pathogenicity process. Two proteins, the proteases NprA (locus identifying tag BA599) and

HtrA (BA3660) have emerged from these studies as representatives of two alternative CO_2/bicarbonate-dependent regulatory pathways: NprA represents proteins which are repressed in the presence of bicarbonate/CO_2 in a manner independent of the presence of the virulence plasmids, while HtrA represents proteins which are induced under high CO_2/bicarbonate tension only in the presence of the virulence plamids. In the present report, we document studies which address the possible significance that the pattern of expression of these two proteases has in *B. anthracis* pathogenicity, and describe the preliminary characterization of the phenotype exhibited by bacteria in which the *htrA* gene was disrupted.

2.2 Results

2.2.1 Differential Proteomic Analyses of the B. anthracis *Secretome*

The proteomic screen of *B. anthracis* secreted proteins was carried out according to the steps depicted in Fig. 2.1 (see legend to figure for details of the various proteomic and serological experimental steps). The study included proteomic inspection of *B. anthracis* cultures of the fully virulent WT Vollum strain and the attenuated plasmid-devoid strain ΔVollum. Secreted proteins were collected at various time points from cultures set up in various rich media (such as BHI [brain/heart infusion]) and low-nutrient media (such as NBY) under aerated or semi-anaerobic bicarbonate/CO_2 supplemented (see Chitlaru et al., 2006) conditions. The later conditions imitate those encountered by the bacteria in the host and are known to be necessary for induction of expression of the bacterial toxin. Taken together, the various secretome proteomic and serological profiles established that *B. anthracis* secretes about 70 abundant proteins which differ in their pattern of expression (see Figs. 2.2 and 2.3 for examples of these different patterns of expression) generating 3 *major* "secretion proteomic signatures": (i) rich media secretome, characterized by the presence of a large number of protein spots (>500, see Fig. 2.4A) (ii) low-nutrient/starvation stress secretome, characterized by the overwhelming (>90% of total) abundance of the protease NprA, and (iii) CO_2/bicarbonate-induced secretome, characterized by the increased abundance of about 15 proteins (besides the toxin components); amongst these, the most pronounced up-regulation is exhibited by the protease HtrA. Notably, while (i) and (ii) are not affected by the presence of the virulence plasmids, the CO_2-induced secretomes (iii) of the WT and the plasmid-cured strains differ dramatically.

2.2.2 The NprA Protease

In rich media, *B. anthracis* secretes a large number of proteins and only minor differences are observed between the Vollum and the ΔVollum strains. However,

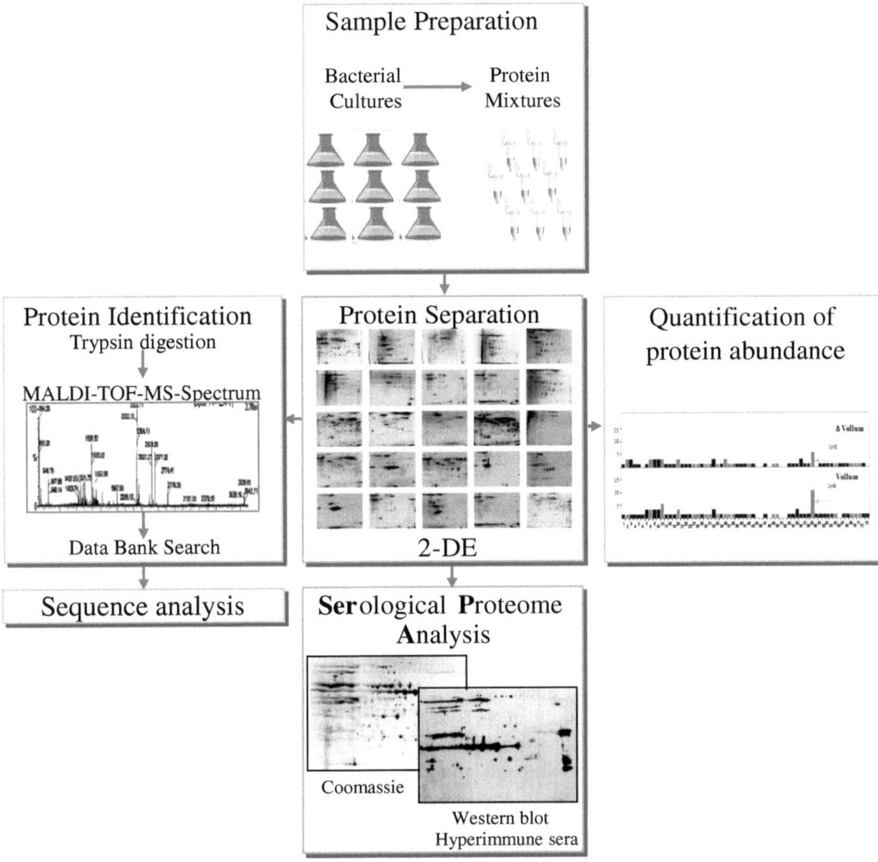

Fig. 2.1 General scheme of proteomic inspection of *B. anthracis* secretomes. Sample Preparation: Bacterial cultures of *B. anthracis* WT or attenuated (virulence-plasmid devoid) strains grown under a variety of physiological conditions served for preparation of secreted protein extracts. Protein Separation: Efficient separation of proteins was achieved by two-dimensional electrophoresis. Protein Identification: Protein spots were identified by MALDI-TOF-MS. Quantification of protein abundance: The relative abundances of individual proteins was determined by the intensity of the respective spots in the 2DE gels (see also Fig. 2.2). Serological Proteome Analysis: the proteomic dissection is complemented by serological analysis (SERPA) in which Western blots are probed with convalescent sera collected from experimental animals infected with *B. anthracis*. Comparison of the Western blots with their twin 2DE maps, enables direct identification of bacterial antigens expressed during infection and eliciting a humoral immune response. Quantification of immunogenicity: The intensity of the immunoreactivity signals generated by a protein spot on the Western blot, compared to the relative abundance of that protein in the mixture, can serve for the rough quantification of the immunogenicity of a protein (see also Fig. 2.3 for exemplification of the serological analysis)

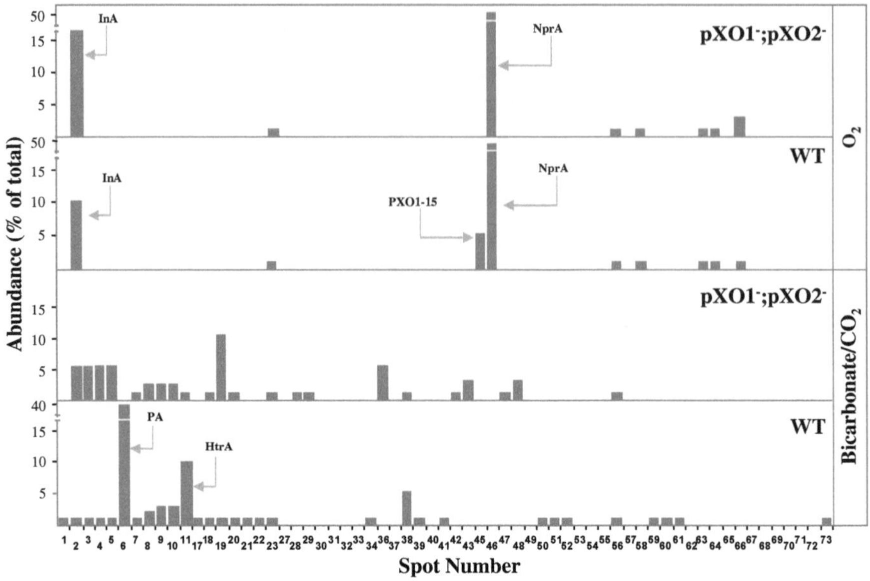

Fig. 2.2 Relative abundance of *B. anthracis* proteins in the low-nutrient secretomes of the wild-type (WT) and virulence-plasmid devoid strains. Secreted proteins were collected from cultures of the wild-type Vollum strain and from the isogenic plasmid devoid non-virulent strain, as indicated. The values represent the averaged abundance (±10%) obtained for individual identified spots obtained in at least three 2-DE duplicate gels representing each protein signature which were scanned and analyzed using the PDQuest software. Absence of spots indicates an abundance lower than 0.1% of the total protein mass. Note the similarity of the proteomic profiles of the secretomes of the two strains cultured aerobically in the low-nutrient media and the differences induced in a plasmid-dependent manner by high bicarbonate/CO_2 concentration. The spot numbers which serve for designation of the proteins were previously detailed (Chitlaru et al., 2006, 2007)

inspection of a late stationary BHI secretome (a stage at which availability of resources may be significantly low), reveals a sharp increase in the level of the protease NprA (Fig. 2.4). More dramatically, in the NBY aerobic media (which contains much lower levels of nutrients compared to the rich BHI media), unusual high amounts of this proteins are observed (Figs. 2.2 and 2.4). In NBY supplemented with CO_2-conditions which simulate the in vivo situation and induce toxin production, NprA is down-regulated (Figs. 2.2 and 2.4). NprA, which in the aerobic media is unusually high, is lowered to a non-detectable level and virtually, disappears completely from the secretome. This pattern of expression was confirmed by the observation that NprA is not immunogenic, namely is not expressed in vivo during infection, despite being highly antigenic (as established by efficient induction of specific antibodies by immunization of mice with a pure preparations of NprA, Chitlaru et al., 2006).

The induction of NprA in the NBY secretome is in line with the suggested role of this protease in the response to starvation stress, in the frame of a quorum sensing mechanism, observed in bacilli of the cereus group in which the PlcR regulatory

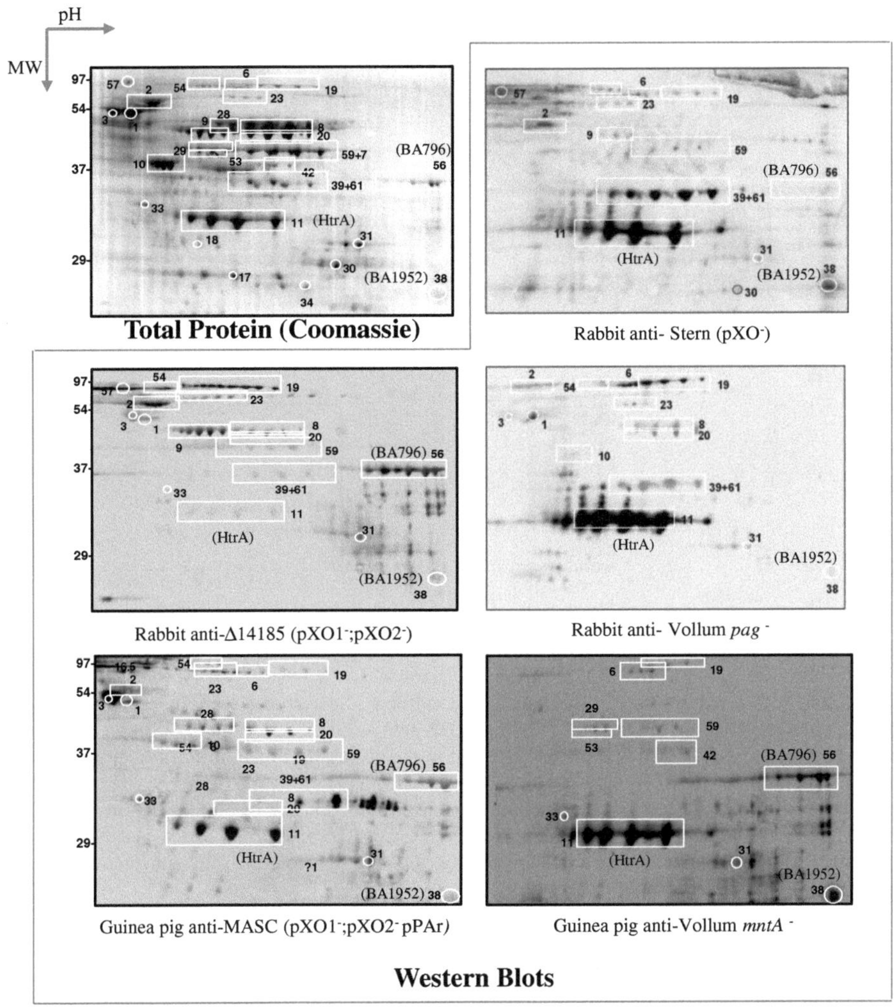

Fig. 2.3 Serologic protein analysis (SERPA) of *B. anthracis* secreted proteins. The coomassie-blue stained 2-DE gel (*upper left pane*, as indicated) of secreted proteins by the *B. anthracis* ΔVollum strain is compared to its corresponding Western blots (all other panels). Each blot was probed with a different hyperimmune sera, generated in rabbits or guinea pigs immunized with different strains of *B. anthracis* (as indicated under each panel). All procedures, details pertaining to the hyperimmune sera as well as the identity of the marked protein spots and a full list of immunogenic proteins were previously documented (Chitlaru et al., 2006, 2007; Gat et al., 2006). Such analyses were carried out with a variety of secretomes. The use of strains and/or conditions in which PA (the most dominant immunogenic product of *B. anthracis*) is absent from the secretome, such as the secretome analysed in this figure, facilitated identification of a large number of immunogenic proteins (the visualization of which is complicated by the masking effect of PA). The immunogenic proteins HtrA, BA1952 and BA796 are indicated. These proteins are potential diagnostic biomarkers owing to their presence in the circulation of infected animals (Sela-Abramovich et al., 2009)

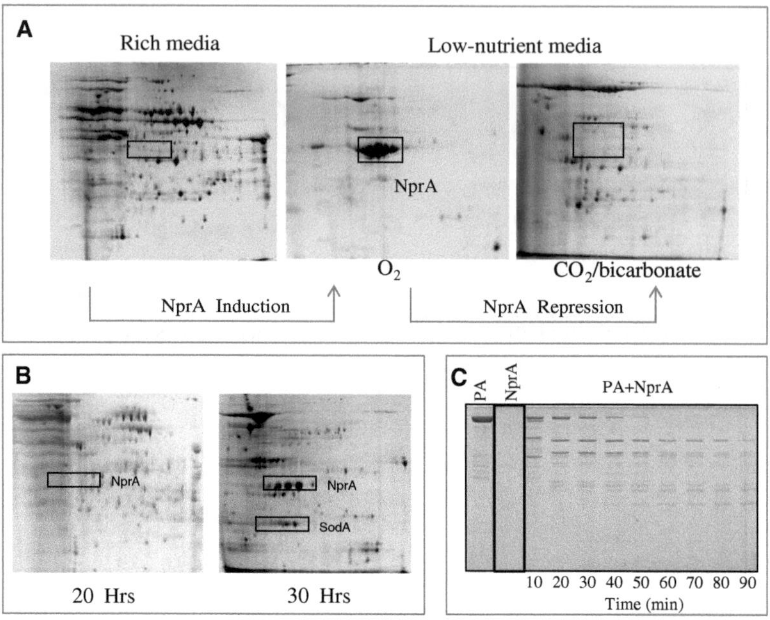

Fig. 2.4 The extracellular protease NprA in induced under starvation conditions and repressed by CO_2/bicarbonate in a pXO1/PXO2 independent manner. (**A**) 2-DE profiles of the secretomes of the Vollum strain, cultured under the indicated conditions. Secretomes were harvested 20 h post inoculation. The position of NprA migration on 2DE gels is indicated by squares. Similar induction and repression of NprA expression was observed in the secretomes of *B. anthracis* strains devoid of the virulence plasmids. (**B**) 2DE profiles of the rich media (BHI media, see Gat et al., 2003) secretome of *B. anthracis*. Note the dramatic increase in the amount of NprA. The protein SodA which is indicative of a state of stress is marked. (**C**) Pure preparations of PA and NprA were incubated for the indicated times and analysed by SDS-PAGE. Note the efficient time-dependent digestion of PA in the presence of NprA

protein is disrupted (such as the situation in *B. anthracis*, Perchat et al., 2007). It should be noted that the unusual high amounts of NprA suggest that this protease may provide additional functions, in addition to quorum sensing signaling, such as exploitation of polypeptides as a nutrient source (Chitlaru et al., 2006). While, these mechanisms may explain the induction of NprA, the dramatic silencing of the NprA expression upon addition of CO_2 to the minimal media, was puzzling and may imply that under conditions in which the anthrax related toxins are expressed, it is beneficial for the bacteria to reduce its proteolitic activity. This possibility was explored by determining the zymography profiles of the secretomes (Chitlaru et al., 2006): we observed that a much higher proteolitic activity acting both on casein and gelatin is secreted in NBY-O_2 compared to the NBY-CO_2 cultures. Interestingly, this proteolitic activity was highly efficient in a zymography test carried out on gels impregnated with PA, strongly indicating that this high proteolitic activity may be deleterious to the bacterium under conditions when its survival depends on the secretion of the toxins, such as those encountered in the host during infection. Here, we bring direct evidence to this concept by showing that pure preparations of NprA digest PA efficiently (Fig. 2.4). Digestion of PA was not observed in similar reactions carried out with pure preparations of other *B. anthracis* proteases such as HtrA or In A (Immune Inhibitor A, not shown).

The fact that the same phenomenon of sharp shut-off of NprA expression occurs equally efficient in the Vollum and ΔVollum strains, demonstrates that CO_2-signalling regulatory circuits independent of the pXO1-encoded *atxA* gene or the presence of pXO2 plasmid are active in *B. anthracis*. This possibility was suggested by Mignot et al. (2003), based on inspection of the CO_2 response of *B. anthracis* strains devoid of either virulence plasmid, yet the present study provides direct evidence that chromosomally encoded CO_2 responsive regulons are involved in determining the virulence of the bacteria.

2.2.3 The HtrA Protease

In the CO_2-supplemented media the pattern of expression of many proteins differ between the Vollum and the ΔVollum strains, demonstrating that regulatory cross-talk mechanisms link the expression of plasmid and chromosomally encoded proteins. Notably, we have recently shown that three proteins among those abundant in the CO_2-induced secretome, can be directly detected in the circulation of infected animals (Sela-Abramovich et al., 2009), in agreement with the notion that high CO_2/bicarbonate tension faithfully recapitulate the conditions encountered in the host. Furthermore, all CO_2-induced exposed proteins are strong immunogens, establishing their in-vivo expression. Amongst these proteins, the protease HtrA is of particular interest. This protein is a well established virulence factor and a potential vaccine candidate in many bacteria, together with its role in the response to stress and secretion process (Chitlaru et al., 2006; Gat et al., 2007). HtrA represents the most abundant (except PA) secreted protein of the Vollum strain in the NBY-CO_2 media (Fig. 2.2), while it is absent from the NBY-aerobic culture. Furthermore, it is one of the strongest immunogens identified in the secretome of *B. anthracis* (Chitlaru et al., 2007; Gat et al., 2007, see also Fig. 2.3). Since HtrA induction depends upon the presence of the virulence plasmids, it is concluded that it is up-regulated in a CO_2-dependent fashion by plasmid-encoded regulatory factors.

To determine whether HtrA is only an accessory to toxin secretion or is by itself involved in aspect of the pathogenesis of the *B. anthracis* infection, such as resilience to stress encountered in the host and/or proteolysis of host tissues, we have generated a mutated strain in which the HtrA gene was disrupted by targeted mutagenesis. This strain exhibits a significant decrease in its ability to withstand a wide variety of stress regimens, including secretion stress (manuscript in preparation). This inability to properly respond to stress is also manifested by the absence of the NprA protease under conditions in which the mutated strain is confronted with starvation conditions (Fig. 2.5). Most interestingly, preliminary proteomic analysis of the mutated strain, establishes that HtrA is essential in the export

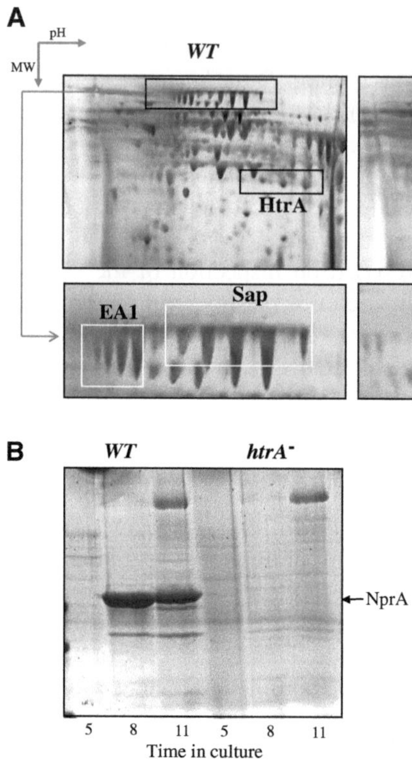

Fig. 2.5 Disruption of the *htrA* gene affects secretion of the S-layer proteins and of the protease NprA. (**a**) 2DE profiles of the rich-media secretomes of an htrA-disrupted strain and of its isogenic parental *B. anthracis* strain. The indicated gel areas spanning the 2DE position of the S-layer proteins Sap and EA1, absent from the secretome of the disrupted strain, are enlarged, as by squares. The HtrA protein spots are also indicated. (**b**) SDS-PAGE analysis of the NBY secretomes generated by the htrA-disrupted strain and its isogenic parental *B. anthracis* strain. Note the complete abrogation of NprA expression

of the S-layer proteins Sap and EA1 and in the assembly of the S-layer. These unexpected phenotypes strengthen the notion that HtrA is important for virulence manifestations.

2.3 Conclusions

In the course of an infection, the bacterial outer-cell membrane, as well as its secreted proteins represent the interphase for bacteria-host interaction and are exposed to the host immune machinery. Consequently, the bacterial membranal proteome and secreteome represent the sub-cellular fractions of choice for searching proteins possessing virulence and antigenicity potential, towards the development of novel vaccines or antibacterial drugs and/or diagnostic tools. Indeed, the proteomic study of secreted proteins was highly beneficial in identifying immunogenic proteins for further evaluation. Furthermore, the proteomic inspection was carried out in a differential manner, involving several bacterial strains, a multitude of culturing conditions, and serological analysis for determining in-vivo expression of secreted proteins; therefore, this analysis enabled dissection of bacterial regulons possibly involved in virulence manifestation. In this context, the study presented here, focusing on the extracellular proteases NprA and HtrA represents an important addition to our current understanding of virulence-related circuits of *B. anthracis*, other than those related to the classic bacterial toxins.

References

Ariel, N., Zvi, A., Grosfeld, H., Gat, O., Inbar, Y., Velan, B., Cohen, S. & Shafferman, A. (2002) Search for potential vaccine candidate ORFs in the *B. anthracis* virulence plasmid pXO1-*in silico* and in vitro screening. *Infect Immun*, 70(12), 6817–27.

Ariel, N., Zvi, A., Makarova, K., Chitlaru, T., Elhanay, E., Velan, B., Cohen, S., Friedlander, A. & Shafferman, A. (2003) Genome-based bioinformatic selection of chromosomal *B. anthracis* putative vaccine candidates coupled with proteomic identification of surface-associated antigens. *Infect Immun*, 71(8), 4563–79.

Aronson, A. I., Bell, C. & Fulroth, B. (2005) Plasmid-encoded regulator of extracellular proteases in *Bacillus anthracis*. *J Bacteriol,* 187(9), 3133–38.

Bradley, K., Mogridge, J., Mourez, M., Collier, B. & Young, J. (2001) Identification of the cellular receptor for anthrax toxin. *Nature*, 414(6860), 225–29.

Brossier, F., Levy, M. & Mock, M. (2002) Anthrax spores make an essential contribution to vaccine efficacy. *Infect Immun* 70(2), 661–64.

Cendrowsky, S., Macarthur, W. & Hanna, P. (2004) *Bacillus anthracis* requires siderophore biosynthesis for growth in macrophages and mouse virulence. *Mol Microbiol* 51(2), 407–17.

Chitlaru, T., Ariel, N., Zvi, A., Lion, M., Velan, B., Shafferman, A. & Elhanany, E. (2004) Identification of chromosomally encoded membranal polypeptides of *Bacillus anthracis* by a proteomic analysis: prevalence of proteins containing S-layer homology domains. *Proteomics*, 4(3), 677–91.

Chitlaru, T., Gat, O., Gozlan, Y., Ariel, N. & Shafferman, A. (2006) Differential proteomic analysis of the *Bacillus anthracis* secretome: distinct plasmid and chromosome CO_2-dependent cross-talk mechanisms modulate extracellular proteolytic activities. *J Bacteriol* 188(10), 3551–71.

Chitlaru, T., Gat. O., Gozlan, Y., Grosfeld, H., Inbar, I. & Shafferman, A. (2007) Identification of in vivo expressed immunogenic proteins by serological proteome analysis of *Bacillus anthracis* secretome. *Infect Immun* 75(6), 2841–52.

Chitlaru, T. & Shafferman, A. (2009) Proteomic studies of *Bacillus anthracis*. *Future Microbiol*, 4(8), 983–98.

Collier, R. J. (2009) Membrane translocation by anthrax toxin. *Mol Aspects Med*, 30, 413–22.

Fellows, P. F., Linscott, M. K., Ivins, B. E., Pitt, M. L. M., Rossi, Gibbs, P. H. & Friedlander, A. M. (2001) Efficacy of a human anthrax vaccine in guinea pigs, rabbits, and rhesus macaques against challenge by *Bacillus anthracis* isolates of diverse geographical origin. *Vaccine,* 19(23–24), 3241–7.

Fisher, N., Sheraton-Rama, L., Herring-Palmer, A., Heffernan, B., Bergman, N. & Hanna, P. (2006) The dltABCD operon of *Bacillus anthracis* Sterne is required for virulence and resistance to peptide, enzymatic, and cellular mediators of innate immunity. *J Bacteriol,* 188(4), 1301–9.

Fouet, A. & Mock, M. (2006) Regulatory networks for virulence and persistence of *Bacillus anthracis*. *Curr Opin Microbiol* 9(2), 160–6.

Fouet, A. (2009) The surface of *Bacillus anthracis*. *Mol Aspect Med*, 30, 374–85.

Friedlander, A., Welkos, S. & Ivins, B. (2002) Anthrax Vaccines. *Curr Top Microbiol Immun*, 271, 34–60.

Gat, O., Grosfeld, H., Ariel, N., Inbar, I., Zaide, G., Broder, Y., Zvi, A., Chitlaru, T., Altboum, Z., Stein, D., Cohen, S. & Shafferman, A. (2006) Search for *Bacillus anthracis* Potential Vaccine Candidates by a Functional Genomic-Serologic screen. *Infect Immun*, 74(7), 3987–4001.

Gat, O., Grosfeld, H. & Shafferman, A. (2007) In vitro screen of bioinformatically selected *Bacillus anthracis* vaccine candidates by coupled transcription, translation and immunoprecipitation analysis. In vitro transcription and translation protocols. In: Grandi, G. (ed), Methods in molecular biology, Humana, Totowa, NJ, 375, pp. 211–33.

Gat, O., Mendelson, I., Chitlaru, T., Ariel, N., Altboum, Z., Levy, H., Weiss, S., Grosfeld, H., Cohen, S. & Shafferman, A. (2005) The solute-binding component of a putative Mn(II) ABC transporter (MntA) is a novel *Bacillus anthracis* virulence determinant. *Mol Microbiol*, 58(2), 533–51.

Gat, O., Zaide, G., Inbar, I., Grosfeld, H., Chitlaru, T., Levy, H. & Shafferman, A. (2008) Characterization of *Bacillus anthracis* iron-regulated surface determinant (Isd) proteins containing NEAT domains. *Mol Microbiol* 70(4), 983–99.

Gohar M., Faegri, K., Perchat, S., Ravnum, S., Okstad, O. A., Gominet, M., Kolsto, A. B. & Lereclus, D. (2008) The PlcR virulence regulon of *Bacillus cereus*. *PLoS One* 3(7), e2793, doi. 10.1371.

Lacy, T. M., & Collier, R. J. (2002) Structure and Function of Anthrax toxin. *Curr Top Microbiol Immun*, 271, 62–85.

Leppla, S. (1982) Anthrax toxin edema factor: a bacterial adenylate cyclase that increases cyclic AMP concentrations of eukaryotic cells. *Proc Nat Acad Sci USA*, 79(10), 3162–6.

Leppla, S. H. (1995) Anthrax toxins. In Moss, J., Iglewski, B., Vaughan, M. & Tu, A. T. (ed.), Bacterial toxins and virulence factors in disease. Marcel Dekker, New York, NY, pp. 543–72.

Klimpel, K. R., Arora, N. & Leppla, S. H. (1994) Anthrax toxin lethal factor contains a zinc metalloprotease consensus sequence which is required for lethal toxin activity. *Mol Microbiol*, 13(6), 1093–100.

Koehler, T. M. (2009) *Bacillus anthracis* physiology and genetics. *Molecular Aspects of Medicine* 30(6), 386–396.

Mesnage, S., Tosi-Couture, E., Mock, M., Gounon, P. & Fouet, A. (1997) Molecular characterization of the *Bacillus anthracis* main S-layer component: evidence that it is the major cell-associated antigen. *Mol Microbiol* 23(6), 1147–55.

Mignot, T., Mock, M., Robichon, D., Landier, A., Lereclus, D. & Fouet, A. (2001). The incompatibility between the PlcR- and AtxA-controlled regulons may have selected a nonsense mutation in *Bacillus anthracis*. *Mol Microbiol*, 42(5), 1189–98.

Mignot, T., Mock, M. & Fouet, A. (2003) A plasmid-encoded regulator couples the synthesis of toxins and surface structures in *Bacillus anthracis*. *Mol Microbiol*, 47(4), 917–27.

Mock, M. & Fouet, A. (2001) Anthrax. *Ann Rev Microbiol* 55, 647-71.

Perchat, S., Gominet, M., Ramarao, N., Nielsen-Leroux, C., Gohar, M. & Lereclus, D. (2007) Characterization of a new quorum-sensing system in the *Bacillus cereus* group. *Bacillus-ACT 2007*, International Conference on *B. anthracis*, *B. cereus* and *B. thuringiensis*, Oslo, Norway.

Perego, M. & HOCH, J. A. (2008) Commingling regulatory systems following acquisition of virulence plasmids by *Bacillus anthracis*. *Trends Microbiol*, 16(5), 215–21.

Sela-Abramovich, S., Chitlaru, T., Gat, O., Grosfeld, H., Cohen, O. & Shafferman, A. (2009) Novel and Unique Diagnostic Biomarkers for *Bacillus anthracis* Infection. *Appl Environl Microbiol* , 75(19), 6157–67.

Wilson, A. C., Soyer, M., Hoch, J. A. & Perego, M. (2008) The bicarbonate transporter is essential for *Bacillus anthracis* lethality. *PLoS Pathogens* 4(11), e1000210.

Chapter 3
Dynamics of *Bacillus anthracis* Infection and Control Mechanisms

Pierre L. Goossens, Ian J. Glomski, Lhousseine Touqui, and Michèle Mock

Abstract The pathogenicity of *Bacillus anthracis*, a gram-positive spore-forming bacterium, mainly depends on two plasmid-encoded virulence factors: a poly-gamma-D-glutamate capsule and two toxins. Bioluminescence imaging was used to follow in real-time *B. anthracis* population dynamics during cutaneous and inhalational infection with bacteria expressing either toxins or capsule. *B. anthracis* spores were shown to germinate and establish infection at the initial site of inoculation in both routes of infection. Encapsulated *B. anthracis* then progress to the draining lymph node, spleen, lung, and ultimately the blood. In contrast, toxinogenic non-encapsulated bacteria are initially confined for a prolonged period to the initial site of infection, and then progress to the draining lymph node, and late in the infection, to the kidneys, and frequently the gastrointestinal tract; there is minimal colonization of the spleen. Bioluminescence imaging was also applied to real-time visualization of the effects of immunisation with toxin. We identified the group IIA secretory phospholipase A_2 (sPLA$_2$-IIA) as a key anthracidal effector of host innate defense. Transgenic mice expressing human sPLA$_2$-IIA are protected against anthrax and in vivo administration of sPLA$_2$-IIA significantly protects sPLA$_2$-IIA-deficient mice against *B. anthracis* infection. The potential of this enzyme as a natural therapeutic agent in adjunction to current therapy for treating anthrax is discussed.

Keywords Anthrax · Capsule · Group IIA secretory phospholipase A_2 · Innate immunity · Toxins

P.L. Goossens (✉)
Unité des Toxines et Pathogénie Bactérienne, URA 2172 CNRS, Institut Pasteur, 75015 Paris, France
e-mail: pierre.goossens@pasteur.fr

A. Shafferman et al. (eds.), *The Challenge of Highly Pathogenic Microorganisms*,
DOI 10.1007/978-90-481-9054-6_3, © Springer Science+Business Media B.V. 2010

3.1 Virulence Factors

Bacillus anthracis, the etiological agent of anthrax, is a Gram-positive, spore-forming rod-shaped bacterium. Dormant spores are highly resistant to adverse environmental conditions and they are able to survive for long periods in contaminated soils. In a suitable environment spores reestablish vegetative growth (Mock and Fouet, 2001).

Anthrax is primarily a disease of herbivores, but all mammals including humans are susceptible. The disease is initiated by the entry of spores into the host body. This can occur via a minor abrasion, an insect bite, or by eating contaminated meat or inhaling airborne spores. There are three types of human infection: cutaneous, gastrointestinal, and inhalational (Mock and Fouet, 2001). Each form can progress to fatal systemic anthrax.

Anthrax is a toxi-infection: an association of toxemia and rapidly spreading infection progressing to septicemia. The pathogenicity of *B. anthracis* mainly depends on two plasmid-encoded virulence factors: two toxins and a poly-gamma-*D*-glutamate capsule, anchored to the cell wall (Candela and Fouet, 2005) which protects bacilli from the immune system, thus promoting systemic dissemination (Candela and Fouet, 2006).

The toxins are composed of three proteins: edema factor (EF), lethal factor (LF), and protective antigen (PA). PA, the common receptor-binding domain, mediates the entry of EF and LF into the target cells. The combination of PA+EF forms the edema toxin (ET). EF is a calmodulin-dependent adenylate cyclase that increases the intracellular concentration of cyclic AMP. The lethal toxin (LT) is composed of PA+LF. LF is a zinc-binding metalloprotease that cleaves mitogen-activated protein kinase (MAPK) kinases (Mock and Fouet, 2001). These toxins alter host cell signalling, thereby modulating the immune response of the host, and can cause death (Tournier et al., 2007).

Our recent studies show that, in addition to its recognized antiphagocytic and non-immunogenic properties, the capsule mediates close interactions between the encapsulated bacteria and liver endothelium, leading to retention of the bacteria in the liver (Piris-Gimenez et al., 2009). The mechanism underlying this interaction is extremely efficient, as shown by the rapid clearance of the encapsulated bacteria from the blood mainly to the liver and the spleen. It was shown that the bacteria remained in the vascular spaces closely associated with the vascular endothelium and they formed microcolonies containing capsule polyglutamate (Fig. 3.1). This substance could play a role in protecting these bacteria against potentially bactericidal cellular or soluble components of the immune system (Keppie et al., 1953; Scorpio et al., 2007), or in impeding antibiotic diffusion during therapeutic intervention. It could also provide a favorable microenvironment for the multiplying bacteria by retaining salts and nutrients.

These results indicate that, in addition to its inhibitory effect on the interaction with the immune system, the capsule surrounding *B. anthracis* plays an active role in mediating the trapping of bacteria within the liver. In wild-type strains of *B. anthracis*, which are both encapsulated and toxin-producing, such

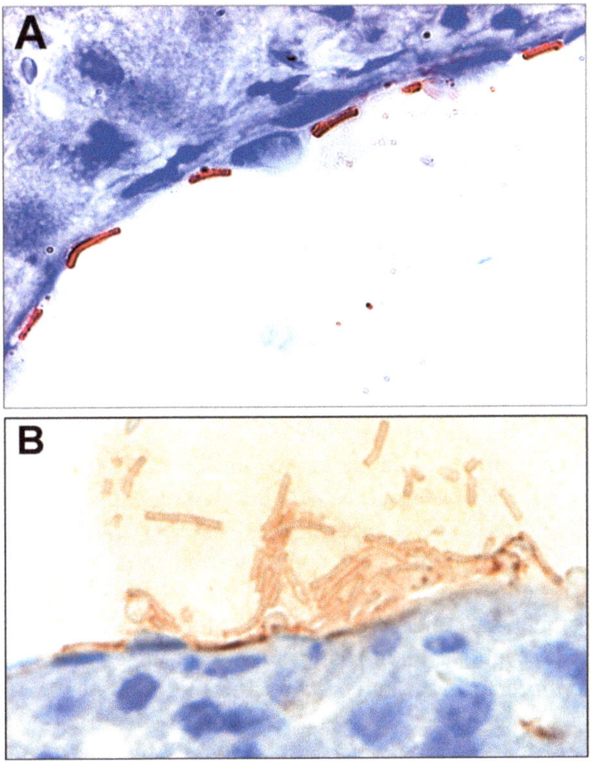

Fig. 3.1 Adhesion of encapsulated *B. anthracis* to the liver endothelium (*Gram staining*) (**a**) and formation of microcolonies imbedded in capsular material (*immunostaining with antipolyglutamate antibodies*) (**b**)

capsule-mediated focalization of toxin secretion would be expected to increase the local concentration of toxins, leading to targeted toxic activity and organ dysfunction. This interaction thus represents a new mechanism by which *B. anthracis* colonizes its host, potentially playing a role in the pathogenesis of anthrax through the localized secretion of toxins or toxic components.

3.2 Interaction with the Host

3.2.1 In Vivo Real Time Monitoring of the Infection

A successful pathogen must overcome a number of host defenses to establish a productive infection. Each stage of the infectious process presents different challenges that may restrict the ability of the bacterium to grow. No pathophysiological studies have yet established precisely how *B. anthracis* infection progresses from one corporal compartment to another and the respective role of each major virulence factor, i.e. toxins and capsule.

The inhalational route and cutaneous inoculation have been extensively used for investigations in animal models. To follow *B. anthracis* population dynamics during infection, we used bioluminescent bacteria and the corresponding imaging technology. The *lux* operon, fused to the promoter of the *pagA* gene encoding PA (a promoter highly expressed in *B. anthracis* in vivo), was introduced into *B. anthracis* strains with various genetic backgrounds (Glomski et al., 2007a, b). The bioluminescent bacteria produce photons, and their location and progression through the host is monitored with an ultra-sensitive digital camera. This powerful technology allows a single mouse to be monitored throughout the course of infection in real-time and eliminates the need to kill mice at each time point.

The bioluminescence imaging was used to follow *B. anthracis* population dynamics during cutaneous and inhalational infection with bacteria expressing either toxins or capsule to analyse their relative contribution in bacterial growth and dissemination characteristics.

We first followed the infection with an encapsulated non-toxinogenic ($Cap^{+}Tox^{-}$) strain to understand how *B. anthracis* interacts *per se*, without the modulating effects of the toxins (Glomski et al., 2007a). We observed that *B. anthracis* spores germinate and establish infections at the initial site of inoculation in both inhalational and cutaneous (in the mouse ear) infections without needing to be transported to draining lymph nodes, and that inhaled spores establish the initial infection in nasal-associated lymphoid tissues. All routes of infection progress to the draining lymph node, spleen, lung, and the blood, leading to death in 48–72 h.

In contrast, toxinogenic non-encapsulated (Tox^{+}, Cap^{-}) bacteria are initially confined for a prolonged period to the initial site of infection (for 72–96 h in the cutaneous model), and then progress to the draining lymph node, and late in the infection, to the kidneys, and frequently the gastrointestinal tract; there is minimal colonization of the spleen (Fig. 3.2b) (Glomski et al., 2007b).

Real-time in vivo detection of bioluminescent *B. anthracis* spread has thus allowed the detection of previously unrecognized portals of bacterial entry and of marked differences in dissemination pattern and dynamics between non-encapsulated and encapsulated strains.

The bioluminescence imaging was also applied to real-time visualization of the effects of immunisation. PA-based vaccines protect mice from toxinogenic non-encapsulated *B. anthracis* infection (Welkos and Friedlander, 1988; Mock and Fouet, 2001), however, where and when adaptive immunity exerts its protective function in vivo remains poorly defined.

Therefore PA-immunized A/J mice or control mice, were infected cutaneously with the bioluminescent Tox^{+}, Cap^{-} strain. Within 8 h, luminescence in the PA-immunized mice was detected at the site of injection in the ear, remained at a plateau until 32 h, and then diminished in intensity until no longer detectable. Luminescence was not detected beyond the ear in PA-immunized mice. In contrast, luminescence in control mice initiated by 8 h at the site of infection, intensified until death and spread to deeper tissues as described above.

These data imply that PA-immunization limits bacterial growth to the site of inoculation and prevented bacterial dissemination in other tissues, implying that immunity is immediately functional in situ. The humoral immune response through neutralization of toxin activity is considered to be the primary effector of PA-based

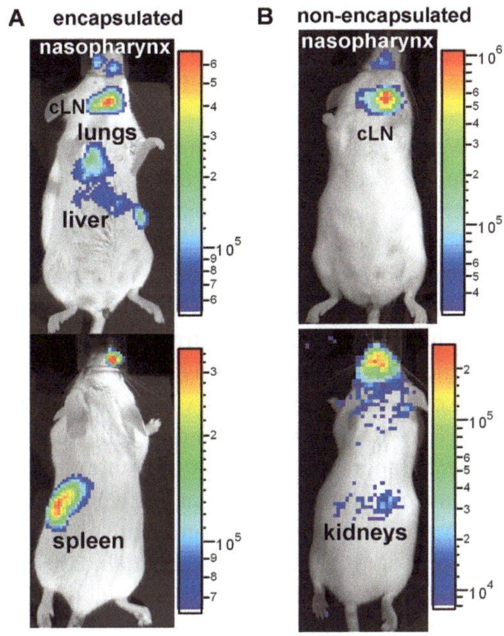

Fig. 3.2 Dissemination pattern of encapsulated (**a**) and non-encapsulated (**b**) *B. anthracis* during inhalational infection. After initial multiplication in the nasopharynx, bacilli disseminate to the draining cervical lymph node (cLN), before colonising (**a**) the spleen, liver and lungs for encapsulated bacteria, or (**b**) kidneys for non-encapsulated bacteria

anti-anthrax immunity (Leppla et al., 2002). This observation thus supports the conclusion that preformed effectors, such as antibodies or PA-reactive CD4 T lymphocytes (Laughlin et al., 2007), are acting to restrict early bacterial growth.

Another successful application of the bioluminescence imaging was in testing the potential therapeutic effects of new bactericidal compounds. No luminescence was detected at the site of injection in the ear in mice treated with the compound and surviving the infection.

3.2.2 Host Control Mechanisms and Subversion

Anthrax is an acute disease, so adaptive immunity cannot be induced rapidly enough to protect naive hosts. Innate immunity is thus the first line of defense for controlling *B. anthracis* infection at an early stage in non-immunised hosts. Little is known about the innate immune response triggered upon infection by *B. anthracis* spores.

The mortality rate of cutaneous anthrax is much lower than those of inhalatory and gastrointestinal anthrax, showing that control mechanisms exist. Our knowledge of host control mechanisms and their subversion by *B. anthracis* has greatly increased over the last few years.

3.2.2.1 Anthracidal Activity and Therapeutic Use of sPLA$_2$-IIA

Substances with bactericidal activity (anthracidal) against *B. anthracis* have been partially purified from anthrax cutaneous lesions in the 1950s. The first-line effectors of the innate immune system that may be involved in this bactericidal activity include group IIA secreted phospholipase A_2 (sPLA$_2$-IIA), an antibacterial acute-phase protein produced in many tissues and secreted by inflammatory cells such as macrophages;

We have demonstrated that sPLA$_2$-IIA is highly bactericidal for *B. anthracis* through hydrolysis of the bacterial membrane phospholipids. This anthracidal effect occurs both against germinated spores and encapsulated bacilli (Piris-Gimenez et al., 2004).

Using in vivo models, we demonstrated that sPLA$_2$-IIA effectively controls *B. anthracis* infection (Piris-Gimenez et al., 2005). C57BL/6 mice are naturally deficient in sPLA$_2$-IIA; administration of human recombinant sPLA$_2$-IIA to these mice significantly protects them against *B. anthracis* infection. Furthermore, transgenic mice expressing human sPLA$_2$-IIA are fully resistant to infection with *B. anthracis*.

sPLA$_2$-IIA is thus a potential therapeutic agent and could be combined with the currently approved treatment of anthrax. sPLA$_2$-IIA treatment appears particularly relevant as we have shown that the two *B. anthracis* toxins inhibit the transcription of sPLA$_2$-IIA in macrophages and thus subvert the host innate immune response (Piris-Gimenez et al., 2004; Raymond et al., 2007, 2009). Delivery of exogenous sPLA$_2$-IIA as an adjunct to therapy should thus overcome this potentially deleterious effect and complement any local and temporary deficiency of sPLA$_2$-IIA that may be a risk factor favoring the development of anthrax. Moreover, a sPLA$_2$-IIA treatment could be particularly valuable because the bactericidal efficacy is independent of any antibiotic resistance expressed by *B. anthracis*.

3.2.2.2 Other Control Mechanisms

Spores have the ability to stimulate the innate and adaptive immune responses. Elucidation of these mechanisms and their use could pave the way to enhancing the host immune response and better control bacterial development. We have thus shown that immunisation with spores leads to the induction of protective T lymphocytes of the CD4 subset that secrete gamma-interferon (Glomski et al., 2007c).

References

Candela, T. & Fouet, A. (2005) *Bacillus anthracis* CapD, belonging to the gamma-glutamyltranspeptidase family, is required for the covalent anchoring of capsule to peptidoglycan. *Mol Microbiol,* 57, 71726.

Candela, T. & Fouet, A. (2006) Poly-gamma-glutamate in bacteria. *Mol Microbiol,* 60, 1091–8.

Glomski, I. J., Piris-Gimenez, A., Huerre, M., Mock, M. & Goossens, P. L. (2007a) Primary involvement of pharynx and peyer's patch in inhalational and intestinal anthrax. *PLoS Pathog,* 3, e76.

Glomski, I. J., Corre, J. P., Mock, M. & Goossens, P. L. (2007b) Noncapsulated toxinogenic *Bacillus anthracis* presents a specific growth and dissemination pattern in naive and protective antigen-immune mice. *Infect Immun,* 75, 4754–61.

Glomski, I. J., Corre, J. P., Mock, M. & Goossens, P. L. (2007c) Cutting edge: IFN-gamma-producing CD4 T lymphocytes mediate spore-induced immunity to capsulated *Bacillus anthracis*. *J Immunol,* 178, 2646–50.

Keppie, J., Smith, H. & Harris-Smith, P. W. (1953) The chemical basis of the virulence of *Bacillus anthracis*. II. Some biological properties of bacterial products. *Br J Exp Pathol,* 34, 486–96.

Laughlin, E. M., Miller, J. D., James, E., Fillos, D., Ibegbu, C. C., Mittler, R. S., Akondy, R., Kwok, W., Ahmed, R. & Nepom, G. (2007) Antigen-specific CD4+ T cells recognize epitopes of protective antigen following vaccination with an anthrax vaccine. *Infect Immun,* 75, 1852–60.

Leppla, S. H., Robbins, J. B., Schneerson, R. & Shiloach, J. (2002) Development of an improved vaccine for anthrax. *J Clin Invest,* 110, 141–4.

Mock, M. & Fouet, A. (2001) Anthrax. *Annu Rev Microbiol,* 55, 647–71.

Piris-Giménez, A., Corre, J. P., Jouvion, G., Candela, T., Khun, H. & Goossens, P. L. (2009) Encapsulated *Bacillus anthracis* interacts closely with liver endothelium. *J Infect Dis,* 200, 1381–9.

Piris-Giménez, A., Paya, M., Lambeau, G., Chignard, M., Mock, M., Touqui, L. & Goossens, P. L. (2005). *In vivo* protective role of human group IIA phospholipase A_2 against experimental anthrax. *J Immunol,* 175, 6786–91.

Piris-Gimenez, A., Wu, Y. Z., Paya, M., Delclaux, C., Touqui, L. & Goossens, P. L. (2004) High bactericidal efficiency of type IIA phospholipase A_2 against *Bacillus anthracis* and inhibition of its secretion by the lethal toxin. *J Immunol,* 173, 521–30.

Raymond, B., Leduc, D., Ravaux, L., Le Goffic, R., Candela, T., Raymondjean, M., Goossens, P. L., & Touqui, L. (2007) Edema toxins impairs anthracidal phospholipase A2 expression by alveolar macrophages. *PLoS Pathog,* 3, e187.

Raymond, B., Ravaux, L., Mémet, S., Wu, Y. Z., Sturny-Leclère, A., Leduc, D., Denoyelle, C., Goossens, P. L., Payá, M., Raymondjean, M. & Touqui, L. (2009) Anthrax lethal toxin down-regulates type-IIA secreted phospholipase A_2 expression through MAPK/NF-kappaB inactivation. *Biochem Pharmacol,* http://dx.doi.org/10.1016/j.bcp.2009.11.023

Scorpio, A., Chabot, D. J., Day, W. A. et al. (2007) Poly-gamma-glutamate capsule-degrading enzyme treatment enhances phagocytosis and killing of encapsulated *Bacillus anthracis*. *Antimicrob Agents Chemother,* 51, 215–22.

Tournier, J. N., Quesnel-Hellmann, A., Cleret, A. & Vidal, D. R. (2007) Contribution of toxins to the pathogenesis of inhalational anthrax. *Cell Microbiol* 9, 555–65.

Welkos, S. L. & Friedlander, A. M. (1988) Comparative safety and efficacy against *Bacillus anthracis* of protective antigen and live vaccines in mice. *Microb Pathog,* 5, 127–39.

Chapter 4
Survival and Trafficking of *Yersinia pestis* in Non-acidified Phagosomes in Murine Macrophages

James B. Bliska

Abstract *Yersinia pestis* is a facultative intracellular bacterial pathogen. Survival of *Y. pestis* in macrophages is thought to play an important role in pathogenesis during the earliest stages of plague. Recent studies have identified several bacterial genes important for survival in macrophages, and determined that *Y. pestis* inhabits a compartment called the *Yersinia*-containing vacuole (YCV), which acquires markers of late endosomes or lysosomes. Furthermore, studies have shown the ability of *Y. pestis* to survive in macrophages activated with the cytokine IFNγ. Some vacuolar pathogens appear to co-opt the process of autophagy for survival in host cells. Alternatively, xenophagy is an autophagic process that is upregulated in activated macrophages and functions to kill bacteria in acidic autophagosomes. Studies were undertaken to investigate the mechanism of *Y. pestis* survival in phagosomes of naïve and activated macrophages, and to determine if the pathogen avoids or co-opts autophagy. Co-localization of the YCV with markers of autophagosomes or acidic lysosomes, and the pH of the YCV, was determined by microscopic imaging of infected macrophages. Results showed that YCVs could contain double membranes characteristic of autophagosomes and co-localized with a marker of autophagic membranes. Interestingly, YCVs failed to acidify below pH of 7. In addition, *Y. pestis* survived equally well in macrophages proficient or deficient for autophagy, showing that the bacterium does not co-opt autophagy for intracellular survival. It is concluded that although *Y. pestis* can reside in autophagosomes, the pathogen avoids destruction by xenophagy by preventing vacuole acidification. *Y. pestis* may actively prevent phagosome acidification in either naïve or activated macrophages by preventing delivery of the vacuolar ATPase (vATPase) to the YCV or direct inactivation of the vATPase.

Keywords *Yersinia pestis* · Macrophage · Phagosome · Autophagy · Vacuolar ATPase

---

J.B. Bliska (✉)
Center for Infectious Diseases and Department of Molecular Genetics and Microbiology, Stony Brook University, Stony Brook, NY 11794, USA
e-mail: jbliska@ms.cc.sunysb.edu

A. Shafferman et al. (eds.), *The Challenge of Highly Pathogenic Microorganisms*,
DOI 10.1007/978-90-481-9054-6_4, © Springer Science+Business Media B.V. 2010

4.1 Introduction

4.1.1 Y. pestis *Pathogenesis*

The agent of plague, *Y. pestis,* is a Gram-negative bacterium in the family Enterobaceriaceae. Plague in humans can manifest in three different forms: bubonic, pneumonic or septicemic (Perry and Fetherston, 1997; Prentice and Rahalison, 2007). *Y. pestis* infections are most commonly transmitted to humans by infected fleas, and these infections typically develop into bubonic plague, or less frequently into septicemic plague (Perry and Fetherston, 1997; Prentice and Rahalison, 2007). Contact with bodily fluids or inhalation of respiratory droplets from infected animals or humans can also result in plague infections in humans. Primary pneumonic plague is initiated following inhalation of droplets bearing *Y. pestis* into the lungs. Pneumonic plague follows a more rapid course and has a higher probability of mortality than bubonic or septicemic plague.

The Centers for Disease Control and Prevention has classified *Y. pestis* as a category A agent due to characteristics that would facilitate its use as a biological weapon (http://www.bt.cdc.gov/agent/plague/). These characteristics include the possibility for dissemination from human to human in respiratory droplets and the high mortality rate of pneumonic plague (Inglesby et al., 2000). In addition, naturally occurring isolates of *Y. pestis* are readily accessible to humans since zoonotic foci of plague exist in many parts of the world, including North America. Frontline antibiotics, such as aminoglycosides, can significantly reduce mortality of pneumonic plague if treatment is initiated within 20 hr of the onset of disease (Inglesby et al., 2000). Although strains of *Y. pestis* encountered in natural environments are generally sensitive to frontline antibiotics, antibiotic resistant strains of *Y. pestis* have been identified (Prentice and Rahalison, 2007). In addition, there is no safe and effective vaccine currently available that can prevent the pneumonic form of the disease (Smiley, 2008; Titball and Williamson, 2004). Therefore, to provide a foundation for the development of new strategies to prevent or treat pneumonic plague, it is important to increase our understanding of plague pathogenesis.

A number of studies have documented that *Y. pestis* can survive and replicate in murine macrophages in vitro (Cavanaugh and Randall, 1959; Charnetzky and Shuford, 1985; Janssen and Surgalla, 1968; Pujol and Bliska, 2003; Pujol et al., 2005, 2009; Straley and Harmon, 1984a; Noel et al., 2009; Grabenstein et al., 2006). *Y. pestis* has been detected by microscopy inside phagocytes in tissues of experimentally infected animals (Finegold, 1969; Meyer, 1950; Welkos et al., 2002). It has also been determined using the technique of flow cytometry that significant numbers of *Y. pestis* bacteria can exist inside spleen macrophages in mice over the first several days of infection (Lukaszewski et al., 2005). However, the primary niche for replication of *Y. pestis* in tissues is extracellular, especially at later stages of infection (Lathem et al., 2005; Sebbane et al., 2005; Welkos et al., 1997).

Y. pestis synthesizes a large number of virulence factors that are important for resistance to phagocytosis (antiphagocytosis) and for extracellular growth (Perry and Fetherston, 1997; Prentice and Rahalison, 2007). These include factors encoded on the plasmids pCD1, pPCP1, pMT1 and the chromosomal *pgm* locus (Perry and Fetherston, 1997; Carniel, 2002). Extended growth at 37°C provides an environmental signal for upregulation of factors important for extracellular growth. For example, expression of a type III secretion system (T3SS) encoded on the plasmid pCD1 and the virulence factors secreted by this pathway, *Yersinia* Outer Proteins (Yops), is upregulated at 37°C (Viboud and Bliska, 2005). Additionally, a pilus surface appendage composed of the F1 protein is maximally expressed after extended growth at 37C and promotes resistance to phagocytosis (Du et al., 2002; Runco et al., 2008). Because the T3SS and F1 are not highly expressed when *Y. pestis* is grown at ambient temperatures (e.g. 26–28°C) the bacteria are efficiently internalized by macrophages or neutrophils (Cavanaugh and Randall, 1959; Pujol and Bliska, 2005). The bacteria are sensitive to killing within neutrophils but are able to proliferate within macrophages (Cavanaugh and Randall, 1959; Pujol and Bliska, 2005). These results indicate that factors sufficient for survival of *Y. pestis* in macrophages are synthesized when the bacteria are grown at ambient temperatures. Thus, when *Y. pestis* is introduced into a host following flea-bite or inhalation of aerosol droplets, it may be internalized by, and transiently survive within, resident tissue macrophages (Cavanaugh and Randall, 1959; Pujol and Bliska, 2005). Infected macrophages would in essence protect the pathogen from bactericidal neutrophils recruited to the locus of infection. In addition, anti-phagocytic determinants of *Y. pestis* may be upregulated within the intracellular environment of the macrophage (Cavanaugh and Randall, 1959). *Y. pestis* may be able to escape from macrophages without killing them or induce their own release by induction of cell death (Cavanaugh and Randall, 1959). Upon leaving the macrophage *Y. pestis* would be conditioned for extracellular growth and primed for resistance to uptake by neutrophils (Cavanaugh and Randall, 1959).

The above "early transient intracellular phase" model of plague pathogenesis as originally proposed by Cavanaugh and Randall (Cavanaugh and Randall, 1959) and Burrows and Bacon (Burrows and Bacon, 1956) was originally invoked with respect to flea-borne *Y. pestis,* infection. However, Sebanne et al. (Sebbane et al., 2005) failed to detect intracellular forms of the bacteria in rat lypmphatic vessels or primary lymph nodes draining the site of intradermal inoculation with *Y. pestis.* Moreover, recent studies indicate that *Y. pestis* bacteria deposited during actual flea-bite may be inherently resistant to phagocytosis (Jarrett et al., 2004; Hinnebusch, 2005;, Hinnebusch and Erickson, 2008). This is because during a blood meal some fleas appear to deposit large fragments of *Y. pestis*-containing biofilms into the dermis (Jarrett et al., 2004; Hinnebusch, 2005; Hinnebusch and Erickson, 2008). Large biofilm fragments may be difficult for macrophages to internalize. In summary, the extent to which *Y. pestis* exists within macrophages during infection in vivo may depend on the nature of the host, the route of inoculation, the temporal stage of the infection, as well as the specific tissue infected.

4.1.2 Role of Macrophages in Innate and Adaptive Host Immunity

Macrophages are important cells of the innate immune system and their functions can influence development of adaptive immunity (Adams and Hamilton, 1984;

Aderem and Underhill, 1999). A major function of macrophages is to internalize and kill invading pathogens. Macrophages express several surface receptors that enable them to bind and internalize a variety of foreign particles. Resident macrophages are present within many different tissues (e.g. alveolar macrophages in the lung). In addition to functioning as professional phagocytes, macrophage can coordinate the ensuing immune response by expression of cytokines and chemokines. The inherent ability of naive macrophages to internalize and kill invading microorganisms is greatly enhanced by activating stimuli, for example the cytokine IFNγ. Macrophages can also function as antigen presenting cells, although less efficiently than dendritic cells.

4.1.3 Trafficking and Acidification of Phagosomes in Macrophages

Following internalization of a bacterium into a macrophage it initially resides within a nacent membrane bound phagosome (Vieira et al., 2002; Meresse et al., 1999). The phagosome is not a static organelle, but one that matures rapidly, through a series of fusion events with endocytic vesicles, into a less habitable compartment termed a phagolysosome (Vieira et al., 2002; Meresse et al., 1999). Phagosomes fuse with early endosomes within 2–5 min and with late endosomes or lysosomes within 10–30 min of their formation (Vieira et al., 2002). A number of markers including lysosome associated membrane proteins (LAMPs) and lysosomal proteases (cathepsin B, D and L) are characteristic of late phagosomes or phagolysosomes. De novo lysosomal components such as cathepsin D are delivered into late phagosomes from Golgi-derived vesicles. In phagolysosomes proteases such as cathepsin D become activated. The pH of the phagolysosomes is significantly lower (~pH4.5) than the pH found in late endosomes (pH5.5–6) due to the action of the vacuolar proton ATPase (vATPase). Recent data suggests that the vATPase, in addition to its role in decreasing phagosome acidity, plays a role in regulating phagosome trafficking. For instance, it appears that phagosomes do not efficiently fuse with lysosomes until the phagosomal environment is partially acidified (Vieira et al., 2002; Huynh and Grinstein, 2007).

4.1.4 Bacterial Strategies for Surviving in Phagosomes

Many species of pathogenic bacteria in addition to *Y. pestis* subvert macrophage function in order to survive and replicate within these cells. Examples include *Brucella, Legionella, Mycobacterium*, and *Salmonella*. These bacteria remain within the vacuole and modify phagosome maturation to prevent destruction. Pathogenic bacteria that survive within phagosomes of macrophages have evolved a number of different strategies to prevent phagosome maturation (Duclos and Desjardins, 2000; Meresse et al., 1999). One of the best-studied examples is *Salmonella enterica* serovar Typhimurium (*S.* Typhimurium). *S.* Typhimurium resides in a macrophage vacuole (the *Salmonella*-containing vacuole (SCV)) that appears to initially fuse with lysosomes (Cuellar-Mata et al., 2002; Oh et al., 1996). The SCV rapidly acquires the late endosomal/lysosomal markers Lamp1 and Lamp2 (Knodler and Steele-Mortimer, 2003; Brumell and Grinstein, 2004). The vATPase is delivered to the SCV, causing the pH within to decrease (Cuellar-Mata et al., 2002). However, within several hours, the SCV is modified into a specialized compartment that is secluded from interaction with lysosomes (Knodler and Steele-Mortimer, 2003; Brumell and Grinstein, 2004). *S.* Typhimurium encodes a T3SS on its chromosome, within *Salmonella* pathogenicity island-2 (SPI-2). The SPI-2 T3SS is required for survival and replication of *S.* Typhimurium in macrophages (Ochman et al., 1996; Hensel et al., 1998). A multitude of effectors of the SPI-2 system have been identified (Knodler and Steele-Mortimer, 2003; Brumell and Grinstein, 2004). Precisely how the effectors of the SPI-2 T3SS interfere with phagosome maturation is only partially understood and is currently an area of intense investigation (Knodler and Steele-Mortimer, 2003; Brumell and Grinstein, 2004).

M. tuberculosis survives within macrophages by stalling maturation of its vacuole at an early stage. The *Mycobacterium*-containing phagosome (MCP) is characterized by the absence of a number of markers that usually associate with late endosomes (Vergne et al., 2004; Rohde et al., 2007). For example, *M. tuberculosis* inhibits acidification of its vacuole by preventing the delivery of vATPase to this compartment. Mature lysosomal proteases and Lamp1 are also excluded from the MCP. Rab5, a small GTPase found on early endosomes, but not late endosomes, is retained by the MCP. Rab7 on the other hand, a small GTPase present on late endosomes, is not detected at the times expected for normal recruitment to maturing phagosomes. Recent publications have presented evidence for the involvement of two mycobacterial lipids in the phagosome maturation arrest and the altered trafficking of the mycobacterial phagosome (Chua et al., 2004). In addition, current work is focusing on the role of a specialized secretion system (ESX-1) (DiGiuseppe Champion and Cox, 2007) in allowing *M. tuberculosis* to inhibit phagosome maturation.

4.1.5 Role of IFNγ in Upregulating the Microbicidal Activities of Macrophages Through NO Production and Autophagy

IFNg, which is secreted by activated natural killer (NK) cells or T lymphocytes, can stimulate increased bactericidal activity in macrophages. Exposure to IFNγ increases expression of the inducible nitric oxide synthase (iNOS) in macrophages. The NO produced by iNOS plays an important role in decreasing survival of intracellular pathogens, including *M. tuberculosis*, in macrophages (Shenoy et al., 2007). IFNγ activated macrophages also exhibit increased rates of macroautophagy (hereafter referred to as autophagy, i.e. "self-eating"), a membrane trafficking process in eukaryotic cells that surrounds cytoplasmic material (e.g. defective organelles) and seals it in a vacuole (the autophagosome). Cargo delivered to autophagosomes is routed for destruction in an autolysosome (Klionsky et al., 2008; Ogawa and Sasakawa, 2006; Levine and Deretic, 2007). Several recent studies have shown that autophagy plays an key role in a protective innate immune response to intracellular pathogens (Ogawa and Sasakawa, 2006; Levine and Deretic, 2007). In the case of *M. tuberculosis*, upregulation of autophagy by IFNγ overrides the MCP maturation block, allowing for delivery of the pathogen to a destructive autolysosome-like phagosome (Ogawa and Sasakawa, 2006; Levine and Deretic, 2007). This process has been termed xenophagy ("foreign-eating") (Levine and Deretic, 2007). Conversely, the process of autophagy has been co-opted by other intracellular bacteria to enhance survival within host cells (Ogawa and Sasakawa, 2006; Mizushima et al., 2008).

The process of autophagy involves a stepwise pathway of membrane movement, enclosure and maturation. The process begins with the phagophore, a crescent-shaped membrane that engulphs cytoplasmic material, or in the case of xenophagy, a phagosome containing a vacuolar pathogen. Following sealing of the phagophore, an autophagosome is formed, which matures through fusions with late endosome and lysosome compartments into an autophagolysosome, where degradation of the organelle or pathogen takes place. The rate of autophagy is regulated by IFNγ as well as by at least two types of signaling enzymes, the mTOR kinase and class I and class II PI3 kinases (Klionsky et al., 2008; Ogawa and Sasakawa, 2006; Levine and Deretic, 2007). The progress of the autophagy pathway is executed by the ATG proteins, which comprise two protein conjugation systems. For example, the ATG5 protein is essential for the formation of the phagophore (Klionsky et al., 2008; Levine and Deretic, 2007). ATG5-deficient macrophages have been generated, validated for defects in autophagy, and used to examine the role of autophagy for pathogen replication (Zhao et al., 2007). The ATG8 protein (also referred to as light chain 3 (LC3) becomes conjugated to phosphatidylethanolamine (PE) during autophagy, and specifically associates with autophagic membranes. LC3 its unconjugated form is called LC3-I and in its PE-conjugated form LC3-II. Turnover of LC3-II occurs when the protein is trapped within autolysosomes (Klionsky et al., 2008). A useful method for measuring rates of autophagy within eukaryotic cells involves monitoring conversion of LC3-I to LC3-II under conditions in which turnover of LC3-II in autolysosomes is prevented. This conversion can be followed by immunoblotting due to a mobility difference between the modified and unmodified forms of LC3 (Klionsky et al., 2008). LC3-II has also beed used extensively as a specific cytological marker for autophagic membranes in eukaryotic cells (Klionsky et al., 2008).

4.2 Results

4.2.1 Genetic Analysis of Y. pestis *Survival in Macrophages*

Prior to our work that is summarized here, very little was known about the nature of the macrophage vacuole inhabited by *Y. pestis*. Early work yielded evidence that *Y. pestis*-containing vacuoles fused with lysosomes in murine macrophages

(Charnetzky and Shuford, 1985; Straley and Harmon, 1984b). Based upon these findings it was suggested that *Y. pestis* survives in a phagolysosome (Straley and Harmon, 1984b). Several studies had investigated the genetic basis for the ability of *Y. pestis* to survive and replicate within macrophages. Straley and Harmon showed that pCD1, pPCP1 and the *pgm* locus were not required for replication of *Y. pestis* in murine macrophages (Straley and Harmon, 1984a). The *phoP* gene, which encodes a transcriptional activator, was shown to be important for survival of *Y. pestis* in macrophages and for virulence in mice (Oyston et al., 2000). In *S.* Typhimurium PhoP-regulated genes important for survival of *S.* Typhimurium in the phagosome include genes of the *pmr* operon. The products of the *pmr* operon function to modify lipid A with aminoarabinose; this increases bacterial resistance to antimicrobial peptides (Ernst et al., 1999; Groisman, 2001). The genes under control of PhoP in *Y. pestis* that are important for survival in macrophages had not been identified.

Several genes important for survival of *Y. pestis* in macrophages were identified in our laboratory. One set of genes that was identified is regulated by PhoP. Products of these genes allow *Y. pestis* to adapt to the environment of the YCV (Grabenstein et al., 2006). The identification of these genes was accomplished by mutagenesis of *Y. pestis* with a Tn5*lacZ* element followed by screening for mutants with a $lacZ^+$ phenotype under conditions of *phoP* overexpression. Mutants with the desired insertions were subsequently examined for the ability to survive in naïve macrophages. This analysis resulted in the identification of three PhoP-regulated genes, involved in modification of LPS (*ugd, pmrK*) or resistance to low Mg^{2+} levels (*mgtC*). Results obtained with *ugd*, *mgtC*, and *ugd mgtC* mutants revealed that the products of *ugd* and *mgtC* function independently to promote early survival of *Y. pestis* in macrophage phagosomes (Grabenstein et al., 2006). The functional characterization of these gene products was not further pursued since they had been characterized in *S.* Typhimurium (Ernst et al., 1999).

A second set of genes specifically required for *Y. pestis* to survive in macrophages activated with IFNγ were also identified (Pujol et al., 2005). This was made possible by the discovery that a *Y. pestis* strain missing the 102 kilobase *pgm* locus from the chromosome was unable to survive in activated macrophages (Pujol et al., 2005). The *pgm* locus encodes well-characterized genes important for virulence, including the *ybt* operon, which encodes for a siderophore synthesis and retrieval pathway (Perry and Fetherston, 1997; Carniel, 2002). Because the *pgm* locus is bounded by IS elements, it can spontaneously recombine out of the chromosome at high frequency, resulting in an attenuated *pgm* mutant (Carniel, 2002). A *Y. pestis* strain containing the *pgm* locus (*pgm*$^+$) could replicate in macrophages stimulated with IFNγ (Pujol et al., 2005). To identify which of the approximately 100 genes in the *pgm* locus were important for survival in activated macrophages, a complementation approach was used (Pujol et al., 2005). This lead to the identification of a three-gene operon, *ripCBA*, that promotes survival of *Y. pestis* in activated macrophages. The *ripCBA* genes appear to encode novel metabolic enzymes. Interestingly, bioinformatics analysis indicates that homologs of the Rip proteins are encoded in operons in a number of other bacteria, including other facultative intracellular pathogens such as *M. tuberculosis*. Work on the *M. tuberculosis* homolog of RipC (CitE) suggests that these enzymes may be involved in a novel pathway of fatty acid biosynthesis (Goulding et al., 2007). To gain additional insight into the functions of these proteins, it will be important to determine the structures of the Rip proteins, to determine if these proteins form a complex, and to identify their role in *Y. pestis* metabolism.

4.2.2 Analysis of Trafficking of the YCV in Macrophages

Trafficking of the YCV in macrophages was characterized by immunofluorescence and thin section electron microscopy (Grabenstein et al., 2006). Results of experiments that utilize lysosomal tracers or antibodies to the LAMP1 or cathepsin D proteins suggested that that the YCV fuses with late-endosomes or lysosomes. Results of thin section EM showed that the YCV adopts a spacious morphology beginning around 8 hr post infection, at which time bacterial replication begins. In more recent work, pH sensitive fluorescent probes (Lysotracker and FITC-labeled *Y. pestis*) were to determine if the YCV acidifies (Steinberg and Swanson, 1994). The results of these studies showed that the YCV fails to acidify and in fact remains at or above neutral pH (Pujol et al., 2009). In contrast, phagosomes containing formaldehyde fixed or heat-killed *Y. pestis* rapidly acidified (Pujol et al., 2009). We now hypothesize that YCVs interact with late endosomes but not lysosomes, because, as discussed above, phagosome acidification is required for efficient fusion of vacuoles with lysosomes. *Y. pestis* appears to prevent acidification of the YCV in cis, as a co-infection experiment with live and fixed bacteria showed that live *Y. pestis* were unable to prevent acidification of phagosomes containing fixed *Y. pestis* even when there were in the same macrophage (Pujol et al., 2009). In addition, de novo protein synthesis does not appear to be required for *Y. pestis* to inhibit phagosome acidification, since *Y. pestis* was able to prevent phagosome acidification in the presence of the bacteriostatic antibiotic chloramphenicol (Pujol et al., 2009). A large number of *Y. pestis* strains and mutants have been tested for the ability to prevent phagosome acidification, and so far no mutant with a deficiency in this activity has been identified. Genes encoded on pCD1, pPCP1, the *pgm* locus or under control of PhoP are not required to prevent phagosome acidification (Pujol et al., 2009). Thus, although a *phoP* mutant is defective for intracellular survival, it remains competent to inhibit phagosome acidification.

4.2.3 Role of Autophagy in Survival of Y. pestis *in Macrophages*

Activation by IFNγ is known to alter the trafficking and environment of phagosomes in macrophages (Rohde et al., 2007; Shenoy et al., 2007). For example, as discussed above, IFNγ treatment is known to activate autophagy in macrophages, which can override the block of phagosome maturation imposed by *M. tuberculosis* (Levine and Deretic, 2007; Ogawa and Sasakawa, 2006). On the other hand, other intracellular pathogens exploit autophagy for establishment of a replicative niche within host cells (Ogawa and Sasakawa, 2006). We therefore considered the possibility that the replication of *Y. pestis* in macrophages exposed to IFNγ resulted from an increase in autophagy, a process that might be exploited by the pathogen for intramacrophage survival (Ogawa and Sasakawa, 2006). To examine this possibility, we looked for evidence of autophagy in macrophages infected with *Y. pestis*. Analysis by thin section EM showed the presence of double membrane structures surrounding about half of the *Y. pestis* bacteria observed inside macrophages (Pujol et al., 2009). Immunofluorescence microscopy was used to detect the presence of a GFP-LC3-II fusion protein on YCVs. Approximately 40% of YCVs in naïve or activated macrophages showed a clear labeling of GFP LC3-II (Pujol et al., 2009). Immunoblotting was used to detect levels of LC3-II under different infection conditions. Interestingly, in macrophages infected with *Y. pestis* there was a significant increase in the steady state level of LC3-II (Pujol et al., 2009). This increase was not seen in macrophages infected with a *phoP* mutant or fixed *Y. pestis,* indicating that bacterial survival was needed for the increase in LC3-II. These results are consistent with the possibility that the YCV is fusing with autophagosomes, and turnover of LC3-II delivered to the YCV is prevented due to a lack of acidification. Finally, use of ATG5-deficient macrophages showed that autophagy was not required for survival of *Y. pestis* in macrophages, and the absence of autophagy did not prevent the block in phagosome acidification (Pujol et al., 2009).

4.3 Discussion

4.3.1 Bacterial Strategies for Subverting Acidification of Phagosomes

The ability to prevent phagosome acidification appears to be a common attribute of several pathogenic bacteria (Huynh and Grinstein, 2007) (Fig. 4.1). As discussed above for *M. tuberculosis*, the arrest in maturation of the MCP appears to preclude the acquisition of a sufficient density of vATPase, and the failure to acidify is most likely a consequence of the block in membrane trafficking (Huynh and Grinstein, 2007). *H. pylori* appears to prevent phagosome acidification by two complementary mechanisms (Huynh and Grinstein, 2007). First, it induces the formation of NH_3 by the action of a potent bacterial urease. The NH_3 serves to sequester incoming protons by forming NH_4. Second, *H. pylori* secretes a cytotoxin called VacA that oligomerizes to form a Cl^- channel. When inserted into the phagosomal membrane, VacA contributes to the permeation of counterions that serve to collapse the electrical potential generated by the vATPase. Dissipation of the electrical potential seems to promote proton accumulation. Together with Cl^-, the accumulated NH4 acts to increase the osmotic content of the phagosomal lumen. This, in turn, drives the influx of H_2O, causing swelling of phagosomes into megasomes (Huynh and Grinstein, 2007). Recently, *Listeria monocytogenes* has been shown to be capable of replicating in macophage phagosomes that expand and do not acidify (Birmingham

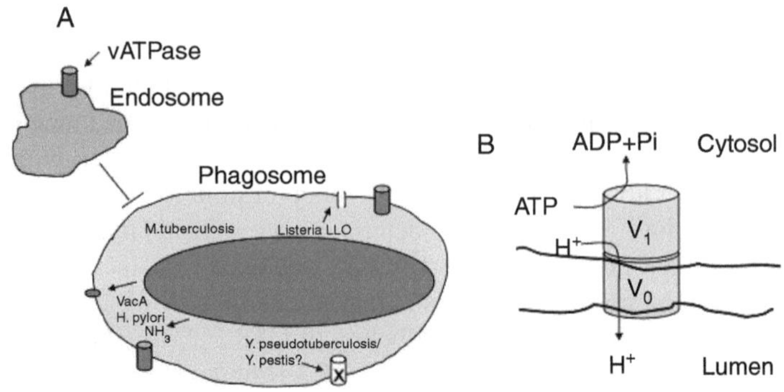

Fig. 4.1 Mechanisms by which pathogens subvert phagosome acidification and model of vATPase. (**a**) Shown are mechanisms by which different pathogens prevent phagosome acidification. *M. tuberculosis* blocks phagosome maturation and therefore prevents delivery of the vATPase from an endosome vesicle. *L. monocytogenes* dissipates the pH gradient by LLO-dependent pore formation. *H. pylori* allows Cl^- transport via VacA, and neutralizes protons via production of NH_3. It is suggested that *Y. pseudotuberculosis* can directly inhibit the action of the vATPase (Tsukano et al., 1999). (**b**) The vATPase shown in membrane of a phagosome is made up of two sectors, each composed of multiple subunits. ATP hydrolysis leads to proton import into the phagosome lumen. The Vo sector is exposed to the lumen of the phagosome and contains targets of molecules that can inhibit the vATPase, including bafilomycin, which inactivates subunit c

et al., 2008). The pore forming hemolysin LLO appears to be required for this process, and it is suggested that pore formation by LLO allows dissipation of the pH gradient.

4.4 Conclusions

The mechanism by which *Y. pestis* survives in macrophages remains poorly understood as compared to other facultative intracellular pathogens such as *S.* Typhimurium or *M. tuberculosis.* As described above, recent work has yielded new insights into mechanisms of *Y. pestis* survival in macrophages. Although these studies have answered many questions, they have also raised new questions. It has been determined that the *Yersinia*-containing vacuole (YCV) can fuse with late endosomes or lysosomes, but fails to acidify and therefore it is unlikely that the YCV represents a true phagolysome as initially proposed (Straley and Harmon, 1984b). We do not know the mechanism by which *Y. pestis* prevents acidification of the YCV, nor do we fully understand how a lack of acidification affects trafficking of this compartment. A previous study reported that *Y. pseudotuberculosis*, which is closely related to *Y. pestis*, could be found inside YCVs that contained the vATPase, but the vATPase was inactive (Tsukano et al., 1999) (Fig. 4.1). However, preliminary data indicate that vacuoles containing *Y. pestis* do not colocalize with the vATPase in macrophages (K. Klein and J. Bliska, unpublished data). Regardless of the mechanism used by *Y. pestis* to prevent phagosome acidification, it is hypothesized that this process allows *Y. pestis* to survive in naïve macrophages or activated macrophages and to avoid the bactericidal process of xenophagy. Finally, although preliminary evidence has been obtained that survival in macrophages is important for plague disease in a mouse model (Bliska and Casadevall, 2009; Oyston et al., 2000), we do not know the precise stage of pathogenesis at which survival in macrophages is important. Additional studies will be required to address the questions outline above. A better understanding of the mechanism of *Y. pestis* survival in macrophages and the role of this process in pathogenesis will provide a foundation of knowledge on which the development of novel countermeasures against pneumonic plague can proceed.

Acknowledgements I thank Kate Klein for permission to cite unpublished data. The work summarized in this chapter was supported by grants from the NIH awarded to J.B.B. (PO1-AI055621 and the Northeast Biodefense Center U54-AI057158-Lipkin).

References

Adams, D. O. & Hamilton, T. A. (1984) The cell biology of macrophage activation. *Annu Rev Immunol,* 2, 283–318.

Aderem, A. & Underhill, D. M. (1999) Mechanisms of phagocytosis in macrophages. *Annu Rev Immunol,* 17, 593–623.

Birmingham, C. L., Canadien, V., Kaniuk, N. A., Steinberg, B. E., Higgins, D. E. & Brumell, J. H. (2008) Listeriolysin O allows Listeria monocytogenes replication in macrophage vacuoles. *Nature,* 451, 350–4.

Bliska, J. B. & Casadevall, A. (2009) Intracellular pathogenic bacteria and fungi – a case of convergent evolution? *Nat Rev Microbiol,* 7, 165–71.

Brumell, J. H. & Grinstein, S. (2004) Salmonella redirects phagosomal maturation. *Curr Opin Microbiol,* 7, 78–84.

Burrows, T. W. & Bacon, G. A. (1956) The basis of virulence in *Pasteurella pestis*: an antigen determining virulence. *Br J Exp Pathol,* 37, 481–93.

Carniel, E. (2002) Plasmids and pathogenicity islands of *Yersinia. Curr Top Microbiol Immunol,* 264, 89–108.

Cavanaugh, D. C. & Randall, R. (1959) The role of multiplication of *Pasteurella pestis* in mononuclear phagocytes in the pathogenesis of fleaborne plague. *J Immunol,* 85, 348–63.

Charnetzky, W. T. & Shuford, W. W. (1985) Survival and growth of *Yersinia pestis* within macrophages and an effect of the loss of the 47-megadalton plasmid on growth in macrophages. *Infect Immun,* 47, 234–41.

Chua, J., Vergne, I., Master, S. & Deretic, V. (2004) A tale of two lipids: Mycobacterium tuberculosis phagosome maturation arrest. *Curr Opin Microbiol,* 7, 71–7.

Cuellar-Mata, P., Jabado, N., Liu, J., Furuya, W., Finlay, B. B., Gros, P. & Grinstein, S. (2002) Nramp1 modifies the fusion of *Salmonella typhimurium*-containing vacuoles with cellular endomembranes in macrophages. *J Biol Chem,* 277, 2258–65.

Digiuseppe Champion, P. A. & Cox, J. S. (2007) Protein secretion systems in Mycobacteria. *Cell Microbiol,* 9, 1376–84.

Du, Y., Rosqvist, R. & Forsberg, A. (2002) Role of fraction 1 antigen of *Yersinia pestis* in inhibition of phagocytosis. *Infect Immun,* 70, 1453–60.

Duclos, S. & Desjardins, M. (2000) Subversion of a young phagosome: the survival strategies of intracellular pathogens. *Cell Microbiol,* 2, 365–77.

Ernst, R. K., Guina, T. & Miller, S. I. (1999) How intracellular bacteria survive: surface modifications that promote resistance to host innate immune responses. *J Infect Dis,* 179, S326–30.

Finegold, M. J. (1969) Pneumonic plague in monkeys. An electron microscopic study. *Am. J. Pathol.,* 54, 167–85.

Goulding, C. W., Bowers, P. M., Segelke, B., Lekin, T., Kim, C. Y., Terwilliger, T. C. & Eisenberg, D. (2007) The structure and computational analysis of Mycobacterium tuberculosis protein CitE suggest a novel enzymatic function. *J Mol Biol,* 365, 275–83.

Grabenstein, J. P., Fukuto, H. S., Palmer, L. E. & Bliska, J. B. (2006) Characterization of phagosome trafficking and identification of PhoP-regulated genes important for survival of Yersinia pestis in macrophages. *Infect Immun,* 74, 3727–41.

Groisman, E. A. (2001) The pleiotropic two-component regulatory system PhoP-PhoQ. *J Bacteriol,* 183, 1835–42.

Hensel, M., Shea, J. E., Waterman, S. R., Mundy, R., Nikolaus, T., Banks, G., Vazquez-Torres, A., Gleeson, C., Fang, F. C. & Holden, D. W. (1998) Genes encoding putative effector proteins of the type III secretion system of *Salmonella* pathogenicity island 2 are required for bacterial virulence and proliferation in macrophages. *Mol Microbiol,* 30, 163–74.

Hinnebusch, B. J. (2005) The evolution of flea-borne transmission in Yersinia pestis. *Curr Issues Mol Biol,* 7, 197–212.

Hinnebusch, B. J. & Erickson, D. L. (2008) Yersinia pestis biofilm in the flea vector and its role in the transmission of plague. *Curr Top Microbiol Immunol,* 322, 229–48.

Huynh, K. K. & Grinstein, S. (2007) Regulation of vacuolar pH and its modulation by some microbial species. *Microbiol Mol Biol Rev,* 71, 452–62.

Inglesby, T. V., Dennis, D. T., Henderson, D. A., Bartlett, J. G., Ascher, M. S., Eitzen, E., Fine, A. D., Friedlander, A. M., Hauer, J., Koerner, J. F., Layton, M., Mcdade, J., Osterholm, M. T., O'toole, T., Parker, G., Perl, T. M., Russell, P. K., Schoch-Spana, M. & Tonat, K. (2000) Plague as a biological weapon: medical and public health management. Working Group on Civilian Biodefense. *JAMA,* 283, 2281–90.

Janssen, W. A. & Surgalla, M. J. (1968) Plague bacillus: survival within host phagocytes. *Science,* 163, 950–2.

Jarrett, C. O., Deak, E., Isherwood, K. E., Oyston, P. C., Fischer, E. R., Whitney, A. R., Kobayashi, S. D., Deleo, F. R. & Hinnebusch, B. J. (2004) Transmission of Yersinia pestis from an infectious biofilm in the flea vector. *J Infect Dis,* 190, 783–92.

Klionsky, D. J., Abeliovich, H., Agostinis, P., Agrawal, D. K., Aliev, G., Askew, D. S., Baba, M., Baehrecke, E. H., Bahr, B. A., Ballabio, A., Bamber, B. A., Bassham, D. C., Bergamini, E., Bi, X., Biard-Piechaczyk, M., Blum, J. S., Bredesen, D. E., Brodsky, J. L., Brumell, J. H., Brunk, U. T., Bursch, W., Camougrand, N., Cebollero, E., Cecconi, F., Chen, Y., Chin, L. S., Choi, A., Chu, C. T., Chung, J., Clarke, P. G., Clark, R. S., Clarke, S. G., Clave, C., Cleveland, J. L., Codogno, P., Colombo, M. I., Coto-Montes, A., Cregg, J. M., Cuervo, A. M., Debnath, J., Demarchi, F., Dennis, P. B., Dennis, P. A., Deretic, V., Devenish, R. J., Di Sano, F., Dice, J. F., Difiglia, M., Dinesh-Kumar, S., Distelhorst, C. W., Djavaheri-Mergny, M., Dorsey, F. C., Droge, W., Dron, M., Dunn, W. A., JR., Duszenko, M., Eissa, N. T., Elazar, Z., Esclatine, A., Eskelinen, E. L., Fesus, L., Finley, K. D., Fuentes, J. M., Fueyo, J., Fujisaki, K., Galliot, B., Gao, F. B., Gewirtz, D. A., Gibson, S. B., Gohla, A., Goldberg, A. L., Gonzalez, R., Gonzalez-Estevez, C., Gorski, S., Gottlieb, R. A., Haussinger, D., HE, Y. W., Heidenreich, K., Hill, J. A., Hoyer-Hansen, M., Hu, X., Huang, W. P., Iwasaki, A., Jaattela, M., Jackson, W. T., Jiang, X., Jin, S., Johansen, T., Jung, J. U., Kadowaki, M., Kang, C., Kelekar, A., Kessel, D. H., Kiel, J. A., Kim, H. P., Kimchi, A., Kinsella, T. J., Kiselyov, K., Kitamoto, K., Knecht, E., et al. (2008) Guidelines for the use and interpretation of assays for monitoring autophagy in higher eukaryotes. *Autophagy,* 4, 151–75.

Knodler, L. A. & Steele-Mortimer, O. (2003) Taking possession: biogenesis of the *Salmonella*-containing vacuole. *Traffic,* 4, 587–99.

Lathem, W. W., Crosby, S. D., Miller, V. L. & Goldman, W. E. (2005) Progression of primary pneumonic plague: a mouse model of infection, pathology, and bacterial transcriptional activity. *Proc Natl Acad Sci U S A,* 102, 17786–91.

Levine, B. & Deretic, V. (2007) Unveiling the roles of autophagy in innate and adaptive immunity. *Nat Rev Immunol,* 7, 767–77.

Lukaszewski, R. A., Kenny, D. J., Taylor, R., Rees, D. G., Hartley, M. G. & Oyston, P. C. (2005) Pathogenesis of Yersinia pestis infection in BALB/c mice: effects on host macrophages and neutrophils. *Infect Immun,* 73, 7142–50.
Meresse, S., Steele-Mortimer, O., Moreno, E., Desjardins, M., Finlay, B. & Gorvel, J. P. (1999) Controlling the maturation of pathogen-containing vacuoles: a matter of life and death. *Nat. Cell. Biol.,* 1, E183–8.
Meyer, K. F. (1950) Immunity in plague; a critical consideration of some recent studies. *J Immunol,* 64, 139–63.
Mizushima, N., Levine, B., Cuervo, A. M. & Klionsky, D. J. (2008) Autophagy fights disease through cellular self-digestion. *Nature,* 451, 1069–75.
Noel, B. L., Lilo, S., Capurso, D., Hill, J. & Bliska, J. B. (2009) Yersinia pestis can bypass protective antibodies to LcrV and activation with gamma interferon to survive and induce apoptosis in murine macrophages. *Clin Vaccine Immunol,* 16, 1457–66.
Ochman, H., Soncini, F. C., Solomon, F. & Groisman, E. A. (1996) Identification of a pathogenicity island required for *Salmonella* survival in host cells. *Proc Natl Acad Sci U S A,* 93, 7800–4.
Ogawa, M. & Sasakawa, C. (2006) Bacterial evasion of the autophagic defense system. *Curr Opin Microbiol,* 9, 62–8.
Oh, Y. K., Alpuche-Aranda, C., Berthiaume, E., Jinks, T., Miller, S. I. & Swanson, J. A. (1996) Rapid and complete fusion of macrophage lysosomes with phagosomes containing Salmonella typhimurium. *Infect Immun,* 64, 3877–83.
Oyston, P. C. F., Dorrell, N., Williams, K., Li, S.-R., Green, M., Titball, R. W. & Wren, B. (2000) The response regulator PhoP is important for survival under conditions of macrophage-induced stress and virulence in *Yersinia pestis*. *Infect. Immun.,* 68, 3419–25.
Perry, R. D. & Fetherston, J. D. (1997) Yersinia pestis – etiologic agent of plague. *Clin Microbiol Rev,* 10, 35–66.
Prentice, M. B. & Rahalison, L. (2007) Plague. *Lancet,* 369, 1196–207.
Pujol, C. & Bliska, J. B. (2003) The ability to replicate in macrophages is conserved between Yersinia pestis and Yersinia pseudotuberculosis. *Infect Immun,* 71, 5892–9.
Pujol, C. & Bliska, J. B. (2005) Turning Yersinia pathogenesis outside in: subversion of macrophage function by intracellular yersiniae. *Clin Immunol,* 114, 216–26.
Pujol, C., Grabenstein, J. P., Perry, R. D. & Bliska, J. B. (2005) Replication of Yersinia pestis in interferon gamma-activated macrophages requires ripA, a gene encoded in the pigmentation locus. *Proc Natl Acad Sci U S A,* 102, 12909–14.
Pujol, C., Klein, K. A., Romanov, G. A., Palmer, L. E., Cirota, C., Zhao, Z. & Bliska, J. B. (2009) Yersinia pestis can reside in autophagosomes and avoid xenophagy in murine macrophages by preventing vacuole acidification. *Infect Immun.*
Rohde, K., Yates, R. M., Purdy, G. E. & Russell, D. G. (2007) Mycobacterium tuberculosis and the environment within the phagosome. *Immunol Rev,* 219, 37–54.
Runco, L. M., Myrczek, S., Bliska, J. B. & Thanassi, D. G. (2008) Biogenesis of the F1 Capsule and Analysis of the Ultrastructure of Yersinia pestis. *J Bacteriol,* Submitted.
Sebbane, F., Gardner, D., Long, D., Gowen, B. B. & Hinnebusch, B. J. (2005) Kinetics of disease progression and host response in a rat model of bubonic plague. *Am J Pathol,* 166, 1427–39.
Shenoy, A. R., Kim, B. H., Choi, H. P., Matsuzawa, T., Tiwari, S. & Macmicking, J. D. (2007) Emerging themes in IFN-gamma-induced macrophage immunity by the p47 and p65 GTPase families. *Immunobiology,* 212, 771–84.
Smiley, S. T. (2008) Current challenges in the development of vaccines for pneumonic plague. *Expert Rev Vaccines,* 7, 209–21.
Steinberg, T. H. & Swanson, J. A. (1994) Measurement of phagosome-lysosome fusion and phagosomal pH. *Methods Enzymol,* 236, 147–60.
Straley, S. C. & Harmon, P. A. (1984a) Growth in mouse peritoneal macrophages of *Yersinia pestis* lacking established virulence determinants. *Infect. Immun.,* 45, 649–54.
Straley, S. C. & Harmon, P. A. (1984b) *Yersinia pestis* grows within phagolysosomes in mouse peritoneal macrophages. *Infect. Immun.,* 45, 655–59.
Titball, R. W. & Williamson, E. D. (2004) Yersinia pestis (plague) vaccines. *Expert Opin Biol Ther,* 4, 965–73.

Tsukano, H., Kura, F., Inoue, S., Sato, S., Izumiya, H., Yasuda, T. & Watanabe, H. (1999) *Yersinia pseudotuberculosis* blocks the phagosomal acidification of B10.A mouse macrophages through the inhibition of vacuolar H^+-ATPase activity. *Microb. Pathog.,* 27, 253–63.
Vergne, I., Chua, J., Singh, S. B. & Deretic, V. (2004) Cell biology of *Mycobacterium tuberculosis* phagosome. *Annu Rev Cell Dev Biol,* 20, 367–94.
Viboud, G. I. & Bliska, J. B. (2005) Yersinia outer proteins: role in modulation of host cell signaling responses and pathogenesis. *Annu Rev Microbiol,* 59, 69–89.
Vieira, O. V., Botelho, R. J. & Grinstein, S. (2002) Phagosome maturation: aging gracefully. *Biochem. J.,* 366, 689–704.
Welkos, S., Pitt, M. L. M., Martinez, M., Friedlander, A., Vogel, P. & Tammariello, R. (2002) Determination of the virulence of the pigmentation-deficient and pigmentation-/plasminogen activator-deficient strains of *Yersinia pestis* in non-human primate and mouse models of pneumonic plague. *Vaccine,* 20, 2206–14.
Welkos, S. L., Friedlander, A. M. & Davis, K. J. (1997) Studies on the role of plasminogen activator in systemic infection by virulent *Yersinia pestis* strain C092. *Microb. Pathog.,* 23, 211–23.
Zhao, Z., Thackray, L. B., Miller, B. C., Lynn, T. M., Becker, M. M., Ward, E., Mizushima, N. N., Denison, M. R. & Virgin, H. W. T. (2007) Coronavirus replication does not require the autophagy gene ATG5. *Autophagy,* 3, 581–5.

Chapter 5
The Inverse Relationship Between Cytotoxicity of *Y. pestis* and Its Virulence

Emanuelle Mamroud, Avital Tidhar, Yinon Levy, Erez Bar-Haim, Gideon Halperin, Yehuda Flashner, Sara Cohen, Avigdor Shafferman, and Ayelet Zauberman

Abstract Modulation of host cell death during infection is a prevalent virulence strategy developed by many bacterial pathogens. *Yersinia pestis*, the causative agent of the fatal plague disease was found to exert limited cytotoxicity towards target immune cells, mediated by type III secretion system effector YopJ. In contrast, the highly cytotoxic closely related *Y. enterocolitica* O:8 causes a self limited gastrointestinal disease. This phenomenon led us to suggest that the reduced cytotoxic potency of *Y. pestis* is related to its increased virulence potential. Generation of *Y. pestis* strain expressing YopP instead of YopJ, enhanced its cytotoxic potency towards macrophages in-vitro and mouse spleen target cells in-vivo. The highly cytotoxic *Y. pestis* strain demonstrated a reduced ability to colonize internal organs of mice infected subcutaneously, and most strikingly was avirulent in a mouse model of bubonic plague. These results indicate inverse relationship between cytotoxic potency and in-vivo virulence. Still, the same YopP-expressing *Y. pestis* strain remained fully virulent to mice upon intravenous or intranasal infections, indicating retention of virulence potential. In addition, it was found that subcutaneous administration of the highly cytotoxic *Y. pestis* strain activated extremely rapid, potent and systemic protective response against concomitant challenges with a virulent strain via the subcutaneous, intravenous or airway routes. These findings may have important implications on the design of future plague vaccine/therapies and contribute to our understanding of virulence strategies of *Y. pestis* in nature.

Keywords Plague · *Y. pestis* · YopJ · YopP · Cytotoxicity · Protective immunity

---

E. Mamroud (✉)
Department of Biochemistry and Molecular Genetics, Israel Institute for Biological Research, Ness-Ziona 74100, Israel
e-mail: emmym@iibr.gov.il

A. Shafferman et al. (eds.), *The Challenge of Highly Pathogenic Microorganisms*,
DOI 10.1007/978-90-481-9054-6_5, © Springer Science+Business Media B.V. 2010

5.1 Introduction

The genus *Yersinia* includes three pathogenic species, *Y. pestis*, *Y. enterocolitica* and *Y. pseudotuberculosis*. These species are genetically closely related but yet differ significantly in their virulence and pathogenesis. Infection by *Y. pestis*, the etiological agent of plague, occurs by intra-dermal fleabites or by droplet inhalation, leading to fatal bubonic or pneumonic plague. *Y. enterocolitica* and *Y. pseudotuberculosis*, on the other hand, cause mild and self-limiting gastrointestinal syndromes upon infection through the oral route (Brubaker, 1991; Perry and Fetherston, 1997). In spite of the differences in pathogenesis, all three species share similar mechanisms for evading the host innate immunity. These mechanisms are attributed mainly to genes encoded by a ~70 kb plasmid which is essential for virulence (Portnoy and Falkow, 1981; Ben-Gurion and Shafferman, 1981; Gemski et al., 1980; Ferber and Brubaker, 1981). This plasmid encode a type III secretion system (TTSS), comprised of a secretion apparatus, chaperones and several effectors (Yops) (Cornelis and Wolf-Watz, 1997). One of the Yops, YopJ (named YopP in *Y. enterocolitica*) was shown to induce apoptotic cell death in *Yersinia*-infected macrophages in vitro (Mills et al., 1997; Monack et al., 1997; Ruckdeschel et al., 1997a; Weeks et al., 2002; Zauberman et al., 2006) as well as in animal models of infection (Lemaitre et al., 2006; Monack et al., 1998; Brodsky and Medzhitov, 2008). We have previously demonstrated that *Y. pestis* has a limited ability to induce programmed cell death in infected macrophages compared to *Y. enterocolitica* O:8 serotype (Zauberman et al., 2006; Zauberman et al., 2007). This observation was found to correlate with downgraded translocation of YopJ from *Y. pestis* to target cells (Zauberman et al., 2006). Similarly, whereas interactions of *Y. enterocolitica* O:8 with dendritic cells (DCs) led to YopP-mediated induction of apoptotic cell death, infection of DCs with *Y. pestis* failed to affect cell viability (Velan et al., 2006; Erfurth et al., 2004). The role of *Y. pestis* YopJ in pathogenesis was examined in several mouse models in which it was found to be dispensable for virulence (Lemaitre et al., 2006; Zauberman et al., 2007; Straley and Bowmer, 1986). Accumulating evidence seem to indicate that the ability to kill host immune cells is not essential for the manifestation of virulence by *Y. pestis*. Moreover, the presence of viable infected immune cells at the infection site may rather allow the invading pathogen to replicate inside a shielded niche provided by the host cells. Given the fact that *Y. pestis* is found intracellularly during the early stages of infection (Cavanaugh and Randall, 1959; Lukaszewski et al., 2005) and can replicate within macrophages (Pujol and Bliska, 2003; Lukaszewski et al., 2005), one may assume that effective apoptotic activity against immune cells might impair its survival in vivo. The observation that *Y. pestis* demonstrated restricted capacity to induce apoptosis in macrophages and DCs (Velan et al., 2006; Erfurth et al., 2004), along with the finding that YopJ was not essential for *Y. pestis* virulence, led us to suggest that reduced apoptotic potency might have contributed to the highly pathogenic phenotype of *Y. pestis* (Zauberman et al., 2006). In the present study, we examined this assumption using the highly virulent *Y. pestis* Kimberley53, its *yopJ* deletion mutant and progeny derivatives expressing in trans either the *Y. pestis* YopJ or the YopP of *Y. enterocolitica* O:8.

5.2 Results

5.2.1 *The Relationship Between Y. pestis Cytotoxicity and Its Virulence in the Mouse Models of Bubonic and Pneumonic Plague*

In order to explore whether enhanced cytotoxic potency would attenuate *Y. pestis* virulence, we used the highly virulent *yopJ*-deleted strain of *Y. pestis* Kimberley53 (Kim53ΔJ) (Zauberman et al., 2007) to generate recombinant strains expressing in trans either YopP (Kim53ΔJ+P) or YopJ (Kim53ΔJ+J) (Zauberman et al., 2009).

Expression of YopP conferred high cytotoxic potency towards macrophage cells, but expression of YopJ did not (Fig. 5.1a). Consistent with our in vitro observations, the YopP-expressing *Y. pestis* strain induced significantly higher levels of apoptosis in infected spleen cells compared to the virulent Kim53pGFP strain, as determined by staining with anti-active caspase-3 antibodies (Fig. 5.1b).

The impact of increased cytotoxicity on *Y. pestis* virulence was tested by subcutaneous (s.c.) infection of mice. Mice infected with the highly cytotoxic strain, Kim53ΔJ+P, exhibited 100% survival even after infection with 1×10^6 cfu, as opposed to 100% mortality in mice infected with 100 cfu of the control Kim53pGFP or the wild-type Kim53 strains (Fig. 5.2a). Increasing the infection dose of Kim53ΔJ+P to 10^7 cfu resulted in 75% survival, indicating that the LD_{50} value of Kim53ΔJ+P is above 10^7 cfu (while the LD_{50} values for Kim53 and Kim53pGFP are 1 cfu). At the same time, the less cytotoxic strain over-expressing YopJ (Kim53ΔJ+J) remained virulent, as reflected by the mortality of infected

A

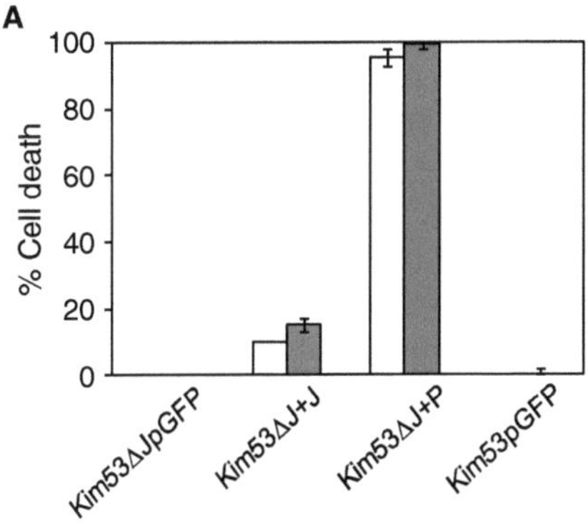

B

Strain	Caspase positive spleen cells in each individual mouse (Avg of 20 fields)			Avg ± SE
Kim53 pGFP	0.1	1.25	0.4	0.58 ± 0.3
Kim53 ΔJ+P	12	9.3	9.5	10.2 ± 0.8

Fig. 5.1 Over-expression of YopP in *Y. pestis* enhanced cytotoxicity to macrophages in vitro and to spleen cells in vivo. (**a**) Cytotoxic effect of *Y. pestis* recombinant strains on J774A.1 macrophage cell line. Cells were infected with Kim53pGFP, Kim53ΔJpGFP, Kim53ΔJ+J and Kim53ΔJ+P at MOI of 10 (*white*) or 50 (*gray*) for 1 h. Cell death was determined at 6 h after infection by LDH release test. (**b**) Number of cells stained by anti-active caspase 3 in spleen sections from mice infected with 1×10^4 cfu of Kim53pGFP or Kim53ΔJ+P strains, isolated at 48 h post infection (p.i)

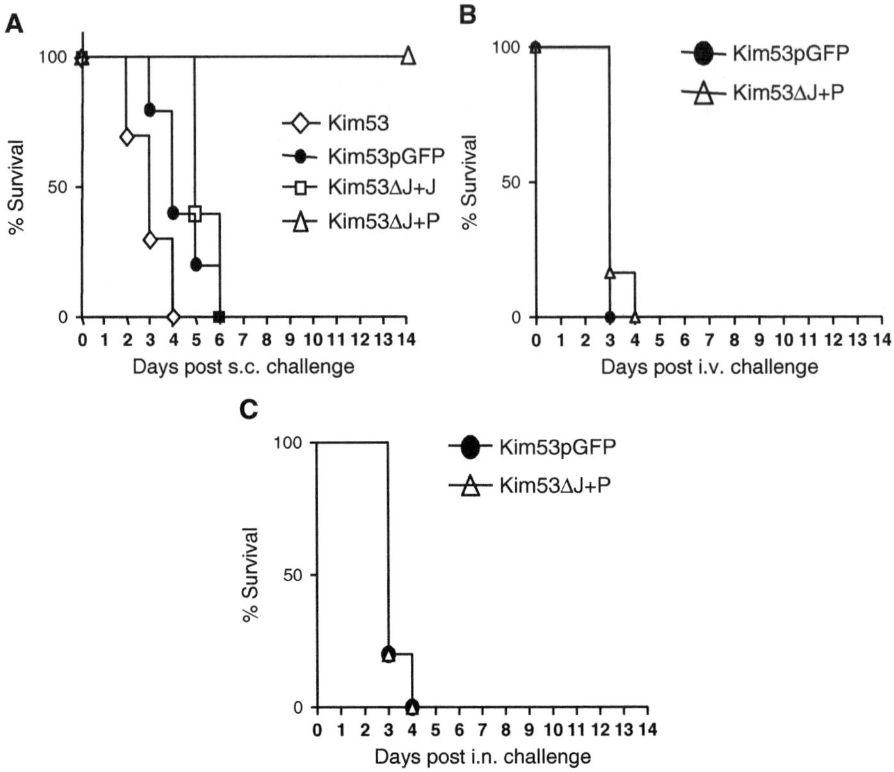

Fig. 5.2 Kim53ΔJ+P is highly attenuated upon subcutaneous infection but is fully virulent via the intravenous and intranasal infection routes. (**a**) Mice were infected subcutaneously with 1×10^2 cfu of Kim53 (*open diamond*), Kim53pGFP (*close circle*) or Kim53ΔJ+J (*open square*), or with 1×10^6 cfu of Kim53ΔJ+P (*open triangle*). (**b**) Mice were infected intravenously with 2×10^3 cfu of Kim53pGFP (*close circle*) or Kim53ΔJ+P (*open triangle*). (**c**) Mice were infected intranasally with 6×10^4 cfu of Kim53pGFP (*close circle*) or Kim53ΔJ+P (*open triangle*)

mice and the calculated LD_{50} value of one cfu (Fig. 5.2a). It therefore appears that concomitant with YopP-mediated enhanced cytotoxicity, the *Y. pestis* bacteria expressing YopP exhibit reduced virulence in the bubonic plague model.

In an attempt to gain deeper insight into mechanisms involved in the observed attenuation of Kim53ΔJ+P under the conditions of s.c. infection, mice were infected with 2,000 cfu intravenously (i.v.), thereby bypassing the subcutaneous barrier. Both *Y. pestis* strains, the YopP-expressing strain and the Kim53pGFP strain, caused 100% mortality with comparable mean time to death (MTTD) of 3.2 and 3.4 days, respectively (Fig. 5.2b).

In view of these results, it became interesting to examine the virulence of Kim53ΔJ+P under the conditions of respiratory infection. Mice were infected intranasally with 6×10^4 cfu of Kim53ΔJ+P (equivalent to ~30 LD_{50} of the wild-type strain). As shown in Fig. 5.2c, mice died within 4 days at the same rate as those infected intranasally with the virulent control strain Kim53pGFP. In addition, both

Table 5.1 Dissemination of *Y. pestis* strains to internal organs and blood following intranasal infection

	Kim53pGFP (cfu/organ or ml)[a]				Kim53ΔJ+P (cfu/organ or ml)[a]			
Mouse	Lung	Spleen	MLN	Blood	Lung	Spleen	MLN	Blood
1	2.1×10^{7}	3.6×10^{5}	4.9×10^{3}	6.5×10^{3}	2.7×10^{9}	1×10^{6}	1×10^{6}	500
2	1.3×10^{9}	2.4×10^{5}	3×10^{7}	6.5×10^{3}	1.3×10^{8}	7×10^{4}	7×10^{5}	<25
3	1×10^{8}	3.4×10^{6}	2.5×10^{7}	1×10^{4}	1×10^{9}	9×10^{5}	4×10^{4}	3.5×10^{3}
4	2×10^{4}	1.4×10^{5}	170	<25	1.5×10^{8}	4×10^{3}	430	120
5	1.5×10^{9}	3×10^{5}	1.3×10^{7}	8×10^{3}	2.4×10^{9}	4×10^{5}	2×10^{7}	620

[a] Mice were infected intranasally with 2×10^5 cfu of Kim53pGFP or Kim53ΔJ+P strains and sacrificed 48 h p.i. Bacterial concentration in blood (cfu/ml) and total bacterial loads in lung, mediastinal lymph node (MLN), and spleens, were determined

strains demonstrated the same i.n. LD_{50} value of 1,800 cfu, which is comparable to that of the parental Kim53 and Kim53ΔJ strains. Moreover, disease progression and systemic bacterial dissemination, as reflected by bacterial cfu counts in the lung, mediastinal lymph node (MLN), spleen and blood, were also comparable.

5.2.2 Overexpression of YopP by Y. pestis Affects Bacterial Colonization of Internal Organs

Host immune cells may serve in vivo as a favorable intracellular niche and a shielding vehicle for plague bacilli (Cavanaugh and Randall, 1959; Pujol and Bliska, 2003; Lukaszewski et al., 2005). It was therefore interesting to determine whether the avirulent phenotype of Kim53ΔJ+P via the s.c. route is associated with the cytotoxic effect of YopP through interference with bacterial propagation and delivery to internal tissues. To this end, we compared colonization of internal organs by Kim53ΔJ+P and the *Y. pestis* Kim53pGFP control strain. As can be seen in Table 5.2, the number of colony-forming units in the draining lymph node (L-ILN) and the spleen of mice infected with Kim53ΔJ+P was about 100-fold lower than with the control *Y. pestis* strain. An even more pronounced disparity was observed in blood. Kim53ΔJ+P could not be detected in blood whereas the *Y. pestis* control strain reached an average concentration of 1×10^6 cfu/ml (Table 5.2). Analysis of Kim53ΔJ+P loads in different lymph nodes (72 h p.i.) indicated that the cytotoxic strain colonized the draining lymph nodes (L-ILN), but was unable to disseminate into the axillary lymph node (axillary-LN) and to the inguinal lymph node located opposite to the site of infection (R-ILN). Interestingly, despite the fact that colonization of the spleen by Kim53ΔJ+P in s.c.-infected mice reached substantial levels (average of 1×10^5 bacteria per spleen), the strain was not virulent via the s.c. route of infection.

Table 5.2 Dissemination of *Y. pestis* strains to blood and internal organs following subcutaneous infection

	Kim53pGFP (60hrs post infection, cfu/organ or ml)[a]			Kim53ΔJ+P (60hrs post infection, cfu/organ or ml)[a]			Kim53ΔJ+P (72hrs post infection, cfu/organ or ml)[a]			
Mouse	Spleen	L-ILN	Blood	Spleen	L-ILN	Blood	Spleen	L-ILN	Axillary-LN	R-ILN
1	$1.4x10^7$	$1.3x10^8$	$4.7x10^5$	$3x10^5$	$4x10^7$	<25	$1x10^7$	$6.5x10^3$	$6.5x10^3$	75
2	$3.6x10^7$	$1.3x10^8$	$2.8x10^6$	$1.8x10^4$	<25	<25	$2.3x10^5$	$3.4x10^5$	400	<25
3	$1.7x10^7$	$4.2x10^8$	$7x10^5$	$1.6x10^5$	$4.7x10^5$	<25	$1.3x10^6$	$1.8x10^6$	<25	<25
4	$1.8x10^6$	$2.5x10^7$	$4.3x10^4$	$1.6x10^4$	<25	<25	75	500	<25	$1.8x10^3$
5	dead	dead	dead	$1x10^4$	$1x10^5$	<25	$1.7x10^5$	220	500	<25
6							$7.7x10^5$	$1.9x10^6$	$5.7x10^4$	<25

[a]Mice were infected at the lower back with 1×10^4 cfu of Kim53pGFP or Kim53ΔJ+P strains and sacrificed at 60 or 72 h p. i. Bacterial concentration in blood (cfu/ml) and total bacterial loads in spleen, draining inguinal lymph nodes (L-ILN), axillary lymph node (axillary-LN) and the lymph node opposite to the site of infection (R-ILN), were determined

5.2.3 Kim53ΔJ+P Generates Effective Systemic Resistance Against Virulent *Y. pestis* Strain

One possible explanation for the reduced infectivity and spread of Kim53ΔJ+P following s.c infection is that this strain kills the transporting immune cells and thus prevents dissemination. An alternative explanation is that these bacteria induce a rapid and efficient immune response that eliminates them from the bloodstream and accounts for the attenuated virulence phenotype.

To clarify this issue, we co-infected mice subcutaneously with Kim53ΔJ+P and the virulent Kim53pGFP strain (one site model). Indeed, co-infection with Kim53ΔJ+P could protect animals from the virulent Kim53pGFP strain (Fig. 5.3a). Such protection could not be achieved by co-infection with other attenuated *Y. pestis* strains such as Kim53Δp10Δp70 (deleted of pPCP1 and pCD1), EV76 (*pgm*$^-$) or Kim53ΔH (deleted of *yopH*) (Fig. 5.3a).

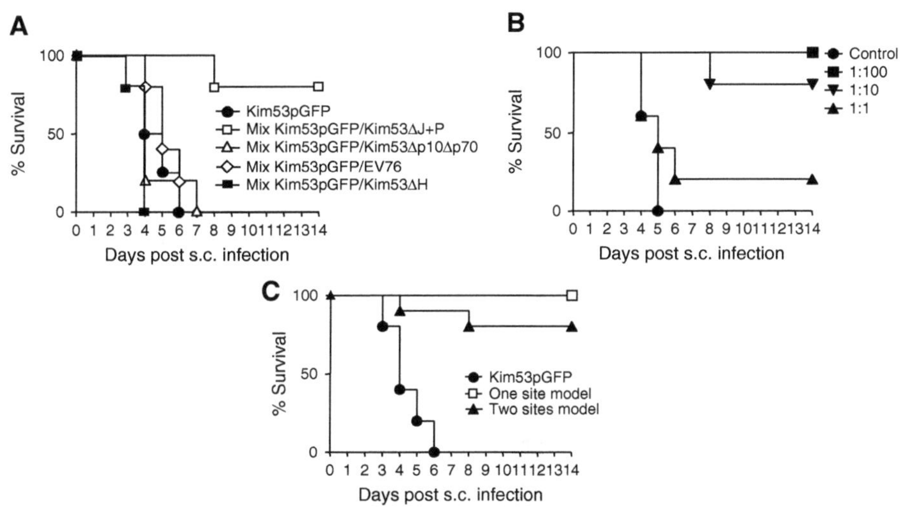

Fig. 5.3 Kim53ΔJ+P induces rapid protection against Kim53pGFP virulence. (**a**) Subcutaneous co-infection at a single site with mixtures of the virulent Kim53pGFP and various *Y. pestis* strains. Mice were infected at the lower back with 1×10^2 cfu of Kim53pGFP alone (*close circle*) or with mixtures of 1×10^2 cfu of Kim53pGFP and 1×10^3 cfu of either Kim53ΔJ+P (*open square*), Kim53Δ10Δ70 (*open triangle*), EV76 (*open diamond*) or Kim53ΔH (*close square*). (**b**) Mice were co-infected with a mixture of Kim53pGFP and Kim53ΔJ+P containing different bacterial ratios (1:1-100, respectively). (**c**) Subcutaneous concomitant infections with Kim53ΔJ+P and Kim53pGFP either at a single site (*one-site model*) or at two separate sites (*two-sites model*). In the one-site model (*open square*), mice were infected at their lower back with mixtures of Kim53pGFP (1×10^2 cfu) and Kim53ΔJ+P (3×10^3 cfu). In the two-sites model (*close triangle*) mice were injected at the upper back with either 3×10^3 cfu of Kim53ΔJ+P strain, and immediately thereafter with 1×10^2 cfu of Kim53pGFP strain at the lower back (a ratio of 1:30). Control mice were infected with 1×10^2 cfu of Kim53pGFP (*close circle*)

The protection against the virulent *Y. pestis* induced by the cytotoxic strain was striking. For example, complete protection was achieved when 10 cfu (~10 LD_{50}) of the virulent Kim53pGFP strain were co-infected with 1×10^3 cfu of Kim53ΔJ+P (a ratio of 1:100) (Fig. 5.3b). Moreover, effective 80% protection against 1×10^2 of virulent *Y. pestis* strain was achieved when the amount of Kim53ΔJ+P cfu in the bacterial mixture was 10 times higher (a ratio of 1:10). The protective efficacy was further highlighted by the finding that the same survival rate (80%) was achieved even when 1×10^4 cfu of Kim53pGFP (equivalent to 10^4 LD_{50}) were inoculated with 1×10^5 cfu of Kim53ΔJ+P (Zauberman et al., 2009). Lowering the ratio between the two strains to 1:1, provided only 20% protection (Fig. 5.3b). To examine whether the Kim53ΔJ+P-mediated resistance to lethal infection with the *Y. pestis* strain is a systemic or a localized effect; the animals were simultaneously infected at separate sites, with each site inoculated subcutaneously with a different strain (two sites model). As shown in Fig. 5.3c, even under conditions of distant s.c. inoculation sites, the ability of Kim53ΔJ+P to counteract the lethality of Kim53pGFP was substantial. It therefore appeared that s.c infection with Kim53ΔJ+P led to the development of rapid systemic resistance against a subcutaneous challenge that would otherwise kill the animal within 6 days. Further, we analyzed whether this protective effect can also protect against i.n., i.v. and intraperitoneal (i.p.) challenges. We infected mice subcutaneously with 1×10^4 cfu of Kim53ΔJ+P strain and concomitantly or 3 days later we challenged the mice with the wild type strain. As can be seen (Fig. 5.4) when the two strains were injected concomitantly, protection was achieved in all infection models. Moreover, when the cytotoxic strain was administrated 3 days before the challenge, the level of protection was dramatically elevated. These results indicate that a time-dependent systemic protection was developed, yielding full protection when the cytotoxic strain was administrated 3 days before the challenge.

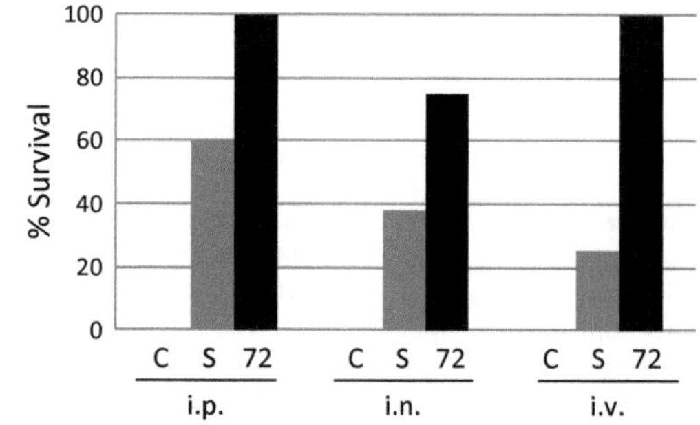

Fig. 5.4 Protection activated by s.c. infection of mice with the cytotoxic strain against lethal intravenous, intranasal or intraperitoneal challenges with virulent *Y. pestis*. Mice were injected subcutaneously with 1×10^4 cfu of Kim53ΔJ+P. Then, either simultaneously (S) or 72 h later (72), mice were exposed intraparentally (i.p.), intranasally (i.n.) or intravenously (i.v.) to 2×10^3 cfu of the virulent Kim53pGFP. Control mice (C) were infected with Kim53pGFP

5.2.4 *The Cross-Priming Mechanism is not Involved in the Protective Effect Induced by the Cytotoxic Kim53ΔJ+P Strain*

The association between the enhanced apoptotic effect of Kim53ΔJ+P and the rapidly acquired protection is reminiscent of the proposed MHC-I-mediated cross-priming mechanism of CD8 T cell activation, which is initiated by pathogen-induced apoptosis (Winau et al., 2006; Winau et al., 2004). However, we found that the attenuation of the cytotoxic strain is not a result of such a cross-priming mechanism since Kim53ΔJ+P strain was also attenuated in the C57BL/6 J mouse strain which is deleted for class I MHC genes ($K^bD^{b-/-}$). Furthermore, inflammatory processes in the spleen did not seem to account for the rapid development of systemic protection, since no significant differences could be demonstrated between the levels of pro-inflammatory cytokines (TNF-α and IL-6) and Th1 cytokines (IFN-γ, IL-2 and IL-12) at 36, 60 and 72 h p.i. with Kim53ΔJ+P or Kim53pGFP (data not shown). In addition, we could not detect development of antibodies against the highly immunogenic F1 antigen of *Y. pestis*.

5.3 Discussion

The ability of bacterial pathogens to induce apoptosis in target immune cells such as macrophages and neutrophils provides an obvious advantage during infection, since these cells would otherwise kill the pathogens. Nevertheless, under particular circumstances, preventing apoptosis can provide a survival advantage by allowing the bacteria to replicate inside a shielded niche and eventually to be transported to target organs while avoiding recognition by the immune system.

To examine the relationship between *Y. pestis* cytotoxicity and its virulence, we generated a cytotoxic *Y. pestis* strain expressing YopP from *Y. enterocolitica* (Kim53ΔJ+P) and assessed its cytotoxic activity and virulence in mouse models of plague disease, in comparison with the control virulent *Y. pestis* strain. We found that the YopP-expressing strain acquired markedly enhanced cytotoxic activity in its interaction with J774A.1 macrophage cells in vitro and in the spleen of infected mice (Fig. 5.1). Concomitant with this enhanced cytotoxicity, the subcutaneously administered YopP-expressing *Y. pestis* strain exhibited a dramatic decrease in virulence, as reflected by elevation of the LD_{50} value from one cfu to more than 10^7 cfu. The colonization of internal organs by the YopP-expressing *Y. pestis* strain was substantially reduced and most significantly, bacteria could not be detected at all in the blood (Table 5.2). It appears therefore that an inverse relationship between YopP-induced cytotoxicity and virulence upon infection via the subcutaneous route does exist. A similar trend was recently reported for a recombinant *Y. pseudotuberculosis* strain expressing *Y. enterocolitica* YopP, which demonstrated impaired virulence in oral mouse infections (Brodsky and Medzhitov, 2008).

It is quite striking however that the virulence of the cytotoxic strain was preserved when it was inoculated via the intravenous and intranasal routes of infection (Fig. 5.2b, c), emphasizing the uniqueness of the subcutaneous mode of infection for the manifestation of the *Y. pestis* attenuated phenotype.

One possible explanation for this site-specific phenotypic attenuation is that the subcutaneous administration of the cytotoxic strain may induce a rapid immune protective response. This notion is indeed supported by the studies of mixed subcutaneous infections of a virulent *Y. pestis* strain and the cytotoxic strain (Fig. 5.3a). The cross-protection provided by the cytotoxic strain could overcome infection with the virulent strain depending on the cfu ratio between the strains (Fig. 5.3b). Two possible mechanisms could account for this finding: a localized destruction of co-infected immune cells leading to disruption of the propagation of *Y. pestis*, or alternatively induction of certain resistance mechanisms. The finding that effective cross-protection could be mediated by subcutaneous inoculation of the cytotoxic strain even when the co-infecting virulent strain was administrated at a separate site supported the assumption of systemic protection (Fig. 5.3c). Most striking was the observation that effective cross-protection against intraperitoneal, intravenous and intranasal lethal challenges of the virulent strain was acquired by subcutaneous administration of the cytotoxic strain (Fig. 5.4).

Pathogens-induced apoptosis may lead to activation of the cross-priming mechanism resulting in CD8 T cell activation (Winau et al., 2006, 2004). However, our results show that the cross-priming mechanism cannot account for the dramatic attenuation of the cytotxic strain. The mechanistic basis underlying the development of the rapid protective response by subcutaneous administration of the cytotoxic strain therefore remains to be determined, and is probably related to a combination of innate and adaptive immunity.

Acknowledgments We thank Mr. Gur David, Mr. Aftalion Moshe, and Mrs. Shirley Lazar for their excellent technical assistance.

References

Ben-Gurion, R. & Shafferman, A. (1981) Essential virulence determinants of different *Yersinia* species are carried on a common plasmid. *Plasmid,* 5, 183–7.

Brodsky, I. E. & Medzhitov, R. (2008) Reduced secretion of YopJ by *Yersinia* limits in vivo cell death but enhances bacterial virulence. *PLoS Pathog,* 4, e1000067.

Brubaker, R. R. (1991) Factors promoting acute and chronic diseases caused by *Yersiniae*. *Clin Microbiol Rev,* 4, 309–24.

Cavanaugh, D. C. & Randall, R. (1959) The role of multiplication of *Pasteurella pestis* in mononuclear phagocytes in the pathogenesis of flea-borne plague. *J Immunol,* 83, 348–63.

Cornelis, G. & Wolf-Watz, H. (1997) The *Yersinia* Yop virulon: a bacterial system for subverting eukaryotic cells. *Mol Microbiol,* 23, 861–7.

Erfurth, S. E., Grobner, S., Kramer, U., Gunst, D. S., Soldanova, I., Schaller, M., Autenrieth, I. B. & Borgmann, S. (2004) *Yersinia enterocolitica* induces apoptosis and inhibits surface molecule expression and cytokine production in murine dendritic cells. *Infect Immun,* 72, 7045–54.

Ferber, D. M. & Brubaker, R. R. (1981) Plasmids in *Yersinia pestis*. *Infect Immun,* 31, 839–41.

Gemski, P., Lazere, J. & Casey, T. (1980) Plasmid associated with pathogenicity and calcium dependency of *Yersinia enterocolitica*. *Infect Immun,* 27, 682–5.

Lemaitre, N., Sebbane, F., Long, D. & Joseph Hinnebusch, B. (2006) *Yersinia pestis* YopJ Suppresses Tumor Necrosis Factor Alpha Induction and Contributes to Apoptosis of Immune Cells in the Lymph Node but Is Not Required for Virulence in a Rat Model of Bubonic Plague. *Infect Immun,* 74, 5126–31.

Lukaszewski, R. A., Kenny, D. J., Taylor, R., Rees, D. G., Hartley, M. G. & Oyston, P. C. (2005) Pathogenesis of *Yersinia pestis* infection in BALB/c mice: effects on host macrophages and neutrophils. *Infect Immun,* 73, 7142–50.

Mills, S., Boland, A., Sory, M., Van Der Smissen, P., Kerbourch, C., Finlay, B. & Cornelis, G. (1997) *Yersinia enterocolitica* induces apoptosis in macrophages by a process requiring functional type III secretion and translocation mechanisms and involving YopP, presumably acting as an effector protein. *Proc Natl Acad Sci USA,* 94, 12638–43.

Monack, D., Mecsas, J., Ghori, N. & Falkow, S. (1997) *Yersinia* signals macrophages to undergo apoptosis and YopJ is necessary for this cell death. *Proc Natl Acad Sci USA.,* 94, 10385–90.

Monack, D. M., Mecsas, J., Bouley, D. & Falkow, S. (1998) *Yersinia*-induced apoptosis in vivo aids in the establishment of a systemic infection of mice. *J Exp Med,* 188, 2127–37.

Perry, R. & Fetherston, J. (1997) *Yersinia pestis*-etiologic agent of plague. *Clin Microbiol Rev,* 10, 35–66.

Portnoy, D. & Falkow, S. (1981) Virulence-associated plasmids from *Yersinia enterocolitica* and *Yersinia pestis*. *J Bacteriol,* 148, 877–83.

Pujol, C. & Bliska, J. B. (2003) The ability to replicate in macrophages is conserved between *Yersinia pestis* and *Yersinia pseudotuberculosis*. *Infect Immun,* 71, 5892–9.

Ruckdeschel, K., Roggenkamp, A., Lafont, V., Mangeat, P., Heesemann, J. & Rouot, B. (1997a) Interaction of *Yersinia enterocolitica* with macrophages leads to macrophage cell death through apoptosis. *Infect Immun,* 65, 4813–21.

Straley, S. C. & Bowmer, W. S. (1986) Virulence genes regulated at the transcriptional level by Ca2+ in *Yersinia pestis* include structural genes for outer membrane proteins. *Infect Immun,* 51, 445–54.

Velan, B., Bar-Haim, E., Zauberman, A., Mamroud, E., Shafferman, A. & Cohen, S. (2006) Discordance in the effects of *Yersinia pestis* on dendritic cell functions: induction of maturation and paralysis of migration. *Infect Immun,* 74, 6365–76.

Weeks, S., Hill, J., Friedlander, A. & Welkos, S. (2002) Anti-V antigen antibody protects macrophages from *Yersinia pestis* – induced cell death and promotes phagocytosis. *Microb Pathog,* 32, 227–37.

Winau, F., Kaufmann, S. H. & Schaible, U. E. (2004) Apoptosis paves the detour path for CD8 T cell activation against intracellular bacteria. *Cell Microbiol,* 6, 599–607.

Winau, F., Weber, S., Sad, S., De Diego, J., Hoops, S. L., Breiden, B., Sandhoff, K., Brinkmann, V., Kaufmann, S. H. & Schaible, U. E. (2006) Apoptotic vesicles crossprime CD8 T cells and protect against tuberculosis. *Immunity,* 24, 105–17.

Zauberman, A., Cohen, S., Mamroud, E., Flashner, Y., Tidhar, A., Ber, R., Elhanany, E., Shafferman, A. & Velan, B. (2006) Interaction of *Yersinia pestis* with macrophages: limitations in YopJ-dependent apoptosis. *Infect Immun,* 74, 3239–50.

Zauberman, A., Tidhar, A., Levy, Y., Bar-Haim, E., Halperin, G., Flashner, Y., Cohen, S., Shafferman, A. & Mamroud, E. (2009) *Yersinia pestis* endowed with increased cytotoxicity is avirulent in a bubonic plague model and induces rapid protection against pneumonic plague. *PLoS One,* 4, e5938.

Zauberman, A., Velan, B., Mamroud, E., Flashner, Y., Shafferman, A. & Cohen, S. (2007) Disparity between *Yersinia pestis* and *Yersinia enterocolitica* O:8 in YopJ/YopP-dependent functions. *Adv Exp Med Biol,* 603, 312–20.

Chapter 6
The NlpD Lipoprotein of *Yersinia pestis* is Essential for Cell Separation and Virulence

Avital Tidhar, Yehuda Flashner, Sara Cohen, Yinon Levy, Ayelet Zauberman, David Gur, Moshe Aftalion, Shirley Lazar, Eytan Elhanany, Anat Zvi, Avigdor Shafferman, and Emanuelle Mamroud

Abstract The well-characterized virulence factors of *Yersinia pestis* the causative agent of plague are those encoded by the virulence plasmids. Previously we have isolated an attenuated *Y. pestis* transposon insertion mutant in which the chromosomal *pcm* gene was disrupted. The *pcm* gene is located within a putative stress response locus that includes the *surE*, *nlpD*, and *rpoS* genes. In this study, we investigated the expression and the role of *pcm* locus genes in *Y. pestis* pathogenesis by constructing a set of isogenic *surE*, *pcm*, *nlpD* and *rpoS* mutants in the fully virulent Kimberley53 strain. We show that the NlpD lipoprotein is the only factor encoded from the *pcm* locus that is essential for *Y. pestis* virulence. A chromosomal deletion of the *nlpD* gene sequence resulted in a drastic reduction in virulence to an LD_{50} of at least 10^7 cfu for subcutaneous and airway routes of infection. The mutant was unable to colonize mouse organs following infection. The unsegmented morphology of the *nlpD* mutant indicates that NlpD is involved in cell separation; however, deletion of *nlpD* did not affect *in vitro* growth rate. Trans-complementation experiments with the *Y. pestis nlpD* gene restored virulence and all other phenotypic defects. Finally, we demonstrate that the *nlpD* mutant could be used as a very potent live vaccine against bubonic and pneumonic plague.

Keywords Plague · *Y. pestis* · Vaccine · Lipoprotein · *nlpD* · *pcm*

6.1 Introduction

Yersinia pestis is the etiological agent of plague, which has caused millions of deaths in three world pandemics and is still a public health issue in some regions of the world. It is a rapidly progressing disease leading to high mortality rates in untreated

---

A. Tidhar (✉)
Department of Biochemistry and Molecular Genetics, Israel Institute for Biological Research, P.O. Box 19, Ness-Ziona, 74100, Israel
e-mail: avitalt@iibr.gov.il

A. Shafferman et al. (eds.), *The Challenge of Highly Pathogenic Microorganisms*,
DOI 10.1007/978-90-481-9054-6_6, © Springer Science+Business Media B.V. 2010

patients and can spread from person to person (Bearden et al., 1997). These characteristics led to the recognition of *Y. pestis* as a potential threat agent (Inglesby et al., 2000).

The ability of *Y. pestis* to respond to the host environment and to overcome immune systems is attributed to the combined activity of multiple virulence mechanisms. These mechanisms include the type III secretion system (TTSS) (Viboud and Bliska, 2005) the plasminogen activator factor (Lathem et al., 2007; Sodeinde et al., 1992; Welkos et al., 1997) and an iron acquisition system (Bearden et al., 1998; Bearden et al., 1997).

In a previous study, we isolated a highly attenuated mutant designated Kim53-K9 in which the *pcm* gene was disrupted by a mini-Tn5 transposon insertion (Flashner et al., 2004). It was demonstrated in several Gram negative bacteria that the *pcm* gene and its neighboring genes *surE, nlpD* and *rpoS*, are involved in survival under environmental stress conditions (Hengge-Aronis, 2002; Anderson, 1997; Visick et al., 1998).

In the present study, we describe the characterization of the *pcm* genomic locus in *Y. pestis*. The expression pattern of the genes was evaluated along with their respective contributions to *Y. pestis* virulence using mouse models of bubonic and pneumonic plague.

6.2 Results

6.2.1 Comparison of the pcm Locus of Y. pestis, E. coli, S. typhimurium and S. flexneri

Inspection of the genome sequence of *Y. pestis* CO92 (Parkhill et al., 2001) revealed that the *pcm* locus shows high conservation of the gene order with respect to the related enteropathogens. However, whereas the *Y. pestis* SurE, Pcm and RpoS proteins share high levels of identity with their related proteins from *E. coli, S. typhimurium* and *S. flexneri* (88–97%), the NlpD lipoproteins are more divergent (64–67%) (Table 6.1). Examination of the NlpD sequences within the *Yersinia* genus revealed that the *nlpD* gene products in *Y. pseudotuberculosis* and *Y. enterocolitica* have relatively high levels of sequence similarity to the corresponding gene product of *Y. pestis* (98 and 94%, respectively, data not shown).

Table 6.1 Similarity comparison of the *Y. pestis pcm* locus-encoded proteins to the corresponding proteins in enterobacterial pathogens

Y. pestis[a] *vs.*	SurE%	Pcm%	NlpD%	RpoS%
E. coli	88	88	64	97
S. typhimurium	89	90	65	97
S. flexneri	88	88	67	97

[a]CO92 genome

6.2.2 *Assessment of Y. pestis Pcm, NlpD and RpoS Expression in Y. pestis Derivatives*

The virulent *Y. pestis* Kimberley53 strain was used to generate a series of isogenic deletion mutants within the *pcm* locus genes. In each of the newly constructed mutants, a defined region of a single gene was deleted and replaced with a kanamycin resistance cassette resulting in *Y. pestis* Kimberley53 derivatives that were designated Kim53Δ*surE*, Kim53Δ*pcm*, Kim53Δ*nlpD* and Kim53Δ*rpoS*. PCR analysis verified that all of the Kimberley53-derived strains carry the chromosomal *pgm* locus as well as the pMT1, pCD1 and pPCP1 plasmids. To evaluate the expression of Pcm, NlpD and RpoS in these strains, bacterial cultures were grown to stationary phase (24 h) at 37°C and 28°C in heart infusion broth (HIB, Difco) and then subjected to Western blot analysis with highly specific antibodies. The patterns of expression were independent of the growth temperature (data not shown). In the Kim53Δ*surE* mutant, Pcm was not detected, whereas the levels of NlpD and RpoS were comparable to the wild type strain (Fig. 6.1). In the Kim53-K9 strain, Pcm and NlpD were not detected, whereas the level of RpoS was comparable to the wild type strain (Fig. 6.1). Furthermore, the level of NlpD was reduced in Kim53Δ*pcm,* whereas that of RpoS was not altered. The *rpoS* gene resides downstream of *nlpD*. In bacterial enteropathogens, the major *rpoS* promoter has been identified within the *nlpD* gene (Navarro-Llorens et al., 2002 and references therein, Hirsch and Elliott, 2005). We therefore constructed two *nlpD*-null mutants: KimΔ$nlpD_L$, in which the putative *rpoS* promoter elements were deleted, and KimΔ*nlpD,* in which these sequences were preserved. As expected, the level of RpoS was decreased in KimΔ$nlpD_L$ but not in KimΔ*nlpD* (Fig. 6.1). The last gene in the *pcm* locus that was analyzed was *rpoS*, which is predicted to encode an alternative sigma factor expressed during stress conditions (Hengge-Aronis, 2002). The Pcm and NlpD expression levels in Kim53Δ*rpoS* were comparable to the levels found in the wild type strain (Fig. 6.1). To further characterize the expression of

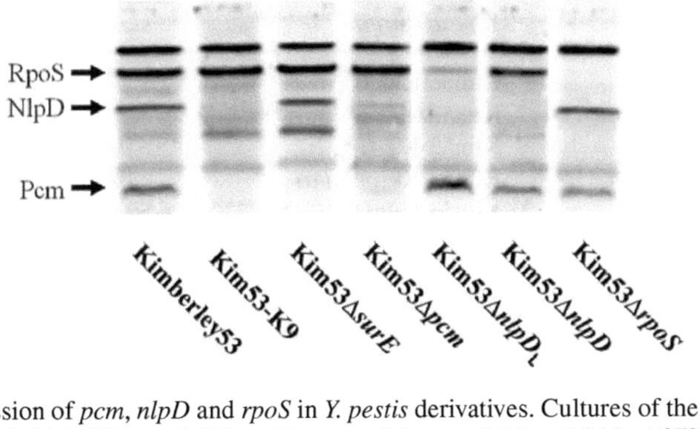

Fig. 6.1 Expression of *pcm, nlpD* and *rpoS* in *Y. pestis* derivatives. Cultures of the *Y. pestis* strains were inoculated (initial $OD_{660} = 0.01$) and incubated for an additional 24 h at 37°C. Western blot analysis was performed with anti-Pcm, anti-NlpD and anti-RpoS antibodies

the genes in the *pcm* locus, primer extension analysis and RT-PCR were preformed, which demonstrated a complex regulation pattern (Tidhar et al., 2009). It appears that transcription of the *pcm* gene is driven by control elements within the *surE* gene and that transcription of *nlpD* can be initiated from two sites within the *surE* and *pcm* genes. The latter can be responsible for *rpoS* transcription which can be initiated as well from a site within *nlpD* gene. These results are compatible with expression patterns of the various deletion mutants and can also explain why the insertion mutation in Kim53-K9 abolished NlpD expression.

6.2.3 Within the pcm Locus, NlpD is the Only Essential Factor for Development of Plague

The virulence of Kimberley53-deletion mutants in mouse models of bubonic [subcoetaneous infection (s.c.)] and pneumonic plague [intranasal infection (i.n.)] was evaluated in comparison to the wild-type Kimberley53 strain. In both infection models the Kim53Δ*pcm* strain (Table 6.2) as well as Kim53Δ*rpoS* (Tidhar et al., 2009) were found to be highly virulent as the wild-type strain. In contrast Kim53-K9 and Kim53Δ*nlpD* strains were avirulent with LD_{50} values higher then 10^7cfu (Table 6.2) and infected mice did not show disease symptoms. To substantiate our observation that the attenuated phenotype of the NlpD-null mutants (Kim53Δ*nlpD* and Kim53-K9) resulted from the loss of NlpD expression, we examined the ability of a trans-complemented *Y. pestis nlpD* gene to restore the mutant's virulence. The complete coding sequence of the *Y. pestis* Kimberley53 *nlpD* gene was cloned into pBR322 and the plasmid was introduced into Kim53Δ*nlpD* and Kim53-K9 to give the complemented strains Kim53Δ*nlpD*(p*nlpD*) and Kim53-K9(p*nlpD*). Western blot analysis confirmed that the NlpD expression level in both strains was

Table 6.2 Virulence of *Y. pestis* strains in mouse models of bubonic and pneumonic plague

	LD_{50} value [a,b] (cfu)	
Y. pestis strain	s.c. route	i.n. route
Kimberley53	1–3	5.5×10^2
Kim53Δ*pcm*	3	3.3×10^3
Kim53-K9	$>1\times10^7$	ND
Kim53-K9(p*nlpD*)	$<1\times10^2$	ND
Kim53Δ*nlpD*	$>2\times10^7$	$>4\times10^7$
Kim53Δ*nlpD*(p*nlpD*)	$<1\times10^2$	$<1.5\times10^3$

ND = not determined
[a]The "<" symbol indicates that the calculated LD_{50} value is the minimal infection dose tested, under which more than 50% of the animals died
[b] The ">" symbol indicates that the calculated LD_{50} value is the maximal infection dose tested under which less than 50% of the animals died

comparable to the wild type strain (data not shown). The *nlpD*-complemented strains regained the virulent phenotype (Table 6.2), indicating that NlpD is an essential *Y. pestis* virulence factor.

6.2.4 Kim53ΔnlpD has Impaired Cell Morphology and Cannot Colonize Internal Organs

Bacterial lipoproteins are components of the cell envelope of Gram-negative bacteria and are usually localized at the periplasmic space anchored to either the outer or the inner membrane (Seydel et al., 1999). *Y. pestis* NlpD has a threonine residue immediately after the fatty-acylated cysteine and is therefore predicted to reside in the outer membrane similarly to the *E. coli* NlpD (Robichon et al., 2005; Seydel et al., 1999). Microscope analyses indicated that *in vitro* culturing of Kim53Δ*nlpD* leads to formation of unsegmented chains (Fig. 6.2a) as opposed to the single cell morphology of Kimberley53 (Fig. 6.2b). Complementation of *nlpD* expression eliminated completely the abnormal filamented cell morphology and restored the wild type morphology (Fig. 6.2c).These results indicate that NlpD is involved in bacterial cell separation. In spite of this change, NlpD is not essential for *Y. pestis* cell growth, at least under laboratory conditions, as shown by the normal growth rate of the *nlpD* mutant (Fig. 6.3a).

Monitoring the colonization of internal organs by Kim53Δ*nlpD* and the wild type strain Kimberley53, revealed a dramatic disparity between the two strains in both bubonic and pneumonic plague models. In contrast to the wild type strain, which reached high bacterial loads in internal organs (lymph nodes, spleen, lungs) and blood at late stages of the disease, Kim53Δ*nlpD* could not be detected in these organs at 24 hours post infection and up to 10 days post infection (data not shown).

These observations indicate that NlpD is required for *Y. pestis* propagation and dissemination to target organs, and corroborate the non-virulent phenotype of Kim53Δ*nlpD* in both pneumonic and bubonic infection models.

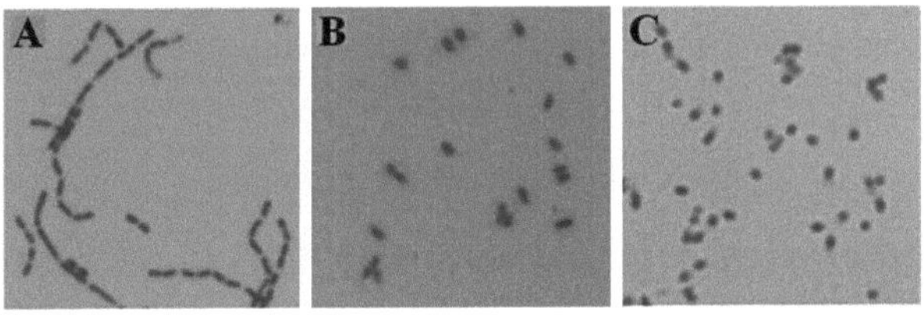

Fig. 6.2 Morphology of *Y. pestis* bacilli. Gram staining of *Y. pestis* strains. Kimberley53Δ*nlpD* (**a**), Kim53 (**b**) and Kim53Δ*nlpD*(p*nlpD*) (**c**) were grown at 28°C in HIB for 24 h. Bacilli were stained and observed by light microscopy at a magnification of ×1000

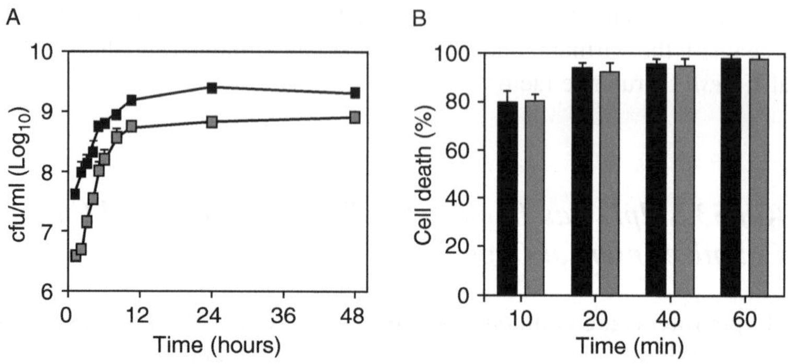

Fig. 6.3 Survival during environmental stress conditions. (**a**) Prolonged growth. Kimberley53 and Kim53Δ*nlpD* were grown in HIB for 48 h at 37°C. (**b**) Resistance to oxidative stress. Bacterial cultures (HIB, 37°C, $OD_{660} = 0.1$) of Kimberley53 (*black*) and Kim53Δ*nlpD* (*light gray*) were exposed to 25 mM of H_2O_2 for the indicated time. Viable cell counts were determined by plating dilutions on BHI agar and incubating at 28°C for 48 h

6.2.5 Evaluation of T3SS Activity and Stress Resistance of the Kim53ΔnlpD Mutant

It can be assumed that deletion of *nlpD* in *Y. pestis* may influence membrane-related functions, such as TTSS activity, which is known to be essential for *Y. pestis* virulence, and thus could lead to the attenuated phenotype. However, preservation of TTSS functionality, at least *in vitro*, was demonstrated by retention of calcium-dependent growth at 37°C, expression and secretion of Yop effectors such as YopE and YopJ into the culture medium following exposure to inducing conditions, and suppression of cytokine secretion from infected macrophages (Tidhar et al., 2009).

The genes flanking *nlpD* were found in many enteropathogens to be involved in survival during stationary phase and under other environmental stress conditions (Hengge-Aronis, 2002). In search of a possible explanation for the attenuation of Kim53Δ*nlpD*, we evaluated the involvement of *Y. pestis* NlpD in resistance to prolonged growth in culture (HIB), and in oxidative conditions that simulate the intra-phagosomal milieu. Exposure of Kim53Δ*nlpD* and Kimberley53 to prolonged *in vitro* growth (up to 96 h), and to oxidative stress (10–50 mM H_2O_2) indicated that NlpD is not essential for resistance to these conditions (Fig. 6.3a, b). In contrast, Kim53Δ*nlpD* was slightly more susceptible than wild type Kimberley53 to acidic pH (Tidhar et al., 2009).

6.2.6 The Potential of Kim53ΔnlpD as a Vaccine Strain

The high level of attenuation of Kim53Δ*nlpD* motivated us to evaluate the potential of this defined mutant to serve as a vaccine. In a model of bubonic plague, mice were injected subcutaneously with Kim53Δ*nlpD* or *Y. pestis* EV76 prototype vaccine

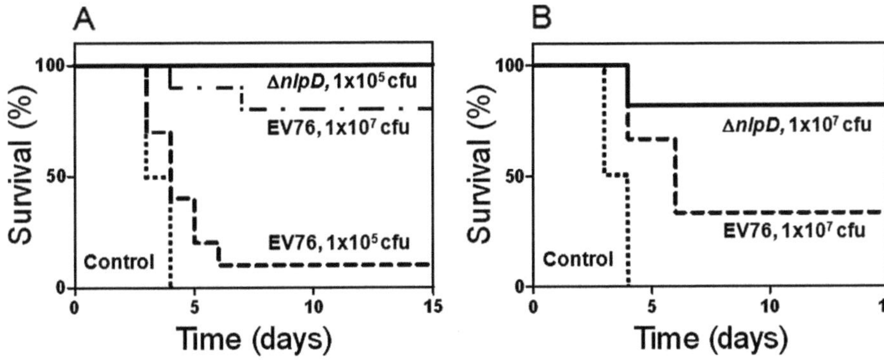

Fig. 6.4 Evaluation of the potential of Kim53Δ*nlpD* as a vaccine strain (**a**) Bubonic plague model. Mice were administrated subcutaneously with 10^5 cfu of Kim53Δ*nlpD* (*strait line*), 10^5 and 10^7cfu of EV76 (*dashed line* and *dash-dot line*, respectively) or with PBS (control, *dotted line*). Fifty days later, mice were challenged subcutaneously with 10^5 LD_{50} of the virulent *Y. pestis* Kimberley53 strain. (**b**) Pneumonic plague model. Mice were administrated subcutaneously with 10^7 cfu of Kim53Δ*nlpD* (*strait line*), EV76 (*dashed line*) or with PBS (control, *dotted line*). Fifty days later, mice were challenged intranasally with $10LD_{50}$ of the virulent *Y. pestis* Kimberley53 strain

strain (1×10^5 cfu). Fifty days later, mice were challenged subcutaneously with 1×10^5 LD_{50} of the fully virulent Kimberley53 strain. All mice vaccinated with Kim53Δ*nlpD* survived s.c. challenge with 1×10^5 LD_{50} of Kimberley53 without showing signs of illness, whereas a similar dose of EV76 failed to elicit significant protective immunity (Fig. 6.4a). EV76 was able to protect mice only at 100 fold higher vaccination dose of 10^7 cfu (Fig. 6.4a).

In the mouse model of pneumonic plague, a single s.c. immunization with 1×10^7 cfu of Kim53Δ*nlpD* or EV76 was followed 50 days later by i.n. challenge of $10LD_{50}$ (5,500 cfu) of the virulent *Y. pestis* Kimberley53 strain. While all control mice died within 4 days, a protection level of 82% was obtained following immunization with 1×10^7 cfu of Kim53Δ*nlpD* (Fig. 6.4b). In contrast, EV76 was able to elicit a protection level of only 33% (Fig. 6.4b). These findings suggest that Kim53Δ*nlpD* may be a suitable platform for a live vaccine.

6.3 Discussion

In the present study, we have characterized the *Y. pestis pcm* locus genes and analyzed their expression and involvement in the pathogenesis of *Y. pestis* using mouse models of bubonic and pneumonic plague. Systematic deletion mutagenesis of the *surE, pcm, nlpD* and *rpoS* genes and complementation studies allowed us to identify the NlpD lipoprotein as the only essential factor for *Y. pestis* pathogenesis in this locus. Subcutaneous and intranasal administration of a *Y. pestis nlpD*-null strain to mice demonstrated that this strain is severely attenuated ($LD_{50}>10^7$cfu, Table 6.2), and is impaired in its ability to colonize internal organs. Trans-complementation experiments verified that NlpD is an essential *Y. pestis* virulence factor (Table 6.2). To the best of our knowledge, this is the first demonstration of a single chromosomal *Y. pestis* factor that is essential for development of both bubonic and pneumonic plague.

The *nlpD*-null mutant did not differ significantly from the wild type strain in the ability to survive during prolonged growth in rich broth and during exposure to oxidative conditions (Fig. 6.3a, b). Yet, we found a slight increase in the mutant's sensitivity to *in vitro* acidic conditions. While the latter does not seem to account for the dramatic loss of virulence, further studies are needed to understand the importance of the sensitivity of the mutant to acidic conditions (Tidhar et al., 2009).

The aberrant shape of the *Y. pestis* Δ*nlpD* bacilli (Fig. 6.2b) suggests that NlpD is important for cell separation. Consistent with this assumption is the structure of NlpD, which contains an N-terminal LysM domain found in a variety of enzymes involved in bacterial cell wall degradation (Bateman and Bycroft, 2000; Buist et al., 2008; Lai et al., 2006). NlpD also contains a C-terminal M23 metallopeptidase region found in proteins that are involved in bacterial cell separation (Bernhardt and de Boer, 2004). The observed phenotype of the *nlpD*-null mutant suggests a linkage between impairment of cell separation and attenuation of virulence. Similar links have already been described for other bacterial pathogens (Kajimura et al., 2005; Pilgrim et al., 2003).

Interestingly, the highly attenuated phenotype of the *nlpD* mutant and its inability to colonize host organs did not seem to prevent the development of immunity against plague following s.c. infection. Immunization of mice with low doses of Kim53Δ*nlpD* resulted in remarkably higher protection levels against bubonic and pneumonic plague than did immunization with the *Y. pestis* EV76 vaccine strain (Fig. 6.4). The observed development of protective immunity could have practical implications in the design of future *Y. pestis* vaccines or therapies against both bubonic and pneumonic plague.

Acknowledgments We thank Mrs. Brut P. and Mrs. Stein D. for excellent technical assistance. We also thank Dr. François Norel from the Institute Pasteur Paris, France for the generous gift of RpoS antibodies.

References

Anderson, P. (1997) Kinase cascades regulating entry into apoptosis. *Microbiol Mol Biol Rev,* 61, 33–46.

Bateman, A. & Bycroft, M. (2000) The structure of a LysM domain from E. coli membrane-bound lytic murein transglycosylase D (MltD). *J Mol Biol,* 299, 1113–9.

Bearden, S. W., Fetherston, J. D. & Perry, R. D. (1997) Genetic organization of the yersiniabactin biosynthetic region and construction of avirulent mutants in Yersinia pestis. *Infect Immun,* 65, 1659–68.

Bearden, S. W., Staggs, T. M. & Perry, R. D. (1998) An ABC transporter system of Yersinia pestis allows utilization of chelated iron by Escherichia coli SAB11. *J Bacteriol,* 180, 1135–47.

Bernhardt, T. G. & De Boer, P. A. (2004) Screening for synthetic lethal mutants in Escherichia coli and identification of EnvC (YibP) as a periplasmic septal ring factor with murein hydrolase activity. *Mol Microbiol,* 52, 1255–69.

Buist, G., Steen, A., Kok, J. & Kuipers, O. P. (2008) LysM, a widely distributed protein motif for binding to (peptido)glycans. *Mol Microbiol,* 68, 838–47.

Flashner, Y., Mamroud, E., Tidhar, A., Ber, R., Aftalion, M., Gur, D., Lazar, S., Zvi, A., Bino, T., Ariel, N., Velan, B., Shafferman, A. & Cohen, S. (2004) Generation of Yersinia pestis attenuated strains by signature-tagged mutagenesis in search of novel vaccine candidates. *Infect Immun,* 72, 908–15.

Hengge-Aronis, R. (2002) Signal transduction and regulatory mechanisms involved in control of the sigma(S) (RpoS) subunit of RNA polymerase. *Microbiol Mol Biol Rev,* 66, 373–95.

Hirsch, M. & Elliott, T. (2005) Stationary-phase regulation of RpoS translation in Escherichia coli. *J Bacteriol,* 187, 7204–13.

Inglesby, T. V., Dennis, D. T., Henderson, D. A., Bartlett, J. G., Ascher, M. S., Eitzen, E., Fine, A. D., Friedlander, A. M., Hauer, J., Koerner, J. F., Layton, M., Mcdade, J., Osterholm, M. T., O'toole, T., Parker, G., Perl, T. M., Russell, P. K., Schoch-Spana, M. & Tonat, K. (2000) Plague as a biological weapon: medical and public health management. Working Group on Civilian Biodefense. *Jama,* 283, 2281–90.

Kajimura, J., Fujiwara, T., Yamada, S., Suzawa, Y., Nishida, T., Oyamada, Y., Hayashi, I., Yamagishi, J., Komatsuzawa, H. & Sugai, M. (2005) Identification and molecular characterization of an N-acetylmuramyl-L-alanine amidase Sle1 involved in cell separation of Staphylococcus aureus. *Mol Microbiol,* 58, 1087–101.

Lai, X., Weng, J., Zhang, X., Shi, W., Zhao, J., Wang, H. & Wang, H. (2006) MSTF: a domain involved in bacterial metallopeptidases and surface proteins, mycobacteriophage tape-measure proteins and fungal proteins. *FEMS Microbiol Lett,* 258, 78–82.

Lathem, W. W., Price, P. A., Miller, V. L. & Goldman, W. E. (2007) A plasminogen-activating protease specifically controls the development of primary pneumonic plague. *Science,* 315, 509–13.

Navarro-Llorens, J. M., Martinez-Garcia, E. & Tormo, A. (2002) Enterobacter cloacae rpoS promoter and gene organization. *Arch Microbiol,* 179, 33–41.

Parkhill, J., Wren, B. W., Thomson, N. R., Titball, R. W., Holden, M. T., Prentice, M. B., Sebaihia, M., James, K. D., Churcher, C., Mungall, K. L., Baker, S., Basham, D., Bentley, S. D., Brooks, K., Cerdeno-Tarraga, A. M., Chillingworth, T., Cronin, A., Davies, R. M., Davis, P., Dougan, G., Feltwell, T., Hamlin, N., Holroyd, S., Jagels, K., Karlyshev, A. V., Leather, S., Moule, S., Oyston, P. C., Quail, M., Rutherford, K., Simmonds, M., Skelton, J., Stevens, K., Whitehead, S. & Barrell, B. G. (2001) Genome sequence of Yersinia pestis, the causative agent of plague. *Nature,* 413, 523–7.

Pilgrim, S., Kolb-Maurer, A., Gentschev, I., Goebel, W. & Kuhn, M. (2003) Deletion of the gene encoding p60 in Listeria monocytogenes leads to abnormal cell division and loss of actin-based motility. *Infect Immun,* 71, 3473–84.

Robichon, C., Vidal-Ingigliardi, D. & Pugsley, A. P. (2005) Depletion of apolipoprotein N-acyltransferase causes mislocalization of outer membrane lipoproteins in Escherichia coli. *J Biol Chem,* 280, 974–83.

Seydel, A., Gounon, P. & Pugsley, A. P. (1999) Testing the '+2 rule' for lipoprotein sorting in the Escherichia coli cell envelope with a new genetic selection. *Mol Microbiol,* 34, 810–21.

Sodeinde, O. A., Subrahmanyam, Y. V., Stark, K., Quan, T., Bao, Y. & Goguen, J. D. (1992) A surface protease and the invasive character of plague. *Science,* 258, 1004–7.

Tidhar, A., Flashner, Y., Cohen, S., Levi, Y., Zauberman, A., Gur, D., Aftalion, M., Elhanany, E., Zvi, A., Shafferman, A. & Mamroud, E. (2009) The NlpD lipoprotein is a novel Yersinia pestis virulence factor essential for the development of plague. *PLoS One,* 4, e7023.

Viboud, G. I. & Bliska, J. B. (2005) Yersinia outer proteins: role in modulation of host cell signaling responses and pathogenesis. *Annu Rev Microbiol,* 59, 69–89.

Visick, J. E., Cai, H. & Clarke, S. (1998) The L-isoaspartyl protein repair methyltransferase enhances survival of aging Escherichia coli subjected to secondary environmental stresses. *J Bacteriol,* 180, 2623–9.

Welkos, S. L., Friedlander, A. M. & Davis, K. J. (1997) Studies on the role of plasminogen activator in systemic infection by virulent Yersinia pestis strain C092. *Microb Pathog,* 23, 211–23.

Chapter 7
Dissection of the Functions of the IglC Protein of *Francisella tularensis*

Jeanette E. Bröms, Moa Lavander, and Anders Sjöstedt

Abstract *Francisella tularensis* harbors genes with similarity to genes encoding components of a type VI secretion system (T6SS). These include *iglA* and *iglB*, the homologues of which are conserved in T6SSs. They are part of the *igl* operon, also encompassing the *iglC* and *iglD* genes. We have used a yeast two-hybrid system to study the interaction of the Igl proteins of *F. tularensis* LVS. Previously, we identified a region of IglA necessary for efficient binding to IglB as well as for IglAB protein stability and intra-macrophage growth with an essential role for a conserved α-helical region. Thus, IglA-IglB complex formation is clearly crucial for *Francisella* pathogenicity and the same interaction is conserved in other human pathogens. Herein, the interaction of IglC with other members of the operon was investigated. It showed no binding to the other members in the yeast two-hybrid assay and we found also that two cysteine residues, C191 and C192, predicted to be putative prenylation sites, played no role for the important contribution of IglC to the intracellular replication of *F. tularensis* although C191 was important for the stability of the protein.

Keywords *Francisella* · IglC · Yeast two-hybrid system · Cysteine residues

7.1 Introduction

Francisella tularensis is a highly virulent, gram-negative facultative intracellular bacterium causing the zoonotic disease tularemia, in many mammals. While little is known about the molecular mechanisms of *Francisella* pathogenesis, a key strategy

---

A. Sjöstedt (✉)
Department of Clinical Microbiology, Clinical Bacteriology, and Laboratory for Molecular Infection Medicine Sweden (MIMS), Umeå University, SE-901 85 Umeå, Sweden
e-mail: anders.sjostedt@climi.umu.se

A. Shafferman et al. (eds.), *The Challenge of Highly Pathogenic Microorganisms*,
DOI 10.1007/978-90-481-9054-6_7, © Springer Science+Business Media B.V. 2010

appears to be its ability to survive and replicate within macrophages. *F. tularensis* enters a phagosome but evades phagosome–lysosome fusion and subsequently escapes into the cytoplasm. Several genes necessary for intramacrophage survival as well as growth within the amoeba *Acanthamoebae castellanii* have been identified, many of which are encoded within the 34-kb *Francisella* Pathogenicity Island (FPI) (Nano and Schmerk, 2007). Almost all of the proteins of the FPI are essentially conserved across subspecies. Much focus has been on the members of the *iglABCD* operon. Studies have shown that IglC and IglD are required for the ability of *F. tularensis* to replicate within the cytosol of macrophages and essential for the bacterial escape from the phagosome into the cytoplasm (Lindgren et al., 2004; Santic et al., 2007; Bönquist et al., 2008). In contrast to IglC and IglD that appear to be unique to *F. tularensis*, homologues of IglA and IglB exist in many bacterial species, most of which are either pathogenic to animals or plants or plant symbionts (Das and Chaudhuri, 2003; Nano et al., 2004). Together with several of the FPI encoded proteins, IglA and IglB show homology to proteins thought to be involved in type VI protein secretion (T6S) (de Bruin et al., 2007; Bingle et al., 2008). Functional T6S was recently demonstrated in pathogens like *Vibrio cholerae*, *Pseudomonas aeruginosa*, enteroaggregative *Escherichia coli* (EAEC), *Aeromonas hydrophila*, *Burkholderia mallei* and *Edwardsiella tarda*, and in many of these, a direct link between protein secretion and virulence has been established (Dudley et al., 2006; Mougous et al., 2006; Pukatzki et al., 2006; Schell et al., 2007; Zheng and Leung, 2007; Suarez et al., 2008). IglA has been found to be a cytoplasmic protein required for intramacrophage growth and virulence of *F. novicida* and an interaction with IglB was demonstrated by immunoprecipitation (de Bruin et al., 2007). We have found using a yeast two-hybrid system that an α-helical domain of IglA required for the interaction with IglB and the stability of the IglAB complex, as well as for intracellular growth and virulence of LVS. A similar domain was identified in the IglA homologues of other pathogens such as *Yersinia pseudotuberculosis*, *P. aeruginosa*, *V. cholerae*, *Salmonella typhimurium* and *E. coli* and was found to be essential for the complex formation of the respective IglAB homologues. We now investigated if IglC showed any interaction in the yeast two-hybrid system with other Igl proteins and also addressed if the presence of specific cysteine residues contributed to the function of the protein.

7.2 Materials and Methods

7.2.1 Bacterial Strains, Plasmids and Growth Conditions

F. tularensis strains were grown on modified GC-agar base or in liquid Chamberlain's medium (Chamberlain, 1965) at 37°C. When necessary, kanamycin (Km; 10 μg/ml) or chloramphenicol (2.5 μg/ml) was added.

7.2.2 Yeast Plasmid Construction

To investigate protein-protein interaction studies in yeast, PCR amplified fragments encoding Igl proteins or mutant IglC derivatives constructed by overlap PCR were introduced into pGADT7 and pGBKT7.

7.2.3 The Yeast Two-Hybrid Assay

Transformation of the *Saccharomyces cerevisiae* reporter strains AH109 and Y187, protein expression analysis of yeast lysates and analysis of protein-protein interactions were performed according to established methods (Bröms et al., 2009). Specifically, interactions were determined by growth of yeast on synthetic dropout minimal agar (Clontech Laboratories) devoid of tryptophan, leucine (SD-LT) and adenine resulting from *ADE2* reporter gene activation. The interactive potential was confirmed by comparative growth at 25°C, 30°C and 37°C to provide an insight into the relative energy required for each interaction, and by induction of two independent reporter genes, *HIS3* and *lacZ*, by growing yeast on SD-LT agar lacking histidine and in liquid culture using ONPG (o-Nitrophenyl-beta-D-Galactopyranoside) (Sigma-Aldrich, St. Louis, MO, USA) as substrate respectively. Due to an intrinsic leakiness with the *HIS3* reporter, 3 mM 3-aminotriazole was added to histidine dropout media to suppress false positives (James et al., 1996). Protein expression was verified using antibodies recognizing the activation or DNA-binding domain of GAL4 (Clontech Laboratories). Strain AH109 was used for all yeast analysis, with the exception of the β-galactosidase assay where strain Y187 was used.

7.2.4 Igl Protein Production

Levels of Igl proteins in pellet fractions of *F. tularensis* grown on modified GC-agar base were analyzed by Western blot using polyclonal antibodies recognizing IglA (BEI Resources, Manassas, VA, USA) or IglD (Agrisera, Vännäs, Sweden) or monoclonal antibodies specific for IglB (BEI Resources) or IglC (Bönquist et al., 2008). Proteins were visualized using the Enhanced Chemiluminescence system (ECL) (Amersham Biosciences, Uppsala, Sweden).

7.2.5 Protein Stability

The intrabacterial protein stability assay was adapted from Feldman and colleagues (Feldman et al., 2002) with some modifications. In short, *F. tularensis* was grown overnight at 37°C in liquid Chamberlain medium, diluted 2 × in fresh medium and grown for 1 h before protein synthesis was stopped by addition of 10 μg/ml chloramphenicol (corresponds to time zero). Samples were taken out at different time points and analyzed by Western blot using antisera recognizing IglA or IglB in combination with ECL.

7.2.6 Cultivation and Infection of Macrophages

To determine the ability of *F. tularensis* to grow within macrophages, J774A.1 cells were infected according to our established methods, and bacteria were plated on modified GC-agar base plates for determination of viable counts (Golovliov et al., 2003).

7.3 Results and Discussion

It has been found that there is a mutual dependency on IglA and IglB for their stability suggesting that IglA and IglB may interact in *F. tularensis* and this was recently shown to be the case in *F. novicida* (de Bruin et al., 2007). Previously, we have used the yeast two-hybrid assay, to study the IglA-IglB interaction in more detail (Bröms et al., 2009). We expressed *iglA* from the GAL4 activation domain contained within pGAD7 and *iglB* from the GAL4 DNA-binding domain expressed by pGBKT7. When either plasmid was transformed into the reporter AH109, it did not result in *ADE2* or *HIS3* reporter gene activation (data not shown). In contrast, when the plasmids were co-transformed, the *ADE2* and *HIS3* reporter genes were activated and they enabled the growth of AH109 on minimal media devoid of adenine and histidine, respectively. The reporter genes were activated irrespective of the growth temperature, 25, 30 or 37°C, and this indicated that there was a strong interaction between IglA-IglB. The interaction was directly quantified by use of a β-galactosidase assay. When IglA and IglB were co-expressed, high levels of β-galactosidase were produced (68.8 ± 4.9 units vs. 0.042 ± 0.007 and 0.067 ± 0.012 for IglA or IglB alone), corroborating the strong activation of the *lacZ* reporter gene. We also investigated putative interactions between IglA and IglB and the other gene products of the *Igl* operon, *i.e.*, IglC and IglD, but we observed no interactions between IglA-IglC, IglA-IglD, IglB-IglC, IglB-IglD or IglC-IglD regardless of vector orientation or growth temperature and none of the Igl proteins formed homodimers (data not shown).

It has been proposed that IglC contains cysteine residues that may serve as putative motifs for prenylation, similar to those found in Rab proteins. Unlike other small GTPases, they exhibit a variety of prenylation motifs at their C-termini, containing either one or, more frequently, two cysteine residues, both of which are modified by geranylgeranyl groups. The C-termini of Rab proteins vary in length and sequence and do not have a consensus sequence, such as the CAAX box, which the Rab geranylgeranyl transferase can recognise. There are six different carboxyl-terminal motifs on the 60 known human Rab proteins, *XXX*CC, *X*CC*XX*, *XX*- C*X*C, CC*XXX*,

XXCCX and *XCXXX* (Pereira-Leal and Seabra, 2001). Most Rab proteins contain two carboxyl-terminal cysteines, both of which undergo geranylgeranylation (Farnsworth et al., 1994). IglC contains the C-terminal sequence CCAAS, with the two cysteines at locations 191 and 192 out of the 209 amino acids of the protein.

To analyze the contribution of the cysteine residues, a mutagenesis strategy was used whereby either one or both were replaced with serine. The resulting mutagenized genes were then introduced into plasmid pKK289 Km, to allow constitutive expression from the GroEL promoter (Bönquist, Lindgren et al., 2008). The plasmid was introduced in the *iglC* mutant and the mutated proteins were expressed in *trans*. The expression was assayed by Western blot analysis. We observed that the C191S protein and the C191SC192S protein showed much lower levels compared to the control expressing the native protein (Fig. 7.1). In contrast, the C192S protein was expressed at wild-type levels, suggesting that it was stable.

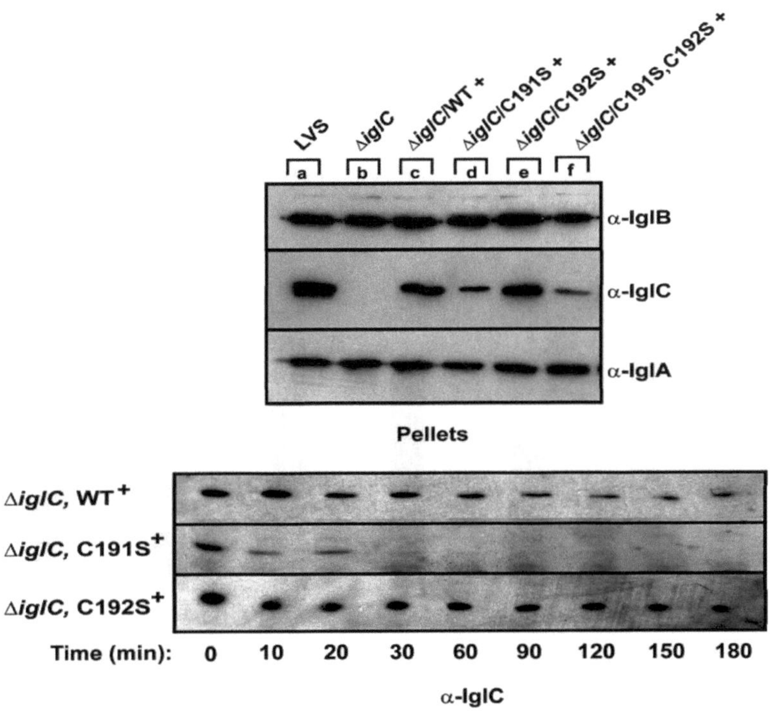

Fig. 7.1 Intrabacterial stability of mutated forms of IglC in *F. tularensis* LVS. The intrabacterial stability of IglC produced by LVS, Δ*iglC* and *trans*-complemented with C191S, C192S, or C191SC192S (*upper panel*) was analyzed by Western blot analysis. The stability or C191S and C192S produced by *trans*-complemented Δ*iglC* (*lower panel*) grown in Chamberlain's medium was examined. At time 0, chloramphenicol was added to stop protein synthesis. Samples from pelleted bacteria were obtained at indicated time points and the amount of proteins detected by Western blot

Using the method by Feldman et al. (2002), we followed the stability of the mutant IglC proteins over time and whereas the wild-type protein and the C192S protein showed no degradation over 180 min, the C191S protein showed a very marked instability and was completely degraded within 30 min (Fig. 7.1).

Since we and others have identified an essential role for all four Igl proteins for intracellular multiplication, we asked whether the mutated IglC proteins affected the phenotype of *F. tularensis*. J774 cells were infected with the null mutant strain Δ*iglC* or the mutants expressing the wild-type *iglC*, C191S, C192S, or C191SC192S genes in *trans*. While Δ*iglC* was essentially unable to grow in J774 cells, the strains complemented with the mutant or wild-type genes grew indistinguishable from parental LVS (Fig. 7.2).

We have previously found that each of the four igl mutants show very similar and highly attenuated phenotypes (Lindgren et al., 2004; Bönquist et al., 2008; Bröms et al., 2009). We also found that the *iglC* and *iglD* mutants of LVS have identical phenotypes and that there seems to be a direct correlation between the incapability of *F. tularensis* to escape from the phagosome and lack of intracellular replication, indicating that the bacterium is incapable of intra-phagosomal replication, at least in most types of macrophages (Bönquist et al., 2008). Moreover, these phenotypic characteristics correlated to a lack of virulence in the mouse model (Golovliov et al., 2003). The function of Igl proteins appear to be conserved across the species since their composition is essentially identical in virulent *F. tularensis* isolates compared to LVS or *F. novicida*. In support of their conserved

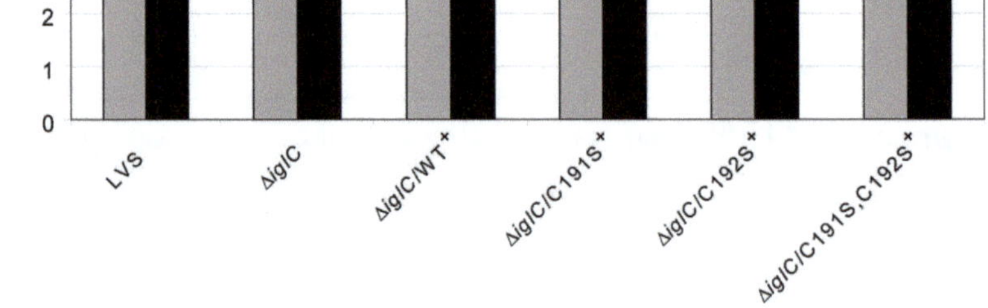

Fig. 7.2 Intracellular growth of strains of *F. tularensis*. J774 cells were infected by various strains of *F. tularensis* at an MOI of 200 for 2 h. After gentamicin treatment, cells were allowed to recover for 30 min after which they were lysed immediately (corresponds to 0 h; *grey bars*) or after 24 h (*black bars*) with PBS-buffered 0.1% sodium deoxycholate solution and plated to determine the number of viable bacteria (log_{10}). Each bar represents the mean values and the error bar indicates the standard deviation

roles, we have observed that a SCHU S4 (subsp. *tularensis*) *iglC* mutant is avirulent (Twine et al., 2005) and the phenotype of a SCHU S4 *iglB* and *iglD* mutants appear to be identical (Kadzhaev et al., 2009). Thus, the cumulative evidence indicates that LVS is a relevant model for studies aimed to dissect the roles of the Igl proteins.

Although the present study found no indication of a direct binding of IglC or IglD to any other Igl protein, the evidence collectively implies that there are functional interactions between the four proteins and that the function of each of the proteins is directly or indirectly dependent on the effective expression of all other three proteins.

Previously, we observed a marked instability of IglA and IglB in the Δ*iglB* and Δ*iglA* mutants of LVS, respectively. Similar findings have been observed in an *F. novicida iglB* and *iglA* mutants (de Bruin et al., 2007; Ludu et al., 2008). By using a similar assay, we now investigated the effects of mutating the C-terminal cysteine residues of IglC. Whereas, we found no phenotype of the C192S mutation, the C191S mutation led to a highly unstable protein.

Recently, it was suggested that site-directed mutagenesis of each of the four cysteine residues of IglC led to impaired intracellular multiplication of the mutants (Barker and Klose, 2009). Moreover, it was suggested that one of the cysteines, the equivalent of C191, may serve as a prenylation site. A post-translational modification of IglC has been suggested since proteomic analysis revealed several isoforms of the protein (Lenco et al., 2005). Our findings did not support biologically important roles of C191 and C192 since the serine substitutions did not lead to any impaired intracellular multiplication. However, we did note that the C191S substitution led to a highly unstable protein with a half-life at least 6 times shorter than that of the wild-type. Thus, C191 plays an important role for the stability of IglC but even though the overall level of the protein was significantly decreased in the mutant, obviously the level was still such that the important function for the virulence of *F. tularensis* was preserved.

7.4 Conclusion

The site-directed mutagenesis strategy used herein is a powerful tool to elucidate the roles of individual amino acids for protein function. The findings imply that the cysteine in position 192 does not play an essential role for the function of IglC whereas C191 is very important for the stability of the protein but not directly involved in any of the biological functions of IglC.

Acknowledgements This work was supported by grants 2006–3426 (JB) and 2006–2877 (AS) from the Swedish Research Council and a grant from the Medical Faculty, Umeå University, and the Umeå Biotech Incubator, Umeå, Sweden. The work was performed in part at the Umeå Centre for Microbial Research (UCMR).

Index words: *Francisella*, type VI secretion system, Igl proteins, yeast two-hybrid system, IglC, cysteine residues.

References

Barker, J. R. & Klose, K. E. (2009) Characterization of the secreted effector protein IglC in *Francisella tularensis* virulence. The 6th International conference on tularemia, Berlin, Robert Koch Institut.

Bingle, L. E., Bailey, C. M. et al. (2008) Type VI secretion: a beginner's guide. *Curr Opin Microbiol*, 11(1), 3–8.

Bröms, J. E., Lavander, M. et al. (2009) A conserved alpha-helix essential for a type VI secretion-like system of *Francisella tularensis*. *J Bacteriol*, 191(8), 2431–46.

Bönquist, L., Lindgren, H. et al. (2008) MglA and Igl proteins contribute to the modulation of *Francisella tularensis* live vaccine strain-containing phagosomes in murine macrophages. *Infect Immun*, 76(8), 3502–10.

Chamberlain, R. E. (1965) Evaluation of live tularemia vaccine prepared in a chemically defined medium. *Appl Microbiol*, 13, 232–5.

Das, S. & Chaudhuri, K. (2003) Identification of a unique IAHP (IcmF associated homologous proteins) cluster in *Vibrio cholerae* and other proteobacteria through in silico analysis. *In Silico Biol*, 3(3), 287–300.

De Bruin, O. M., Ludu, J. S. et al. (2007) The *Francisella* pathogenicity island protein IglA localizes to the bacterial cytoplasm and is needed for intracellular growth. *BMC Microbiol*, 7(1), 1.

Dudley, E. G., Thomson, N. R. et al. (2006) Proteomic and microarray characterization of the AggR regulon identifies a pheU pathogenicity island in enteroaggregative *Escherichia coli*. *Mol Microbiol*, 61(5), 1267–82.

Farnsworth, C., Seabra, M. et al. (1994) Rab geranylgeranyl transferase catalyzes the geranylgeranylation of adjacent cysteines in the small GTPases Rab1A, Rab3A, and Rab5A. *Proc Natl Acad Sci USA*, 91, 11963–7.

Feldman, M. F., Muller, S. et al. (2002) SycE allows secretion of YopE-DHFR hybrids by the *Yersinia enterocolitica* type III Ysc system. *Mol Microbiol*, 46(4), 1183–97.

Golovliov, I., Sjöstedt, A. et al. (2003) A method for allelic replacement in *Francisella tularensis*. *FEMS Microbiol Lett*, 222(2), 273–80.

James, P., Halladay, J. et al. (1996) Genomic libraries and a host strain designed for highly efficient two-hybrid selection in yeast. *Genetics*, 144(4), 1425–36.

Kadzhaev, K., Zingmark, C. et al. (2009) Identification of genes contributing to the virulence of *Francisella tularensis* SCHU S4 in a mouse intradermal infection model. *PLoS One*, 4(5), e5463.

Lenco, J., Pavkova, I. et al. (2005) Insights into the oxidative stress response in *Francisella tularensis* LVS and its mutant DeltaiglC1+2 by proteomics analysis. *FEMS Microbiol Lett*, 246(1), 47–54.

Lindgren, H., Golovliov, I. et al. (2004) Factors affecting the escape of *Francisella tularensis* from the phagolysosome. *J Med Microbiol*, 53(Pt 10), 953–8.

Ludu, J. S., De Bruin, O. M. et al. (2008) The *Francisella* pathogenicity island protein PdpD is required for full virulence and associates with homologues of the type VI secretion system. *J Bacteriol*, 190(13), 4584–95.

Mougous, J. D., Cuff, M. E. et al. (2006) A virulence locus of *Pseudomonas aeruginosa* encodes a protein secretion apparatus. *Science*, 312(5779), 1526–30.

Nano, F. E. & Schmerk, C. (2007) The *Francisella* pathogenicity island. *Ann NY Acad Sci*, 1105, 122–37.

Nano, F. E., Zhang, N. et al. (2004) A *Francisella tularensis* Pathogenicity Island Required for Intramacrophage Growth. *J Bacteriol*, 186(19), 6430–6.

Pereira-Leal, J. & Seabra, M. C. (2001) Evolution of the Rab family of small GTP-binding proteins. *J Mol Biol*, 313, 889–901.

Pukatzki, S., Ma, A. T. et al. (2006) Identification of a conserved bacterial protein secretion system in *Vibrio cholerae* using the *Dictyostelium* host model system. *Proc Natl Acad Sci USA*, 103(5), 1528–33.

Santic, M., Molmeret, M. et al. (2007) A *Francisella tularensis* pathogenicity island protein essential for bacterial proliferation within the host cell cytosol. *Cell Microbiol*, 9(10), 2391–403.
Schell, M. A., Ulrich, R. L. et al. (2007) Type VI secretion is a major virulence determinant in *Burkholderia mallei*. *Mol Microbiol*, 64(6), 1466–85.
Suarez, G., Sierra, J. C. et al. (2008) Molecular characterization of a functional type VI secretion system from a clinical isolate of *Aeromonas hydrophila*. *Microb Pathog*, 44(4), 344–61.
Twine, S., Bystrom, M. et al. (2005) A mutant of *Francisella tularensis* strain SCHU S4 lacking the ability to express a 58-kilodalton protein is attenuated for virulence and is an effective live vaccine. *Infect Immun*, 73(12), 8345–52.
Zheng, J. & Leung, K. Y. (2007) Dissection of a type VI secretion system in *Edwardsiella tarda*. *Mol Microbiol*, 66(5), 1192–206.

Chapter 8
Yersinia pestis Lipopolysaccharide in Host-Pathogen Interactions

Andrey P. Anisimov, Svetlana V. Dentovskaya, Anna N. Kondakova, Buko Lindner, Rima Z. Shaikhutdinova, Nina A. Kocharova, Sof'ya N. Senchenkova, and Yuriy A. Knirel

Abstract Isogenic *Y. pestis* strains with a single mutation in 14 genes of lipopolysaccharide (LPS) biosynthetic pathways were constructed. Using high-resolution electrospray ionization mass spectrometry, the full LPS structures were elucidated in each mutant, and the sequence of monosaccharide transfers in the assembly of the LPS core was inferred. Truncation of the core decreased significantly the resistance of bacteria to normal human serum (NHS) and polymyxin B. Impairing of LPS biosynthesis resulted also in reduction of LPS-dependent enzymatic activities of plasminogen activator. A gradual truncation of the LPS core was accompanied by a decrease of virulence in mice and guinea pigs. However, the reduction in virulence remained behind the decrease of bacterial resistance to innate immunity factors. E.g., *waaQ* mutant deficient in HepIII transferase was highly susceptible to polymyxin B and NHS but was as virulent as the parental strain in both animal models. *Y. pestis* mutants with two or less sugar residues in the core were not only susceptible to antimicrobial cationic peptides and NHS but also avirulent in animal infection models. This finding demonstrated that the LPS structure is crucial for the lethality of plague infection, and *waaC*, *hldE* and *waaA* or their protein products can be considered as promising candidates for targeting *Y. pestis* virulence using specific inhibitors.

Keywords Host-pathogen interactions · Lipopolysaccharide · *Yersinia pestis*

8.1 Introduction

Lipopolysaccharide (LPS, endotoxin) is an important factor of pathogenicity common for most Gram-negative bacteria. The full LPS molecule (S-LPS) consists of three well-defined domains: (i) lipid A composed of sugars, fatty acids and phosphate; it represents the endotoxic principle of the LPS and anchors it in the

A.P. Anisimov (✉)
State Research Center for Applied Microbiology and Biotechnology, Obolensk, 142279, Moscow Region, Russia
e-mail: a-p-anisimov@yandex.ru

A. Shafferman et al. (eds.), *The Challenge of Highly Pathogenic Microorganisms*,
DOI 10.1007/978-90-481-9054-6_8, © Springer Science+Business Media B.V. 2010

outer leaflet of the outer membrane; (ii) a core oligosaccharide containing charged groups; and (iii) an O-chain polysaccharide (O-antigen), which protrudes from the cell surface to the environment and carries immunodeterminant groups eliciting specific antibodies in animals. The Gram-negative bacterium *Yersinia pestis*, the etiological agent of bubonic and pneumonic plague, is a clone evolved from the enteric pathogen *Yersinia pseudotuberculosis* 1,500–20,000 years ago (Achtman et al., 1999). In the course of the divergence *Y. pestis* lost by deletion or inactivation a lot of genes useful for enteropathogenic yersiniae but unnecessary for the vector-borne lifestyle of *Y. pestis*. Particularly, it lost the ability to synthesize O-chain (Skurnik et al., 2000), and, as a result, the LPS of *Y. pestis* is restricted to the core and lipid A moieties (R-LPS).

The O-antigen-deficient LPS plays a role in resistance of *Y. pestis* to serum-mediated lysis (Anisimov et al., 2005; Porat et al., 1995) that is necessary for survival of the bacteria in mammalian blood and for transmission between insects and mammals (Perry and Fetherston, 1997). It determines also bacterial resistance to cationic antimicrobial peptides (Anisimov et al., 2005; Bengoechea et al., 1998), a key component of the innate immunity in both mammals and insects (Dimopoulos, 2003). Reduction in the degree of acylation of lipid A of *Y. pestis* at the mammalian host temperature (37°C) (Kawahara et al., 2002) makes this pathogen-associated molecule invisible for pattern recognition Toll-like receptor 4 (Montminy et al., 2006) and enables uncontrolled multiplication of *Y. pestis* in the host organism resulting in septic shock and death necessary for further transmission of the pathogen by fleas (Brubaker, 1991).

Recently, the fine LPS structures have been reported in representatives of various intraspecies groups of *Y. pestis* grown at 37, 21–28 or 6°C to mimic the conditions in warm-blooded hosts, insects and animals during winter hibernation, respectively (Dentovskaya et al., 2008; Knirel et al., 2006, 2008). Variations in the LPS structure of *Y. pestis* have been documented and found to depend on the growth temperature, composition of the nutrient media and belonging of *Y. pestis* strains to a certain intraspecies group. However, structural elements of *Y. pestis* LPS responsible for different pathogenic properties of this macromolecule are characterized scarcely. This work aims at unraveling the impact of particular LPS components on *Y. pestis* virulence, including resistance of the bacteria to various antimicrobial factors, and to understand better the biological significance of the temperature-dependent LPS structural variations. For these purposes, we identified genes that are involved with biosynthesis of the *Y. pestis* LPS, generated the corresponding non-polar mutants and studied their LPS structures and virulence-related biological properties.

8.2 Materials and Methods

Parental strain *Y. pestis* 231 (bv. antiqua) was isolated in the Aksai focus, Kirghizia. Search for LPS associated genes was performed through annotated genome sequences of *Y. pestis* strains (http://www.ericbrc.org/portal/eric/yersiniapestis?id=enteropathogens&subid=yersiniapestis). When relevant information was absent, genes with already revealed functions from other bacteria were used as templates for search of orthologs in genome sequence of *Y. pestis* strain CO92 by using the BLAST network service at the National Center for Biotechnology Information. The nomenclature proposed by Reeves et al. (1996) is used for designation of bacterial polysaccharide synthesis genes. Mutant strains with impaired LPS biosynthesis pathway were generated by one-step inactivation of the chromosomal genes mentioned above by λ Red recombination technology (Datsenko and Wanner, 2000). Structure of mutant LPSs was determined as described earlier (Knirel et al., 2005, 2008). The sensitivity of *Y. pestis* strains to polymyxin B (PMB) was tested using different doses of this cationic antimicrobial peptide to calculate the minimum inhibitory concentration (MIC) as described previously (Anisimov et al., 2005). A pool of normal human serum (NHS) was obtained from ten non-immunized healthy volunteers. The complement was inactivated by incubating NHS at 56°C for 30 min. Bactericidal properties of NHS were studied by incubation of bacteria with serum for 1 h as described earlier (Anisimov et al., 2005). Coagulation and fibrinolysis were assayed as described previously (Bahmanyar and Cavanaugh, 1976).

BALB/c mice (~20 g each) and guinea pigs (~260 g each) of both sex were used in animal experiments that were approved by the ethical committee of the State Research Center for Applied Microbiology and Biotechnology. Animals were kept in cages in groups at the most of five and three, respectively, and allowed to feed and drink *ad libitum* during experiment terms. Six naïve groups of mice (each of 40) and five naïve groups of guinea pigs (each of 12) were administered subcutaneously (s.c.) in the right inguinal region with serial 10-fold dilutions (10^7 to 1 cfu, five mice for one dose; 10^4 to 10^1 cfu, three guinea pigs for one dose, respectively) of 2-day *Y. pestis* agar cultures grown at 25°C. Humane endpoints were strictly observed. Animals that succumbed to infection were sacrificed and examined bacteriologically. The remaining animals were observed for 21 days. The animals that survived were humanely killed by carbonic gas inhalation. Mortality times were recorded and the LD_{50} and 95% confidence intervals (CI) were determined according to the method of Kärber (Finney, 1978).

8.3 Results and Discussion

8.3.1 Generation of Mutants in LPS Biosynthesis Genes

Each single gene with a predicted function in LPS biosynthesis was successfully inactivated using PCR products with kanamycin and/or chloramphenicol resistance cassettes. Each single mutant was complemented *in trans* by the corresponding wild type gene (data not shown) to rule out the possibility that the phenotype observed is caused by a polar effect or a spontaneous mutation elsewhere in the genome. All generated mutants were shown to be non-polar.

8.3.2 Structures of Mutant LPSs and Confirmation of LPS Biosynthesis Gene Functions

The LPS of each mutant was degraded with mild acid to yield a lipid A precipitate and a core oligosaccharide; the latter was purified by gel chromatography on Sephadex G-50. The isolated core and lipid A samples or/and the whole undegraded LPSs were studied by electrospray ionization Fourier transform ion-cyclotron resonance mass spectrometry in the negative ion mode. Mass spectra of the whole LPS of *arnT* mutant, core oligosaccharides from *waaL, wabC, wabD, waaQ* and *waaF* mutants (Knirel et al., 2007) and lipid A of *lpxM* mutant (Anisimov et al., 2007) have been reported earlier. The data obtained were compared with the data of wild-type *Y. pestis* LPS with known structure (Dentovskaya et al., 2008, Kawahara et al., 2002, Knirel et al., 2005, 2006, 2008), and functions of the inactivated genes were confirmed (Fig. 8.1).

waaL *(YPO0417) mutant.* A sample from *waaL* mutant showed essentially the same MS pattern as the wild type LPS except that 2-acetamido-2-deoxy-β-D-glucose (GlcNAc)-containing compounds were absent from the mutant LPS. As in wild type, there were present oligosaccharides having either β-D-galactose (Gal) or D-*glycero*-α-D-*manno*-heptose-IV (DD-HepIV) at the non-reducing terminus. The absence of GlcNAc from *waaL* mutant suggested that ligase WaaL encoded by the inactivated gene is responsible for adding GlcNAc to the LPS core. The same GlcNAc-lacking LPS chemotype was detected in *wecA* (YPO3866) mutant, which is unable to synthesize undecaprenyl diphosphate derivative of GlcNAc, a donor substrate for WaaL.

wabC *(YPO0186) and* wabD *(YPO0187) mutants.* Core samples from *wabC* and *wabD* mutants were distinguished by the absence of either terminal DD-HepIV

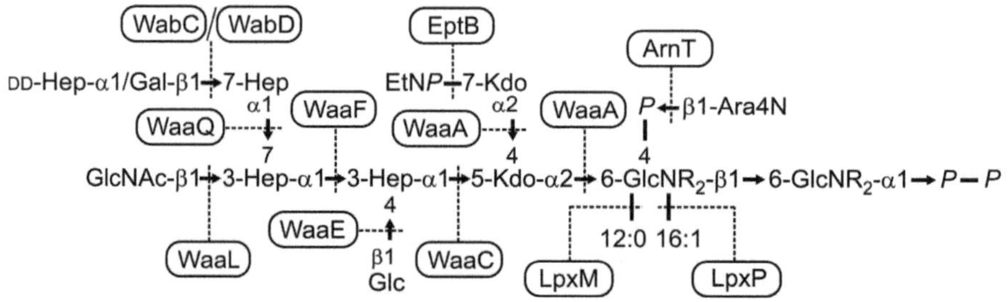

Fig. 8.1 Structure of *Y. pestis* LPS and functional assignment of glycosyl- and acyl-transferases involved with LPS biosynthesis. Ara4N, 4-amino-4-deoxy-L-arabinose; DD-Hep and Hep, D-*glycero*- and L-*glycero*-D-*manno*-heptose, respectively; Kdo, 3-deoxy-D-*manno*-oct-2-ulosonic acid; Ko, D-*glycero*-D-*talo*-oct-2-ulosonic acid; EtNP, phosphoethanolamine; 12:0, lauroyl; 16:1, palmitoleoyl (hexadec-9-enoyl); R stands for 3-hydroxymyristoyl. In cultures grown at 25°C, DD-HepIV and Gal, KdoII and Ko alternate pairwise whereas at 37°C DD-HepIV and KdoII predominate. EtNP in the core is mainly expressed in cultures grown at 6°C. Diphosphate in lipid A is expressed at 37°C and replaced with Ara4NP at 25°C. Glycine linked to HepI (authors' unpublished data) is not shown

or terminal Gal, and, therefore, the encoded enzymes are heptosyltransferase IV and galactosyltransferase, respectively. The inability to incorporate either of these monosaccharides does not interfere with biosynthesis of the remaining part of the LPS core.

waaQ *(YPO0416) mutant.* The core of *waaQ* mutant was devoid of L-*glycero*-α-D-*manno*-heptose-III (HepIII). As expected, neither DD-HepIV nor Gal was present as these monosaccharides are carried by HepIII. GlcNAc that is linked to L-*glycero*-α-D-*manno*-heptose-II (HepII) occurs only in a minority of molecules. Therefore, *waaQ* encodes heptosyltransferase III, and WaaL transfers GlcNAc to the core in the absence of HepIII inefficiently. The inner core region was not affected by inactivation of *waaQ*.

waaE *(YPO0054) mutant.* The mass spectrum of the whole LPS from *waaE* mutant showed the major species with the core moiety containing D-*glycero*-D-*talo*-oct-2-ulososnic acid (Ko), 3-deoxy-D-*manno*-oct-2-ulososnic acid-I (KdoI), L-*glycero*-α-D-*manno*-heptose-I (HepI) and HepII and the lipid A moiety consisting of bisphosphorylated glucosamine disaccharide backbone with four 3-hydroxymyristic (3HO14:0) groups and one 4-amino-4-deoxy-L-arabinose (Ara4N) residue. The lack from the *waaE* mutant LPS of both β-D-glucose (Glc) and HepIII suggested that *waaE* encodes glucosyltransferase and that incorporation of HepIII into the core requires the prior addition of Glc to HepI. As in *waaQ* mutant, only a small proportion of GlcNAc-containing core variants were present in *waaE* mutant.

waaF *(YPO0057) mutant.* A single heptose residue (HepI) was found in the LPS of *waaF* mutant, and, hence, the inactivated gene encodes heptosyltransferase II. The lack of Glc from *waaF* mutant suggested that its addition to the core requires the prior transfer of HepII onto HepI.

waaC *(YPO0056) mutant.* Only Kdo (Ko) mono- and disaccharides were obtained by mild acid degradation of the *waaC* mutant LPS. The mass spectrum of the whole LPS revealed the major compounds having KdoII→KdoI or Ko→KdoI disaccharide core linked to lipid A containing four 3HO14:0 groups and one Ara4N residue. The lack of any heptose from the core suggested that *waaC* encodes heptosyltransferase I. The same LPS chemotype was found in *hldE* (YPO0654) mutant, which is unable to synthesize the activated Hep nucleotide derivative.

waaA *(YPO0055) mutant.* No core component was isolated after mild acid degradation of the *waaA* mutant LPS. The mass spectrum of the whole LPS showed that it is restricted to the lipid A moiety with four (major) and three (minor) 3HO14:0 groups. Therefore, it was suggested that the inactivated gene encodes Kdo transferase WaaA, a bifunctional enzyme that catalyzes transfer of both KdoI to lipid A and KdoII to KdoI (Raetz and Whitfield, 2002).

Only a minority of LPS molecules in *waaA* mutant contain Ara4N. A relatively low content of Ara4N (on the average not more than one residue per molecule) was observed also in lipid A of other *Y. pestis* LPS mutants with deeply truncated core, including *waaE* and *waaC* mutants. In contrast, in wild type LPS (Knirel et al., 2005) as well as in *waaL, wabC* and *wabD* mutants both phosphate groups in lipid A are almost completely glycosylated with Ara4N. These findings indicate a lower efficiency of Ara4N transfer to lipid A in the LPS having an incomplete inner core region or, especially, no core as in *waaA* mutant.

eptB *(YPO4013) mutant.* A core variant containing phosphoethanolamine (EtNP) on terminal Ko has been found to be abundant in the wild-type LPS grown at 6°C (Knirel et al., 2006). No EtNP was present in the LPS of *eptB* mutant cultivated at the same temperature, while the carbohydrate core backbone and lipid A in the mutant and wild type were the same. Therefore, *eptB* encodes EtNP transferase.

arnT *(YPO2421) mutant.* The mass spectrum of the whole LPS showed that *arnT* mutant is fully unable to incorporate Ara4N to lipid A and, therefore, the inactivated gene encodes Ara4N transferase ArnT. Inactivation of *arnT* had no influence on the LPS core structure.

lpxM *(YPO2063) mutant.* No hexaacyl form was found in lipid A from *lpxM* mutant but a pentaacyl form with four 3HO14:0 groups and one palmitoleoyl (16:1) group. None of lower acylated variants present included a lauroyl group either. Other significant changes were observed neither in lipid A nor in the core of the mutant LPS. Therefore, it was concluded that *lpxM* encodes lauroyl transferase LpxM.

lpxP *(YPO3632) mutant.* As in lipid A from *lpxM* mutant, no hexaacyl form was found in lipid A from *lpxP* mutant but in this case the highest acylated variant contained four 3HO14:0 groups and one lauroyl (12:0) group. None of lower acylated variant included 16:1 group either, and, therefore, *lpxP* encodes palmitoleoyl transferase LpxP.

8.3.3 Resistance of Mutants to Bactericidal Action of Polymyxin B and Normal Human Serum

Resistance of *Y. pestis* to cationic antimicrobial peptides depends on the content of Ara4N, whose temperature-dependent incorporation to lipid A is regulated by the two-component PhoP/PhoQ regulatory system (Rebeil et al., 2004). Accordingly, we found that *Y. pestis arnT* mutant deficient in Ara4N was highly susceptible to PMB, as reported for *arnT* mutants of some other bacteria (Nizet, 2006). *waaQ, waaE, waaF, waaC, hldE* and *waaA* mutants with deeply truncated LPS core were 31– 250 times less resistant to PMB (MIC < 20 U mL^{-1}) than the wild-type strain (Table 8.1). The drop of the resistance may be accounted for by a low content of Ara4N in the LPS having an incomplete inner core region. Strains carrying mutations in genes responsible for adding the lateral core monosaccharides (*wabC, wabD, wecA* and *waaL*) were as resistant to PMB as the parental strain (MIC >625 U mL^{-1}).

waaQ, *waaE*, *waaF*, *waaC* and *waaA* deficient strains with an impaired inner core region were highly susceptible to the bactericidal action of normal human serum (NHS) (Table 8.1). The serum killing was complement-mediated as these mutants were resistant to heat-inactivated serum. All other generated mutants, including *arnT* mutant with an Ara4N-lacking LPS, were almost as resistant as the parental strain.

Table 8.1 Biological properties of wild-type strain *Y. pestis* 231 and derived LPS mutants

Mutation in gene	Core oligosaccharide structure	Relevant characteristics
Wild type	Gal/DD-Hep–Hep Glc Ko GlcNAc···Hep–Hep–Kdo–LA	PMB^{R}, NHS^{R}, Fib^{++}, Coa^{++}, mLD$_{50}$ = 2 (1–4) cfu, gpLD$_{50}$ 7 (2–27) cfu
YPO0186 (*wabC*)	Gal–Hep Glc Ko GlcNAc···Hep–Hep–Kdo–LA	PMB^{R}, NHS^{R}, Fib^{++}, Coa^{++}
YPO0187 (*wabD*)	DD-Hep–Hep Glc Ko GlcNAc···Hep–Hep–Kdo–LA	PMB^{R}, NHS^{R}, Fib^{++}, Coa^{++}, mLD$_{50}$ = 8 (1–32) cfu, gpLD$_{50}$ 7 (2–27) cfu
YPO3866 (*wecA*)	Gal/DD-Hep–Hep Glc Ko Hep–Hep–Kdo–LA	PMB^{R}, NHS^{R}, Fib^{+}, Coa^{+}
YPO0417 (*waaL*)	Gal/DD-Hep–Hep Glc Ko Hep–Hep–Kdo–LA	PMB^{R}, NHS^{R}, Fib^{+}, Coa^{+}, mLD$_{50}$ = 13 (2–50) cfu, gpLD$_{50}$ 32 (8–126) cfu
YPO0416 (*waaQ*)	Glc Ko GlcNAc···Hep–Hep–Kdo–LA	**PMB^{S}, NHS^{S}**, Fib^{+}, Coa^{+}, mLD$_{50}$ = 5 (1–16) cfu, gpLD$_{50}$ 15 (4–58) cfu
YPO0054 (*waaE*)	Ko GlcNAc···Hep–Hep–Kdo–LA	**PMB^{S}, NHS^{S}**, Fib^{+}, Coa^{+}, mLD$_{50}$ = 32 (8–126) cfu, gpLD$_{50}$ **> 10^{4}** cfu
YPO0057 (*waaF*)	Ko Hep–Kdo–LA	**PMB^{S}, NHS^{S}, Fib^{-}, Coa^{-}**
YPO0056 (*waaC*)	Ko Kdo–LA	**PMB^{S}, NHS^{S}, Fib^{-}, Coa^{-}**
YPO0654 (*hldE*)	Ko Kdo–LA	**PMB^{S}, NHS^{S}, Fib^{-}, Coa^{-}**, mLD$_{50}$ = **2.0×10^{5}** (5.0×10^{4}–7.9×10^{5}) cfu, gpLD$_{50}$ **> 10^{4}** cfu
YPO0055 (*waaA*)	LA	**PMB^{S}, NHS^{S}, Fib^{-}, Coa^{-}**
YPO2421 (*arnT*)	Gal/DD-Hep–Hep Glc Ko GlcNAc···Hep–Hep–Kdo–LA*	**PMB^{S}**, NHS^{R}, Fib^{++}, Coa^{++}
YPO4013 (*eptB*)	Gal/DD-Hep–Hep Glc Ko* GlcNAc···Hep–Hep–Kdo–LA	PMB^{R}, NHS^{R}, Fib^{+}, Coa^{+}
YPO2063 (*lpxM*)	Gal/DD-Hep–Hep Glc Ko GlcNAc···Hep–Hep–Kdo–LA*	PMB^{R}, NHS^{R}, Fib^{++}, Coa^{++}
YPO3632 (*lpxP*)	Gal/DD-Hep–Hep Glc Ko GlcNAc···Hep–Hep–Kdo–LA*	PMB^{R}, NHS^{R}, Fib^{++}, Coa^{++}

LA, wild-type lipid A; LA∗, lipid A deficient in Ara4N, lauroyl (12:0) or palmitoleoyl (16.1) group; Ko∗, core deficient in PEtN. *Dotted line* indicates a non-stoichiometric amount of GlcNAc. Significant changes in biological properties are indicated in *bold face.* PMB^{R}, minimal inhibitory concentration (MIC) of polymyxin B (PMB) = 625–1250 U ml^{-1}; PMB^{S}, MIC of PMB $\leq$ 20 U ml^{-1}; NHS^{R}, resistant to normal human sera; NHS^{S}, after incubation of mutants grown at 25°C with NHS serum for 1 h the amounts of viable bacteria were three to five lg cfu ml^{-1} lower when compared with the wild type or mutants incubated in heat-inactivated NHS; Fib and Coa, coagulase and fibrinolytic activities of plasminogen activator; Fib^{++} or Coa^{-}, the complete clot lysis or absence of clotting, respectively; Fib^{+}, Coa^{+}, any degree of clot lysis or incomplete clotting (from a loose clot to a solid clot in liquid plasma); Fib^{-}, Coa^{++}, a solid clot; mLD$_{50}$ and gpLD$_{50}$, LD$_{50}$ for mice and guinea pigs, respectively; cfu, colony forming unit.

Serum resistance is necessary for survival and multiplication of causative agents of vector-borne diseases in mammal blood for further transmission by insects to new hosts (Perry and Fetherston, 1997). *Y. pestis* R-LPS incorporated into liposomes was shown to ensure serum resistance in contrast to S-LPS from serum-susceptible yersiniae phenotypes (Porat et al., 1995). On the other hand, serum resistance in *Y. pestis* is mediated by an outer membrane protein Ail (OmpX), and it has been hypothesized that changes in the LPS structure may influence the Ail conformation and specific activity (Bartra et al., 2008). Our data of *Y. pestis* LPS mutants show that the full inner core composed of six sugar residues (KdoI, KdoII or Ko, HepI, HepII, HepIII and Glc) is required for the serum resistance, whereas the outer core monosaccharides (DD-HepIV, Gal and GlcNAc) are of little importance.

8.3.4 The Full LPS Core is Necessary for Enzymatic Activity of Plasminogen Activator

Fibrinolytic and plasmocoagulase activities of Pla were tested in *Y. pestis* strains with mutations in the genes of the LPS core biosynthetic pathway (Table 8.1). It was found that both activities are first decreased (in *wecA, waaL, waaQ* and *waaE* mutants) and then totally abolished (in *waaF, waaC, hldE* and *waaA* mutants) with a gradual reduction of the core.

Plasminogen activator Pla, a transmembrane protease belonging to omptins, is one of the recognized *Y. pestis* pathogenicity factors responsible for systemic spread of bacteria in the host organism (Kukkonen and Korhonen, 2004). In some *Y. pestis* strains, Pla expression is obligatory for full subcutaneous virulence. In contrast, in all representatives of subspecies *caucasica*, which are naturally deficient in plasmid pPst (pPla, pPCP1 or pYP) coding for Pla, or in a number of artificial Pla^- mutants of subspecies *pestis* strains, the absence of Pla does not affect the high subcutaneous virulence, which is comparable to that of classical wild-type Pla^+ strains (for review see Anisimov et al., 2004). Recently, it has been demonstrated that the presence of an S-type LPS with a long-chain O-antigen inhibits the action of *Y. pestis* Pla, whereas R-LPS is necessary for proper folding of Pla and manifestation of its enzymatic activities (Pouillot et al., 2005).

Our data on Pla fibrinolytic and plasmocoagulase activities in *Y. pestis* strains with gradual truncation of the core suggest that all eight wild-type *Y. pestis* core constituent sugars are necessary for maximal enzymatic activities of Pla. A smaller core consisting of seven to five sugar residues was still able to ensure the fibrinolytic and coagulase activities, though at noticeably reduced levels, whereas a deeper truncation resulted in inactivation of Pla.

8.3.5 Virulence of Mutants in Mice and Guinea Pigs

A gradual truncation of the LPS core was accompanied by a decrease of bacterial virulence in mice and guinea pigs (Table 8.1). In all cases the animal death was dose-related. However, the reduction in virulence remained behind the decrease of the bacterial resistance to the innate immunity factors. For instance, *waaQ* mutant deficient in HepIII transferase synthesis was highly susceptible to PMB and NHS but was as virulent as the parental strain in both animal models. In mice, further reduction of the core, e.g. as in *waaE* mutant deficient in synthesis of glucosyltransferase, was accompanied by a slight but reliable increase in LD_{50} values ($2 \rightarrow 32$ cfu) and mean times to death ($4.7 \rightarrow 7.9$ days). In any case such LD_{50} is extremely low and *waaE* mutant was still highly virulent (for mice). The least virulent for mice was *hldE* mutant deficient in synthesis of Hep and possessing a core restricted to a Kdo→Kdo or Ko→Kdo disaccharide. In guinea pig, neither *waaE* nor *hldE* mutant caused animal death during 21 days of experiment.

Bacteriological examination of spleen from animals succumbed to infection indicated that they were all positive for *Y. pestis* during experiment terms. Indeed, a heavy growth of bacteria was observed on plates inoculated with homogenized spleen specimens. Organs from survivors infected with strain 231 and derived *wabD*, *waaL*, *waaQ* (mice and guinea pigs) or *waaE* (mice) mutants on the day 22 postinfection were all negative for *Y. pestis*. However, 60– 90% spleen specimens from survivors of both animals challenged with *hldE* mutant and guinea pigs infected with *waaE* mutant on the day 22 postinfection were positive for *Y. pestis*: solitary colonies of bacteria were observed on plates inoculated with homogenized spleen specimens.

wabD, *waaL* and *waaQ* mutants of *Y. pestis* subsp. *pestis* were not attenuated in a biologically significant way as each was still extraordinarily virulent in both mice and guinea pigs (Table 8.1). Only truncation of the core down to five sugar residues caused a noticeable decrease in subcutaneous virulence of *Y. pestis* subsp. *pestis* for guinea pigs (=10^3 cfu). A further truncation of the core to two sugar residues abolished virulence for both mice and guinea pigs.

8.4 Conclusion

To sum up, *Y. pestis* knock-out mutants with two or less sugar residues in the LPS core were not only susceptible to antimicrobial cationic peptides and NHS but also avirulent in murine and guinea pig infection models. This finding demonstrated that the LPS structure is crucial for the lethality of plague infection, and *waaC*, *hldE* and *waaA* or their protein products can be considered as promising candidates for targeting *Y. pestis* virulence using specific inhibitors.

Acknowledgments Authors thank Professor P. Reeves (University of Sydney, Australia) for suggesting the gene names for *wabC* and *wabD*. This work was supported by the Russian Foundation for Basic Research (grants 06-04-49280 for A.P.A. and 08-04-00403 for S.V.D.) and the Russian Federal Service for Consumer Rights and Human Well-being Surveillance (contract # 74-D).

References

Achtman, M., Zurth, K., Morelli, G., Torrea, G., Guiyoule, A. & Carniel, E. (1999) *Yersinia pestis*, the cause of plague, is a recently emerged clone of *Yersinia pseudotuberculosis*. *Proc Natl Acad Sci. USA*, 96, 14043–8.

Anisimov, A. P., Dentovskaya, S. V., Titareva, G. M., Bakhteeva, I. V., Shaikhutdinova, R. Z., Balakhonov, S. V., Lindner, B., Kocharova, N. A., Senchenkova, S. N., Holst, O., Pier, G. B. & Knirel, Y. A. (2005) Intraspecies and temperature-dependent variations in susceptibility of *Yersinia pestis* to bactericidal action of serum and polymyxin B. *Infect Immun*, 73, 7324–31.

Anisimov, A. P., Lindler, L. E. & Pier, G. B. (2004) Intraspecific diversity of *Yersinia pestis*. *Clin Microbiol Rev*, 17, 434–64.

Anisimov, A. P., Shaikhutdinova, R. Z., Pan′kina, L. N., Feodorova, V. A., Savostina, E. P., Bystrova, O. V., Lindner, B., Mokrievich, A. N., Bakhteeva, I. V., Titareva, G. M., Dentovskaya, S. V., Kocharova, N. A., Senchenkova, S. N., Holst, O., Devdariani, Z. L., Popov, Y. A., Pier, G. B. & Knirel, Y. A. (2007) Effect of deletion of the *lpxM* gene on virulence and vaccine potential of *Yersinia pestis* in mice. *J Med Microbiol*, 56, 443–53.

Bahmanyar, M. & Cavanaugh, D. C. (1976) Plague manual, Geneva, WHO.

Bartra, S. S., Styer, K. L., O′bryant, D. M., Nilles, M. L., Hinnebusch, B. J., Aballay, A. & Plano, G. V. (2008) Resistance of *Yersinia pestis* to complement-dependent killing is mediated by the Ail outer membrane protein. *Infect Immun*, 76, 612–22.

Bengoechea, J. A., Lindner, B., Seydel, U., Díaz, R. & Moriyón, I. (1998) *Yersinia pseudotuberculosis* and *Yersinia pestis* are more resistant to bactericidal cationic peptides than *Yersinia enterocolitica. Microbiology*, 144, 1509–15.

Brubaker, R. R. (1991) Factors promoting acute and chronic diseases caused by yersiniae. *Clin Microbiol Rev*, 4, 309–24.

Datsenko, K. A. & Wanner, B. L. (2000) One-step inactivation of chromosomal genes in *Escherichia coli* K-12 using PCR products. *Proc Natl Acad Sci USA,* 97, 6640–5.

Dentovskaya, S. V., Bakhteeva, I. V., Titareva, G. M., Shaikhutdinova, R. Z., Kondakova, A. N., Bystrova, O. V., Lindner, B., Knirel, Y. A. & Anisimov, A. P. (2008) Structural diversity and endotoxic activity of the lipopolysaccharide of *Yersinia pestis. Biochemistry (Mosc.),* 73, 192–9.

Dimopoulos, G. (2003) Insect immunity and its implication in mosquito-malaria interactions. *Cell Microbiol,* 5, 3–14.

Finney, D. J. (1978) Statistical method in biological assay, 3rd Ed., Charles Griffin London, England.

Kawahara, K., Tsukano, H., Watanabe, H., Lindner, B. & Matsuura, M. (2002) Modification of the structure and activity of lipid A in *Yersinia pestis* lipopolysaccharide by growth temperature. *Infect Immun,* 70, 4092–8.

Knirel, Y. A., Dentovskaya, S. V., Bystrova, O. V., Kocharova, N. A., Senchenkova, S. N., Shaikhutdinova, R. Z., Titareva, G. M., Bakhteeva, I. V., Lindner, B., Pier, G. B. & Anisimov, A. P. (2007) Relationship of the lipopolysaccharide structure of *Yersinia pestis* to resistance to antimicrobial factors. *Adv Exp Med Biol,* 603, 88–96.

Knirel, Y. A., Dentovskaya, S. V., Senchenkova, S. N., Shaikhutdinova, R. Z., Kocharova, N. A. & Anisimov, A. P. (2006) Structural features and structural variability of the lipopolysaccharide of *Yersinia pestis*, the cause of plague. *J Endotoxin Res*, 12, 3–9.

Knirel, Y. A., Kondakova, A. N., Bystrova, O. V., Lindner, B., Shaikhutdinova, R. Z., Dentovskaya, S. V. & Anisimov, A. P. (2008) New features of *Yersinia* lipopolysaccharide structures as revealed by high-resolution electrospray ionization mass spectrometry. *Adv Sci Lett*, 1, 192–8.

Knirel, Y. A., Lindner, B., Vinogradov, E. V., Kocharova, N. A., Senchenkova, S. N., Shaikhutdinova, R. Z., Dentovskaya, S. V., Fursova, N. K., Bakhteeva, I. V., Titareva, G. M., Balakhonov, S. V., Holst, O., Gremyakova, T. A., Pier, G. B. & Anisimov, A. P. (2005) Temperature-dependent variations and intraspecies diversity of the structure of the lipopolysaccharide of *Yersinia pestis. Biochemistry,* 45, 1731–43.

Kukkonen, M. & Korhonen, T. K. (2004) The omptin family of enterobacterial surface proteases/adhesins: from housekeeping in *Escherichia coli* to systemic spread of *Yersinia pestis. Int J Med Microbiol,* 294, 7–14.

Montminy, S. W., Khan, N., Mcgrath, S., Walkowicz, M. J., Sharp, F., Conlon, J. E., Fukase, K., Kusumoto, S., Sweet, C., Miyake, K., Akira, S., Cotter, R. J., Goguen, J.D. & Lien, E. (2006) Virulence factors of *Yersinia pestis* are overcome by a strong lipopolysaccharide response. *Nat Immunol,* 7, 1066–73.

Nizet, V. (2006) Antimicrobial peptide resistance mechanisms of human bacterial pathogens. *Curr Issues Mol Biol,* 8, 11–26.

Perry, R. D. & Fetherston, J. D. (1997) *Yersinia pestis*-etiologic agent of plague. *Clin Microbiol Rev*, 10, 35–66.

Porat, R., Mccabe, W. R. & Brubaker, R. R. (1995) Lipopolysaccharide-associated resistance to killing of yersiniae by complement. *J Endotoxin Res*, 2, 91–7.

Pouillot, F., Derbise, A., Kukkonen, M., Foulon, J., Korhonen, T. K. & Carniel, E. (2005) Evaluation of O-antigen inactivation on Pla activity and virulence of *Yersinia pseudotuberculosis* harbouring the pPla plasmid. *Microbiology*, 151, 3759–68.

Raetz, C. R. H. & Whitfield, C. (2002) Lipopolysaccharide endotoxins. *Annu Rev Biochem* 71, 635–700.

Rebeil, R., Ernst, R. K., Gowen, B., Miller, S. I. & Hinnebusch, B. J. (2004) Variation in lipid A structure in the pathogenic yersiniae. *Mol Microbiol*, 52, 1363–73.

Reeves, P. R., Hobbs, M., Valvano, M. A., Skurnik, M., Whitfield, C., Coplin, D., Kido, N., Klena, J., Maskell, D., Raetz, C. R. & Rick. P. D. (1996) Bacterial polysaccharide synthesis and gene nomenclature. *Trends Microbiol*, 4, 495–503.

Skurnik, M., Peippo, A. & Ervelä, E. (2000) Characterization of the O-antigen gene clusters of *Yersinia pseudotuberculosis* and the cryptic O-antigen gene cluster of *Yersinia pestis* shows that the plague bacillus is most closely related to and has evolved from *Y. pseudotuberculosis* serotype O:1b. *Mol Microbiol*, 37, 316–30.

Chapter 9
The Interactions Between Pathogens and Dendritic Cells: From Paralysis of Cells to Their Recruitment for Bacterial Colonization

Baruch Velan, Erez Bar-Haim, Gal Markel, Hila Cohen, Ayelet Zauberman, Sara Cohen, and Avigdor Shafferman

Abstract The encounter between invading microorganisms and dendritic cells (DC) triggers a series of events which include uptake and degradation of the microorganism, induction of a cell maturation process, as well as enhancement of DC migration to the draining lymph nodes. Bacteria of the genera *Yersinia* and *Francisella* have developed different strategies to counteract these events as a measure to evade host defense.

We observed that interaction of the *Yersinia enterocolitica* with DC impairs their maturation into functional immune cells and triggers premature cell death. *Y. pestis*, has no effects on DC viability or maturation, yet impairs DC functions related to cytoskeleton rearrangement in a virulence plasmid dependent way. DC pulsed with *Y. pestis* fails to migrate toward the chemokine CCL19. Moreover, while instillation of a virulence plasmid-cured *Y. pestis* strain into mice airways triggers effective transport of airway DC to the mediastinal lymph node (MdLN), instillation of *Y. pestis* harboring the plasmid fails to do so.

Interaction of *Francisella tularensis* live vaccine strain (LVS) with DC leads to impairment of cell maturation but at the same time allows for intracellular bacterial propagation and effective cell migration, thus paving the way to utilization of DC as vehicles for bacterial dissemination. Indeed, airway infection of mice with LVS results in trafficking of bacteria-carrying DC from the respiratory tract to the draining MdLN. Furthermore, impairment of DC migration in vivo by two independent mechanisms reduces bacterial colonization of the lymph node and delays the onset of morbidity and the time to death of the infected mice.

Taken together, these observations attest to the major role of DC in mounting protection against pathogens, which has in turn led to evolution of various bacterial strategies to counteract and even exploit DC functions.

Keywords *Francisella tularensis* · *Yersinia pestis* · Dendritic cells · Migration · Immune-evasion

---

B. Velan (✉)
Department of Biochemistry and Molecular Genetics, Israel Institute for Biological Research, P.O. Box 19, Ness-Ziona 74100, Israel
e-mail: baruchv@iibr.gov.il

A. Shafferman et al. (eds.), *The Challenge of Highly Pathogenic Microorganisms*, DOI 10.1007/978-90-481-9054-6_9, © Springer Science+Business Media B.V. 2010

9.1 Introduction

Dendritic cells serve as the sentinels of the immune system. DC are found along body surfaces, in-between and under epithelial cells, and accumulate at the sites of pathogen entry. DC are capable of inducing of tolerance or immunity by sampling the body peripheral antigens and transferring them to the draining lymph nodes (LN), in order to present them to T and B cells. DC are able to sense external stimuli by pattern recognition receptors (PRR). Upon encountering such a signal, DC go through a maturation process, switching from peripheral phagocytic cells into professional Antigen Presenting Cells (APC) and regulators of the immune response in the lymph node (Steinman and Hemmi, 2006).

The first step of DC function as professional APC is uptake of antigens and response to extracellular stimuli. Uptake is mediated by a large set of phagocytic receptors present on the surfaces of immature DC. Sensing of external stimuli is mediated mainly by the Toll-like receptor (TLR) family. Triggering of these receptors results in translocation of NF-κB into the nucleus and induction of a set of genes leading to DC maturation (Fig. 9.1) (Niedergang et al., 2004).

Maturation of DC is manifested by processing of protein antigens and their presentation as peptides by MHC molecules. DC maturation also involves up-regulation of the expression of co-stimulatory molecules, which later take part in direct interaction with T cells. Among them are CD80 and CD86 that deliver activation signal to T

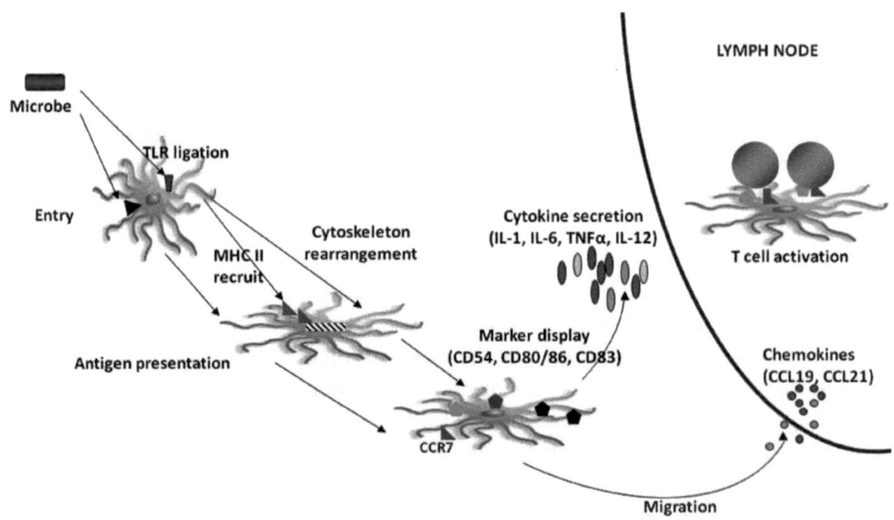

Fig. 9.1 Maturation of DC upon interaction with invading microorganisms. DC are triggered by direct contact with pathogens or their products through TLR. They up-regulate their antigen processing and presentation machinery, as well as display of co-stimulatory (CD80, CD86) and adhesion molecules (CD54, CD83) on their surface. Mature DC secrete pro-inflammatory and Th1-directing cytokines, and migrate to lymph nodes by up-regulation of CCR7 and chemotaxis to its ligands CCL19 and CCL21

cells, adhesion molecules (CD54, CD83) and signaling molecules (CD40) (Savina and Amigorena, 2007; Steinman and Hemmi, 2006). In addition, DC maturation results in early production and secretion of pro-inflammatory cytokines, for example IL-1, IL-6 and TNF-α, as well as IL-12 that promotes the differentiation of non-differentiated helper T cells (Th0) into Th1 type.

A key step for antigen presentation is the migration of DC from the periphery, where they encounter antigens, into the draining LN, where naïve and memory T cells are located. This mobilization is regulated mainly by the chemokine receptor CCR7 on DC, and the chemokines CCL19 and CCL21 that originates from lymphatic vessels (Niedergang et al., 2004; Steinman and Hemmi, 2006).

Different pathogens evolved diverse strategies in order to affect DC function, from direct killing to tightly-regulated mechanisms that exploit DC as a part of the pathogenesis process. Here, we focus on two bacterial pathogens, which are both are classified as a category-A threat agent by the CDC (Darling et al., 2002), *F. tularensis* and *Y. pestis*. Both pathogens are characterized by a low infection dose, paralysis of the immune response, and intracellular growth phase, which is probably more pronounced in *F. tularensis*. We, and others, predicted that DC play a central role in the pathogenicity of *F. tularensis* and *Y. pestis* (Marketon et al., 2005; Bosio et al., 2007), and our work is aimed at examining this assumption.

F. tularensis is a facultative intracellular bacterium that infects and replicates within several cell types. Upon phagocytosis by a target cell, the bacteria blocks the phagosome-lysosome fusion, escapes the phagosome and replicates in the cytosol. In vivo, *F. tularensis* disseminates from the primary infection site into target organs, and it resides mainly at the lung, liver and spleen, irrespectively of the infection site. The basis for *F. tularensis* virulence is not fully resolved as yet. The *F. tularensis* genome contains a pathogenicity island of approximately 17 genes, each one of them is contributing to the bacterial virulence. Most of the studies on *F. tularensis* pathogenesis are carried with the LVS strain, which is non-virulent to humans but retains high virulence in mice (Sjostedt, 2007).

The genus *Yersinia* includes 3 species, *Y. pestis*, *Y. enterocolitica* and *Y. pseudotuberculosis*. Although they are genetically closely related, they differ significantly in their pathogenesis in humans. Only *Y. pestis*, the etiologic agent of plague, causes a severe and fatal disease in men. A major part of the *Yersinia* pathogenesis is governed by a ∼70 kb plasmid that encodes for a type III secretion system (TTSS) and several effector proteins (Yops) that are translocated into a target mammalian cell (Pujol and Bliska, 2005). The work described here was carried-out with the vaccine strain EV76 strain of *Y. pestis*, and WA 0:8 strain of *Y. enterocolitica*.

9.2 Results and Discussion

9.2.1 Killing of Dendritic Cells by Pathogens

Alterations in apoptotic pathways by intracellular pathogens can lead to activation or suppression of dendritic cell death. To examine it we infected bone marrow-derived DC (BMDC) with *Y. pestis* or *Y. enterocolitica* at MOI 20 and followed the DC vitality. *Y. enterocolitica* induced a virulence plasmid dependent cytotoxicity in 65% of the DC within 4 h, and in more than 90% of the cells by 12 h, which was dependent on the virulence plasmid (Velan et al., 2006, and similarly in Erfurth et al., 2004,). On the other hand a lower cytotoxic effect was observed when DC were infected with *Y. pestis*, since more than 50% of the cells were vital 12 and 24 h post infection (Velan et al., 2006; Pujol and Bliska, 2005). One can speculate that delayed apoptosis of infected cells by intracellular pathogens allows the phagocytic cell to serve as a niche for replication, and later-on enables re-infection by apoptotic bodies that carry bacteria and are engulfed by other phagocytic cells. When we examined the *F. tularensis* bacteria (strain LVS) we did not observe a significant cytotoxicity of DC in the early stages of infection (not shown), as also observed in infected macrophages (Lai and Sjostedt, 2003).

Induction of premature cell by more than a few pathogens, such as *Legionella pneumophilia* (Nogueira et al., 2009) *Salmonella* sp., *Shigella* sp., *Streptococcus pneumonia* and *P. aeruginosa* was previously reported (Bhavsar et al., 2007; Finlay and McFadden, 2006; Niedergang et al., 2004; Roy and Mocarski, 2007).

Interestingly, several studies described a reverse correlation between cytotoxicity and virulence in a number of animal models for different pathogens, such as *Y. pestis* (Zauberman et al., 2009; Chapter 5) and *F. tularensis* (Weiss et al., 2007). In the case of *Yersinia*, replacing the YopJ gene of *Y. pestis* with the gene of YopP of *Y. enterocolitica* was sufficient to generate a highly cytotoxic bacterium with lower virulence in mice (Zauberman et al., 2009). Therefore, a well controlled regulation of host cell death may result in a higher pathogenicity.

9.2.2 The Cell as a Niche for Pathogen Proliferation

For intracellular bacteria the strategy of choice is to be taken-up by the suitable cell and utilize it as a replication niche. This requires uptake by cells and avoiding intracellular elimination mechanisms. One of the common strategies for intracellular survival and replication is to block phagosome maturation into a lysosome. That was described for *F. tularensis*, although the molecular mechanism is not yet fully characterized (Sjöstedt, 2007). Although described originally in macrophages, intracellular replication in DC was demonstrated in a similar manner. When we examined it side by side, we observed similar characteristic of LVS intracellular growth between macrophages and bone marrow-derived DC (BMDC). The percent of infected cells was 5–10%, and the growth rate of intracellular bacteria was approximately 100 fold over 24 h post infection (Bar-Haim et al., 2008). We therefore conclude that *F. tularensis* is able to evade the killing mechanisms of both cell types. This is not the case for every pathogen, for example *P. ginaivalis* is efficiently eliminated by polymorphonuclear cells (PMN), while able to survive within DC for more than 24 h (Niedergang et al., 2004), and *Brucella* infects DC more efficiently than macrophages (Ueno et al., 2007).

Several pathogens move from intracellular and extracellular phases following in-vivo infection. Among them are *Y. pestis* and *Y. enterocolitica. Yersinia* spp., which inject effector molecules via their type III secretion system (T3SS) to the target cells, mainly DC, macrophages and neutrophils (Marketon et al., 2005). Among the injected effectors are YopH, a tyrosine phosphatase that dephosphorylates cytoskeletal proteins, and YopE, a Rho-GTPase-activatig protein, which both (as well as other effectors) target membrane re-assembly (Niedergang et al., 2004; Pujol and Bliska, 2005). When we examined it in BMDC we could observe intracellular survival of *Y. pestis* for at least 12 h (Velan et al., 2006).

9.2.3 Manipulation of Immune Induction

Upon stimulation and maturation DC secrets pro-inflammatory cytokines and present antigens to T cells. When we infected BMDC with *F. tularensis* of the LVS strain or *Y. pestis* of the EV76 strain we observed blocking of TNF-α secretion in the infected cells for both bacteria (Fig. 9.2). In the case of EV76 the TNF-α blocking is dependent on the expression of the Yop effectors, *i.e.* presence of the virulence plasmid and growth in 37°C, but not in 28°C (Velan et al., 2006). *Yersinia* spp. targets the MAPK signaling pathway with the YopP/J effector, which among many other effects leads to a reduced TNF-α secretion (Zauberman et al., 2007). Suppression of

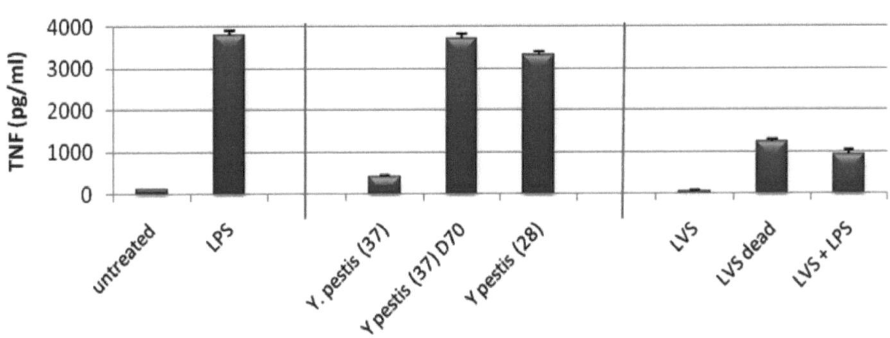

Fig. 9.2 TNF-α secretion by infected DC. BMDC (bone marrow derived DC) were pulsed with *Y. pestis* EV76 bacteria grown in 37 or 28°C, EV76 missing their virulence plasmid (Δ70), live LVS or formalin-inactivated dead LVS. The multiplicity of infection (MOI) of the infecting *Y. pestis* was 5 bacteria per cell and the MOI of LVS was 200 bacteria per cell. Live LVS was also co-administered with 1 μg/ml *E. coli* LPS, and LPS alone served as a positive control for DC induction. Cytokine concentration in the medium was examined 24 h post infection

TNF-α by LVS depends on the viability of the bacteria, since formalin-inactivated LVS trigger TNF-α secretion 50 times higher than the non-infected or LVS-infected cell. When LVS infection was given together with *E. coli* LPS it was able to reduce the stimulatory effect of the LPS, from secretion of 3,800–950 pg/ml.

Bosio *et al.* observed that *F. tularensis* of the highly virulent Schu4 strain do not induce the production of pro- or anti-inflammatory cytokines in human DC, while strains of lower virulence activate the production of pro-inflammatory cytokines and co-stimulatory molecules on the cell surface (Bosio et al., 2007).

Subversion of antigen processing and presentation is evident by several pathogens. This can be clearly demonstrated in DC infected by *Y. enterocolitica* where induction of co-stimulatory molecules and MHC class II is abrogated (Erfurth et al., 2004; Velan et al., 2006). Interestingly, we were able to see effective maturation in *Y. pestis* (EV76) infected DC (Velan et al., 2006).

We have also examined the induction of BMDC maturation following infection with LVS (Fig. 9.3a). Dead LVS bacteria induce DC maturation at a comparable level to that of *E. coli* LPS, as was observed for all of the 6 maturation markers examined. On the other hand, induction of marker display by live LVS was rather ineffective. This was manifested mainly in induction of CD40 and CD54. For both markers the median and mean levels of expression were as low as those of the untreated cells. The effect on MHC class II expression was less pronounced, and the median and mean level of expression of cells pulsed with live or dead LVS were comparable. This could be attributed to the differences in the regulation of the expression of these surface molecules (Savina and Amigorena, 2007; Ueno et al., 2007).

We also examined the potential of live LVS to block the stimulatory effect of *E. coli* LPS by co-pulsing onto BMDC. We observed a significant reduction of the stimulatory effect of the LPS, manifested by decreased expression of CD40 and CD83 (Fig. 9.3b). Therefore, it seems that LVS actively impairs DC potential to present antigens to T cells. Similarly, DC pulsed with dead *L. pneumophila*, but not live bacteria, undergo phenotypic maturation (Kikuchi et al., 2004). Down-regulation of CD1 and class II MHC molecules was described in the case of several pathogens, such as *Mycobacterium tuberculosis* (Merrell and Falkow, 2004), and reduction of surface expression of MHC II in *S. enterica*-infected DC (Bhavsar et al., 2007).

9.2.4 Modulation of DC Migration

Blocking of DC migration to LN can be used to prevent the induction of acquired immunity. On the other hand allowing or inducing DC migration by an intracellular pathogen can serve for bacterial dissemination. These two opposing strategies are illustrated in the cases of *Y. pestis* and *F. tularensis*. In the case of *Y. pestis*, infected DC were paralyzed for migration towards the CCL-19 chemokine, although induction of the chemokine receptor CCR7 was observed to the same level of *E. coli* -stimulated cells (Velan et al., 2006). Moreover, migration of respiratory tract

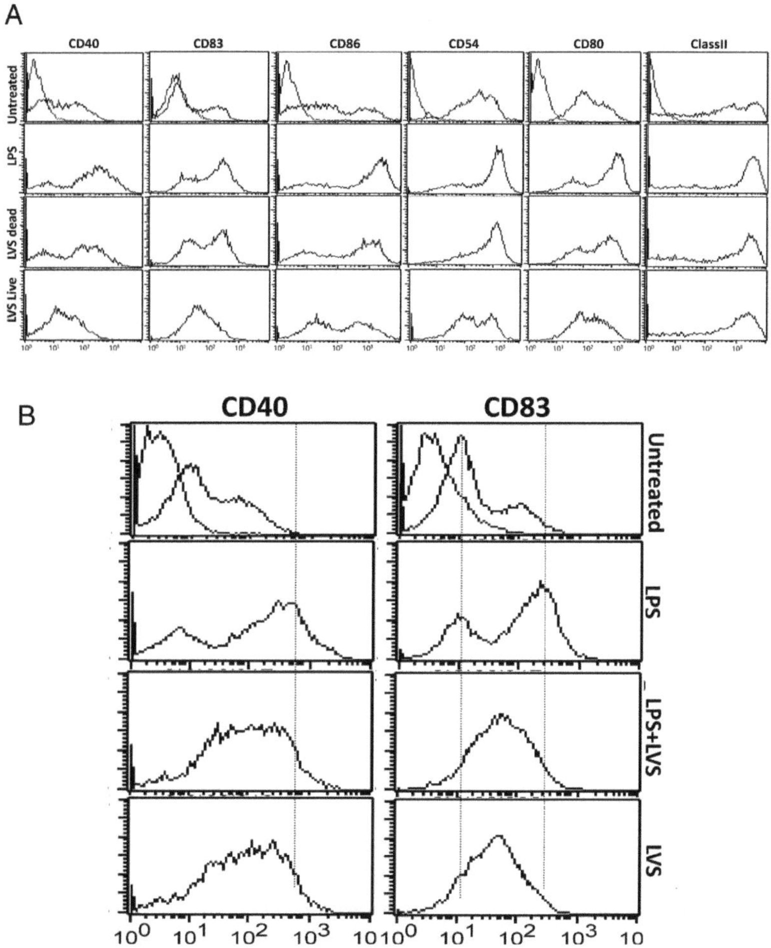

Fig. 9.3 Partial maturation of LVS-infected BMDC (**a**) and block of LPS stimulation (**b**). (**a**) Cells were pulsed with live or formalin-inactivated LVS at MOI of 200 bacteria or bacterial equivalent per cell, or with 1 μg/ml *E. coli* LPS. The cells were incubated for 24 h (37°C, 5%CO_2), washed, and stained for 30 min with the monoclonal antibodies for the maturation markers or their appropriate isotype-matched control antibodies. Stained cells were washed and immediately analyzed by FACS. (**b**) BMDC were pulsed with live LVS (MOI 200), LVS + 0.03 μg LPS or 0.03 μg LPS. One hour later cells were washed and added with fresh medium, incubated for additional 23 h and analyzed for the expression of CD40 and CD83 as described above

DC (RTDC) to the draining lymph node was impaired in *Y. pestis* EV76-infected mice (Velan et al., 2006).

In contrast, when we infected DC with LVS in vitro we observed induction of the CCR7 receptor and active migration to the CCL-19 cytokine in a transwell migration assay (Bar-Haim et al., 2008). Moreover, in LVS-infected mice we observed accumulation of RTDC in the draining LN (mediastinal), and were able to prove that these migratory RTDC were carrying LVS. We next examined the relevance of co-migration of RTDC and LVS bacteria to the pathogenesis. We attempted to block RTDC migration to the draining LN following airways infection with LVS. We utilized two blocking agents: FTY720, a sphingosine analogue that counteracts the functions of the sphingosine-1-phosphate (S1P), or BW245c, an agonist of the prostaglandin D receptor (DP1) (Bar-Haim et al., 2008). With both blockers we observed reduction of DC migration which correlates with reduction in bacterial accumulation. Notably, counts of immigrating bacteria were diminished by up to 100 times in of FTY720-treated mice. Treatment with BW245c also resulted in a delay in the onset of disease signs and eventual animal death.

Hence we conclude that DC migration serve in the early stages of LVS dissemination, and have a role in its pathogenicity. Similarly, it was found that *Mycobacterium tuberculosis* predominantly infects alveolar macrophages and DC, but dissemination to the draining LN is found only in RTDC (Humphreys et al., 2006). In the intestine, *Salmonella typhomurium* are taken-up by sub-epithelial DC and are transported via the bloodstream to the liver and spleen. Another example for dissemination in DC is of *Listeria monocytogenes* that are entering mucosal DC in the intestine and are transferred to the mesenteric lymph node as an early step in the infection process (Niedergang et al., 2004).

9.3 Conclusions

Here we describe various mechanisms used by pathogens to modulate or subvert the function of DC as a part of their pathogenicity. We concentrate on three substantially different strategies by three bacteria (Fig. 9.4). The first is used by *Y. enterocolitica*, we saw that it kills DC early after the encounter between the pathogen and the cell, therefore no other manipulation by the pathogen is required. One should note, however, that the cytotoxic process itself might attract and activate other immune cells and result in induction of effective innate immunity. The second strategy is used by *Y. pestis*, which utilizes the DC as a shelter during the early steps of infection, and paralyses the immune response with its translocated effectors, and thus blocks its movement to the LNs. The third strategy is that of *F. tularensis*, which modulates several aspects of DC function. On one hand, it employs DC as a niche for replication where it is sheltered from interactions with other immune cells. At the same time we observed that *F. tularensis* prevents DC maturation and cytokine secretion, to avoid immune induction. The other level of complexity is provided by modulation of DC mobility, to allow trafficking of bacteria-carrying DC into the LN, thus facilitating *F. tularensis* dissemination (Bar-Haim et al., 2008).

One of the unique characteristics of *F. tularensis* is the low dose required for infection, which in the case of the Schu strain is lower than 100 bacteria in human airways infection (Sjostedt, 2007). Therefore, successful infection could require an early step of intracellular replication sheltered from the immune elimination

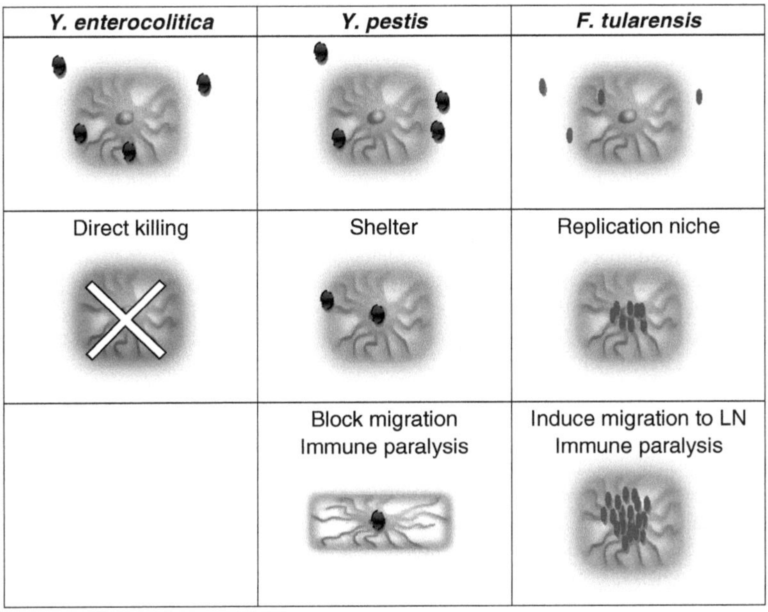

Fig. 9.4 Strategies for DC function subversion. *Y. enterocolitica* directly kill DC in an early time point. *Y. pestis* utilizes DC as a shelter during the early steps of infection, and in parallel blocks its functions. *F. tularensis* invades into DC, replicates intracellularly and utilizes the cells for dissemination into the LN

and avoiding triggering of immune responses. Another fundamental aspect of *F. tularensis* pathogenicity is the very effective spreading from the infection site. Indeed, when we blocked DC migration following airways infection we were able to delay disease appearance and death from *F. tularensis* infection. Altogether our data demonstrates a sophisticated modulation of DC by *F. tularensis*, and those DC modulations have a central role in *F. tularensis* pathogenicity.

References

Bar-Haim, E., Gat, O., Markel, G., Cohen, H., Shafferman, A. & Velan, B. (2008) Interrelationship between Dendritic Cell Trafficking and *Francisella tularensis* Dissemination following Airway Infection. *Plos Pathogens,* 4, e1000211.

Bhavsar, A. P., Guttman, J. A. & Finlay, B. B. (2007) Manipulation of host-cell pathways by bacterial pathogens. *Nature,* 449, 827–34.

Bosio, C. M., Bielefeldt-Ohmann, H. & Belisle, J. T. (2007) Active suppression of the pulmonary immune response by *Francisella tularensis* Schu4. *J Immun,* 178, 4538–47.

Darling, R. G., Catlett, C. L., Huebner, K. D. & Jarrett, D. G. (2002) Threats in bioterrorism I: CDC category A agents. *Emerg Med Clin North Am,* 20, 273–309.

Erfurth, S. E., Grobner, S., Kramer, U., Gunst, D. S. J., Soldanova, I., Schaller, M., Autenrieth, I. B. & Borgmann, S. (2004) *Yersinia enterocolitica* Induces Apoptosis and Inhibits Surface Molecule Expression and Cytokine Production in Murine Dendritic Cells. *Infect Immun,* 72, 7045–54.

Finlay, B. B. & McFadden, G. (2006) Anti-immunology: evasion of the host immune system by bacterial and viral pathogens. *Cell*, 124, 767–82.

Humphreys, I. R., Stewart, G. R., Turner, D. J., Patel, J., Karamanou, D., Snelgrove, R. J. & Young, D. B. (2006) A role for dendritic cells in the dissemination of mycobacterial infection. *Microbes Infect*, 8, 1339–46.

Kikuchi, T., Kobayashi, T., Gomi, K., Suzuki, T., Tokue, Y., Watanabe, A. & Nukiwa, T. (2004) Dendritic cells pulsed with live and dead *Legionella pneumophila* elicit distinct immune responses. *J Immun*, 172, 1727–34.

Lai, X. H. & Sjostedt, A. (2003) Delineation of the molecular mechanisms of *Francisella tularensis*-induced apoptosis in murine macrophages. *Infect Immun*, 71, 4642–6.

Marketon, M. M., Depaolo, R. W., Debord, K. L., Jabri, B. & Schneewind, O. (2005) Plague bacteria target immune cells during infection. *Science*, 309, 1739–41.

Merrell, D. S. & Falkow, S. (2004) Frontal and stealth attack strategies in microbial pathogenesis. *Nature*, 430, 250–6.

Niedergang, F., Didierlaurent, A., Kraehenbuhl, J. P. & Sirard, J. C. (2004) Dendritic cells: the host Achille's heel for mucosal pathogens? *Trends Microbiol*, 12, 79–88.

Nogueira, C. V., Lindsten, T., Jamieson, A. M., Case, C. L., Shin, S., Thompson, C. B. & Roy, C. R. (2009) Rapid Pathogen-Induced Apoptosis: A Mechanism Used by Dendritic Cells to Limit Intracellular Replication of *Legionella pneumophila*. *PLoS Pathog*, 5, e1000478.

Pujol, U. & Bliska, J. B. (2005) Turning Yersinia pathogenesis outside in: subversion of macrophage function by intracellular yersiniae. *Clin Immunol*, 114, 216–26.

Roy, C. R. & Mocarski, E. S. (2007) Pathogen subversion of cell-intrinsic innate immunity. *Nat Immunol*, 8, 1179–87.

Savina, A. & Amigorena, S. (2007) Phagocytosis and antigen presentation in dendritic cells. *Immunol Rev*, 219, 143–56.

Sjostedt, A. (2007) Tularemia: History, epidemiology, pathogen physiology, and clinical manifestations. *Ann NY Acad Sci*, 1105, 1–29.

Steinman, R. M. & Hemmi, H. (2006) Dendritic cells: translating innate to adaptive immunity. *Curr Top Microbiol Immunol*, 311, 17–58.

Ueno, H., Klechevsky, E., Morita, R., Aspord, C., Cao, T., Matsui, T., Di Pucchio, T., Connolly, J., Fay, J. W., Pascual, V., Palucka, A. K. & Banchereau, J. (2007) Dendritic cell subsets in health and disease. *Immunol Rev*, 219, 118–42.

Velan, B., Bar-Haim, E., Zauberman, A., Mamroud, E., Shafferman, A. & Cohen, S. (2006) Discordance in the effects of Yersinia pestis on the dendritic cell functions manifested by induction of maturation and paralysis of migration. *Infect Immun*, 74, 6365–76.

Weiss, D. S., Brotcke, A., Henry, T., Margolis, J. J., Chan, K. & Monack, D. M. (2007) In vivo negative selection screen identifies genes required for Francisella virulence. *Proc Nat Acad Sci USA*, 104, 6037–42.

Zauberman, A., Tidhar, A., Levy, Y., Bar-Haim, E., Halperin, G., Flashner, Y., Cohen, S., Shafferman, A. & Mamroud, E. (2009) *Yersinia pestis* Endowed with Increased Cytotoxicity Is Avirulent in a Bubonic Plague Model and Induces Rapid Protection against Pneumonic Plague. *PLoS ONE*, 4, e5938.

Zauberman, A., Velan, B., Mamroud, E., Flashner, Y., Shafferman, A. & Cohen, S. (2007) Disparity between *Yersinia pestis* and *Yersinia enterocolitica* O : 8 in YopJ/YopP-dependent functions. *Genus Yersinia: Genomics Funct*, 603, 312–20.

Chapter 10
Comparative Approaches to Identify Host Factors Specifically Targeted by *Yersinia pestis* During the Infectious Process

Françoise Guinet, Patrick Avé, Louis Jones, Michel Huerre, and Elisabeth Carniel

Abstract The genus *Yersinia* includes three species pathogenic for humans and animals: *Y. enterocolitica*, *Y. pseudotuberculosis* and *Y. pestis*. The two former species behave like true enteropathogens, i.e. they cause mild intestinal symptoms and are transmitted by the fecal-oral route. In contrast, *Y. pestis* is the etiologic agent of plague, a highly severe and often fatal disease, which is transmitted by flea bites (bubonic plague) or aerosols (pneumonic plague). The plague bacillus is one of the most pathogenic microorganisms of the bacterial kingdom, but the mechanisms it specifically uses to kill its host so efficiently remain largely unknown. Despite drastically different clinical and epidemiological features between enteropathogenic *Yersinia* and the plague agent, it appears that *Y. pestis* is a clone recently emerged from *Y. pseudotuberculosis* (less than 20,000 years ago). Genome comparison indicates that the two species are genetically highly similar. This close genetic relationship is used to carry out comparative pathophysiological studies between *Y. pestis* and its recent ancestor *Y. pseudotuberculosis,* in an attempt to identify the host factors and/or cell lineages specifically targeted by the plague bacillus during the infectious process. Another comparative approach based on the identification of genetic and physiological differences between mouse strains that are either resistant or susceptible to plague is another mean to identify host responses specifically diverted by the plague bacillus.

Keywords Clustering analysis · Plague · Pathophysiology · *Yersinia* · *Y. Pestis* · *Y. pseudotuberculosis*

Abbreviations

cfu	colony-forming units
id	intradermal

F. Guinet (✉)
Unité des *Yersinia*, Institut Pasteur, 28 rue du Docteur Roux, 75724 Paris Cedex 15, France
e-mail: francoise.guinet@pasteur.fr

A. Shafferman et al. (eds.), *The Challenge of Highly Pathogenic Microorganisms*,
DOI 10.1007/978-90-481-9054-6_10, © Springer Science+Business Media B.V. 2010

LD_{50}	lethal dose 50
MCA	multiple correspondence analysis
pi	post-infection

10.1 Introduction

Plague is a bacterial infectious disease that has left an indelible mark in the human mind and is still present in the collective consciousness. Despite considerable progress in its prevention and cure, the plague has made a new comeback in its natural reservoirs (rodents) and accidental hosts (humans) (Stenseth et al., 2008). The steady increase in reported human cases during the past 15 years coupled with its "reappearance" in areas otherwise declared plague-free for several decades, has led to categorize plague as a re-emerging disease (Schrag et al., 1995). The re-emerging threat of the plague has been further exacerbated by the recent identification of a multi-drug resistant strain (Galimand et al., 1997).

Plague is caused by a Gram-negative enterobacterium named *Yersinia pestis* and is primarily a zoonotic disease affecting rats and many other rodents. In humans, bubonic plague is the most common clinical form of the disease and is characterized by the development of a painful lymph node (bubo) draining the site of the fleabite, a high fever and a severe faintness. From the bubo, *Y. pestis* disseminates to deeper tissues (e.g. liver, spleen, lung) and death is a common outcome (40–70% of the patients) when appropriate antibiotherapy is not readily available (Perry and Fetherston, 1997). Pneumonic plague is an extremely severe form of the disease resulting from inter-human aerosol transmission. This pneumopathy is systematically lethal in usually less than 3 days if effective antibiotherapy is not administered at the onset of symptoms. *Y. pestis* is thus considered to be one of the most pathogenic organisms for Humans. In the experimental mouse model, the LD_{50} of *Y. pestis* for laboratory mice is $\approx$10 bacteria by the subcutaneous route and $\approx$1 bacterium by the intravenous route (Perry and Fetherston, 1997).

Despite its extreme pathogenicity, the mechanisms specifically used by *Y. pestis* to kill its host so rapidly and efficiently still remain to be elucidated. A fruitful mean to approach this question consists in comparing two genetically very close organisms with different pathogenicity potentials, in order to identify the tissue lesions and host responses specifying a *Y. pestis* infection.

Y. pestis is a clone recently derived from *Yersinia pseudotuberculosis* with whom its shares most of its genetic material (Achtman et al., 1999). In spite of their close genetic relatedness, the two species exhibit dramatically different life cycles and pathogenic potentials. While *Y. pestis* is the etiologic agent of plague, a highly severe and often fatal disease transmitted by flea bites or aerosols, *Y. pseudotuberculosis* is acquired through consumption of contaminated food or water and causes digestive symptoms that most often subside spontaneously in a few days or weeks (Smego et al., 1999).

Intradermal or subcutaneous inoculation of *Y. pestis* in laboratory rodents causes a bubonic plague very similar to the flea-transmitted disease (Flexner, 1901; Jawetz and Meyer, 1944; Sebbane et al., 2005). In these models, the LD_{50} can be less than 10 cfu. When administered subcutaneously, *Y. pseudotuberculosis* is still less virulent than *Y. pestis*, with LD_{50}s higher than that of *Y. pestis* by several orders of magnitude (Perry and Fetherston, 1997; Une and Brubaker, 1984; Pouillot et al., 2005). However, the pathophysiological mechanisms underlying this virulence difference are not known. Here, we carried out a comparative study of the disease induced by id injection of either *Y. pestis* or *Y. pseudotuberculosis* in mice in order to identify pathophysiological features specific to bubonic plague (Guinet et al., 2008).

10.2 Mortality Studies

Bacterial suspensions of *Y. pestis* CO92 (Parkhill et al., 2001) and *Y. pseudotuberculosis* IP32953 (Chain et al., 2004) were injected in the ear lobe of 8-week female outbred OF1 mice. The id LD_{50} of the two strains was found to be 15 cfu and 5×10^5 cfu, respectively, confirming the known virulence difference between the two *Yersinia* species when injected at peripheral sites. Further mortality studies showed that, in the 2,000–15,000 cfu range of infective loads, *Y. pestis* inoculations consistently led to the animals death while *Y. pseudotuberculosis* infections ended in their recovery. In the next experiments, infective doses within this range were given to the animals, to monitor events differentially associated with lethal versus benign infections.

10.3 Bacterial Progression

Injection sites and draining lymph nodes were collected 24 and 48 h after inoculation of similar doses of *Y. pestis* or *Y. pseudotuberculosis*, and bacterial loads in these organs were determined by cfu enumeration on agar plates. The results show that *Y. pseudotuberculosis* is no less efficient than *Y. pestis* to colonize the injection site. Actually, even higher numbers of *Y. pseudotuberculosis* were recovered from the infected ear at the two time points studied (Fig. 10.1a, b). Moreover, there was no significant difference, between the two *Yersinia* species, in the level of lymph node infection at 24 h pi (Fig. 10.1c), indicating that *Y. pestis* and *Y. pseudotuberculosis* were similarly able to reach this organ. However, *Y. pestis* did achieve higher numbers in the draining lymph node than its closely related species 2 days after injection (Fig. 10.1d). Thus, the higher virulence of *Y. pestis* cells does not appear to result from more efficient proliferation at the injection site, nor to a better ability to translocate to the draining lymph node, but rather to a more efficient multiplication in the draining lymph node.

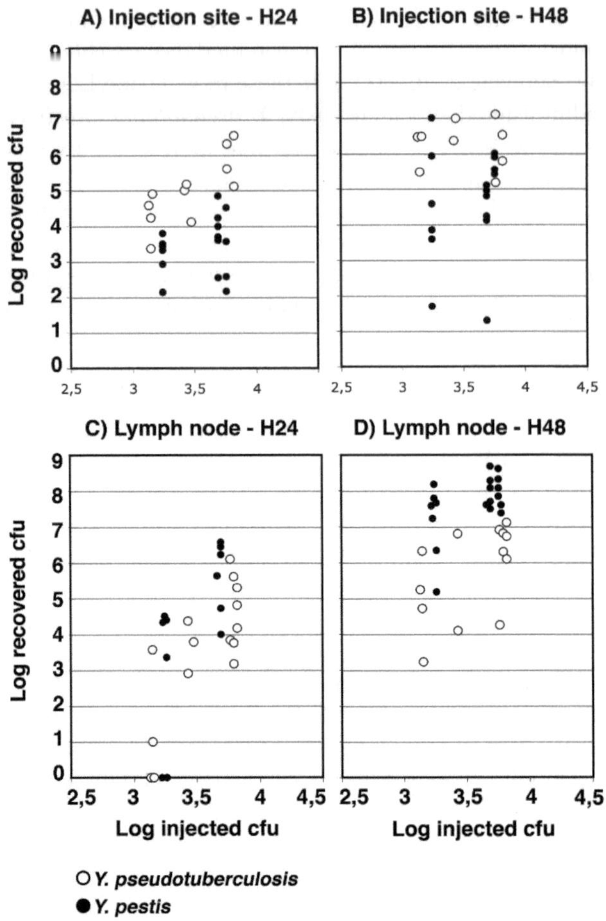

Fig. 10.1 Bacterial counts at the intra dermal site of injection in the ear lobe (**a**, **b**) and in the draining lymph node (**c**, **d**), 24 h (**a**, **c**) and 48 h (**b**, **d**) post-infection

10.4 Pathology

10.4.1 Macroscopic Differences

At the gross pathology level, both *Yersinia* species produced only a slight inflammatory reaction around the injection site, whereas the proximal node started to be enlarged and highly inflammatory between 24 and 48 h pi. *Y. pestis*-infected nodes were more hemorrhagic than those infected with *Y. pseudotuberculosis* (Fig. 10.2) and their consistency was firm.

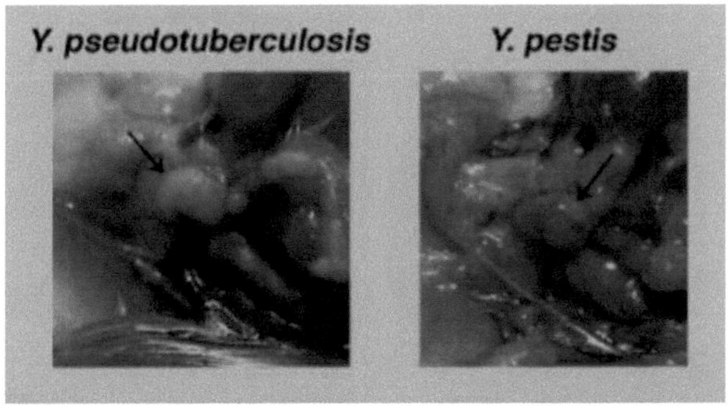

Fig. 10.2 Macroscopical aspect of the draining lymph node 48 h post-infection with *Y. pseudotuberculosis* (the organ is enlarged and frankly purulent) or *Y. pestis* (the node is enlarged, densely infiltrated and hemorrhagic)

10.4.2 Differences in Tissue Lesions

Microscopically, *Y. pestis* generated more destructive lesions than *Y. pseudotuberculosis* in draining lymph nodes at 48–72 h pi (Fig. 10.3a, b). To confirm that the observed difference was statistically significant across a large number of samples, and to determine which elementary lesions were significantly associated with one *Yersinia* species or the other, a detailed and semi-quantitative analysis of the lymph node alterations was performed.

To this end, 42 elementary lesions (criteria) were defined. The criteria were chosen to investigate the following features: extent and organization of inflammatory reaction; amount of bacteria; invasiveness of bacteria in the lymph node tissue; breaching of the inflammatory barrier; alterations in tissue density; signs of final lymph node destruction. Lymph nodes from 88 mice were studied, each one being scored blindly as "+" or "–" for each criterion. Clustering analysis of the resulting +/– table displayed three main clusters: one comprised lymph nodes infected with either *Y. pestis* or *Y. pseudotuberculosis* at an early stage pi and harboring no or minimal lesions (type 1); the second cluster (type 2) contained lymph nodes infected with *Y. pseudotuberculosis* for 48 or 72 h; the third cluster (type 3) grouped the lymph nodes infected with *Y. pestis* and generally collected at 48 h pi (Guinet et al., 2008). Therefore, the second and third cluster defined histological profiles specifically associated with *Y. pseudotuberculosis* and *Y. pestis* infections, respectively.

Another grouping method, Multiple Correspondence Analysis, confirmed that *Y. pestis* and *Y. pseudotuberculosis* lesions formed distinct histopathological patterns (Fig. 10.4). Furthermore, the "test-value" function of the MCA method allows to estimate the relative weight of each criterion in the discrimination between two separate groups (Guinet et al., 2008). In Table 10.1 are listed the criteria most specific to *Y. pseudotuberculosis* or *Y. pestis* infections. For each infection type, the seven criteria with the highest *T*-values are displayed. They show that the most

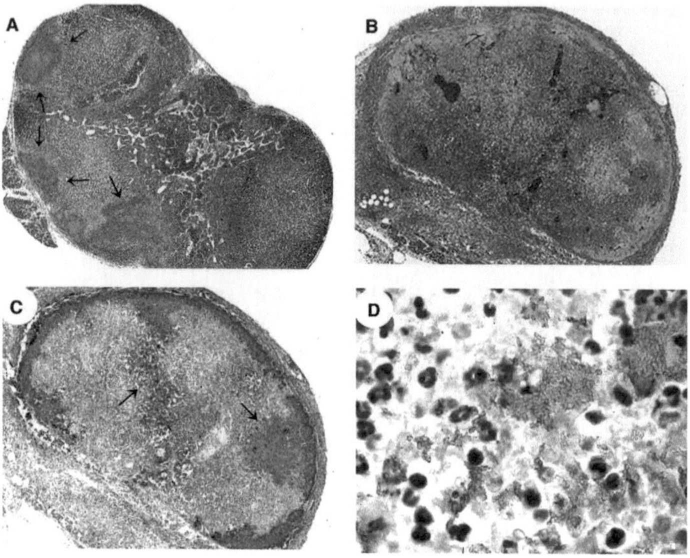

Fig. 10.3 Examples of draining lymph node histopathology. (**a**) *Y. pseudotuberculosis* colonies can be seen as small light areascontained within peripheral abscesses (*arrows*) whereas the rest of the organ has a normal appearance. (**b**) *Y. pestis*-infected lymph node. The normal node architecture is completely obliterated, there are intranodal hemorrhages, vascular congestions and zones of cellular depletion. No organized inflammatory reaction is visible. (**c**) Infiltrating *Y. pestis* infection in the lymph node tissue. The immunostained bacterial infiltrates (arrows) are darker than the rest of the organ. (**d**) High magnification showing bacterial rods detached from nearby bacterial aggregates and making their way between and around host cells. Note the characteristic horseshoe-shaped nuclei of the inflammatory infiltrate *Stainings:* (**a**) and (**b**) hematoxylin-eosin; (**c**) and (**d**) immunostaining with a *Y. pestis* antibody. *Magnifications*: (**a**), (**b**) and (**c**) 4 X objective; (**d**) 100 X objective

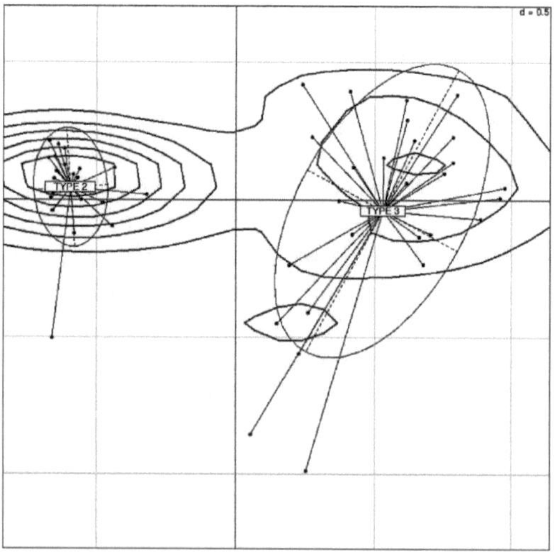

Fig. 10.4 Multiple Correspondence Analysis (MCA) of type 2 and type 3 histopathological profiles. Each dot corresponds to one lymph node. In this figure, only lymph nodes of histotypes 2 or 3, as defined by the dendrogram method (see text), are represented. The figure confirms the type 2/type 3 grouping and shows that the two types, which correspond to advanced *Y. pseudotuberculosis* and *Y. pestis* lesions, respectively, do not overlap

Table 10.1 List of the 14 criteria having the highest power of discrimination between *Y. pestis* or *Y. pseudotuberculosis* lesions in the lymph node

Criteria	*T*-value
Patches of densely packed bacteria, bordered by PMNs[a]	−5.7
Peripheral band of PMNs containing bacterial foci	−5.5
Normal LN[a] tissular density	−5.22
Abscess-type structure > 25% of surface	−4.76
Prominent inflammatory front line	−4.46
Abscess-type structure, polar	−3.34
Abscess-type structure, wedge-shaped	−3.19
Bacterial projections from the subcapsular sinus	+5.22
Free-floating bacteria	+5.74
Bacteria in ≥ 1/3 of sub-capsular sinus	+5.79
Enlarged blood vessels packed with RBCs[a]	+5.93
«Moth eaten » appearence	+5.97
Bacterial infiltration around LN cells	+6
Apparently intact host cells within a bacterial zone	+6.01

T-values represent the discriminating power of each criterion. Negative and positive *T*-values correspond to criteria specifically associated with *Y. pseudotuberculosis* and *Y. pestis* lesions, respectively

[a] *PMN*: polymorphonuclear cells; *LN*: lymph node; *RBC*: red blood cells

specific characteristics of *Y. pseudotuberculosis* infection correspond to the formation of localized abscess-type polymorphonuclear infiltrates that enclose the bacterial foci and preserve the overall structure of the organ. In contrast, *Y. pestis* infection is specified by a diffuse bacterial dissemination within the tissue, extensive lymph node destruction, and high bacterial numbers. Examples of *Y. pestis* invasive and infiltrative behavior are shown in Fig. 10.3c, d.

10.5 Conclusion

Comparative follow-up of *Y. pestis* and *Y. pseudotuberculosis* infections after intradermal delivery of the bacteria provides important clues for the understanding of the pathological processes specifically triggered by *Y. pestis*. Our results indicate that the first organ where noticeable differences are found is the draining lymph node, and not the injection site. Neither does *Y. pseudotuberculosis* appear more impaired in its ability to translocate and settle in the proximal node, than the bubonic plague agent. Features specifically associated, in the lymph node, with lethal *Y. pestis* infections, are the presence of very high numbers of bacteria which display a distinctly invasive behavior between and around the host cells, and are associated with profound alterations of the organ architecture. Further work with defined *Yersinia* mutants should help better understand the links between the observed infection characteristics and the mechanisms underlying the exceptional severity of plague. Another mean of exploring the specific targets of *Y. pestis* is to compare the host responses induced by the plague agent in susceptible and resistant hosts. This other comparative approach is also undergoing.

References

Achtman, M., Zurth, K., Morelli, G., Torrea, G., Guiyoule, A. & Carniel, E. (1999) *Yersinia pestis*, the cause of plague, is a recently emerged clone of *Yersinia pseudotuberculosis*. *Proc Natl Acad Sci USA*, 96, 14043–8.

Chain, P. S., Carniel, E., Larimer, F. W., Lamerdin, J., Stoutland, P. O., Regala, W. M., Georgescu, A. M., Vergez, L. M., Land, M. L., Motin, V. L., Brubaker, R. R., Fowler, J., Hinnebusch, J., Marceau, M., Medigue, C., Simonet, M., Chenal-Francisque, V., souza, B., Dacheux, D., Elliott, J. M., Derbise, A., Hauser, L. J. & Garcia, E. (2004) Insights into the evolution of *Yersinia pestis* through whole-genome comparison with *Yersinia pseudotuberculosis*. *Proc Natl Acad Sci USA*, 101, 13826–31.

Flexner, S. (1901) The pathology of bubonic plague. *Am J Med Sci*, 122, 396–416.

Galimand, M., Guiyoule, A., Gerbaud, G., Rasoamanana, B., Chanteau, S., Carniel, E., & Courvalin, P. (1997). Multidrug resistance in *Yersinia pestis* mediated by a transferable plasmid [see comments]. *N Engl J Med*, 337, 677–680.

Guinet, F., Ave, P., Jones, L., Huerre, M. & Carniel, E. (2008) Defective innate cell response and lymph node infiltration specify Yersinia pestis infection. *PLoS One*, 3, e1688.

Jawetz, E. & Meyer, K. F. (1944) The behaviour of virulent and avirulent *P. pestis* in normal and immune experimental animals. *J Infect Dis*, 74, 1–13.

Parkhill, J., Wren, B. W., Thomson, N. R., Titball, R. W., Holden, M. T., Prentice, M. B., Sebaihia, M., James, K. D., Churcher, C., Mungall, K. L., Baker, S., Basham, D., Bentley, S. D., Brooks, K., Cerdeno-Tarraga, A. M., Chillingworth, T., Cronin, A., Davies, R. M., Davis, P., Dougan, G., Feltwell, T., Hamlin, N., Holroyd, S., Jagels, K., Karlyshev, A. V., Leather, S., Moule, S., Oyston, P. C., Quail, M., Rutherford, K., Simmonds, M., Skelton, J., Stevens, K., Whitehead, S. & Barrell, B. G. (2001) Genome sequence of *Yersinia pestis*, the causative agent of plague. *Nature*, 413, 523–7.

Perry, R. D. & Fetherston, J. D. (1997) *Yersinia pestis*–etiologic agent of plague. *Clin Microbiol Rev*, 10, 35–66.

Pouillot, F., Derbise, A., Kukkonen, M., Foulon, J., Korhonen, T. K. & Carniel, E. (2005) Evaluation of O-antigen inactivation on Pla activity and virulence of *Yersinia pseudotuberculosis* harbouring the pPla plasmid. *Microbiology*, 151, 3759–68.

Schrag, S. J., & Wiener, P. (1995). Emerging infectious diseases: what are the relative roles of ecology and evolution? *Trends Evol Ecol (TREE)*, 10, 319–24.

Sebbane, F., Gardner, D., Long, D., Gowen, B. B. & Hinnebusch, B. J. (2005) Kinetics of disease progression and host response in a rat model of bubonic plague. *Am J Pathol*, 166, 1427–39.

Smego, R. A., Frean, J. & Koornhof, H. J. (1999) Yersiniosis I: Microbiological and clinicoepidemiological aspects of plague and non-plague *Yersinia* infections [Review]. *Eur J Clin Microbiol Infect Dis*, 18, 1–15.

Stenseth, N. C., Atshabar, B. B., Begon, M., Belmain, S. R., Bertherat, E., Carniel, E., Gage, K. L., Leirs, H. & Rahalison, L. (2008) Plague: past, present, and future. *PLoS Med*, 5, e3.

Une, T. & Brubaker, R. R. (1984) In vivo comparison of avirulent Vwa- and Pgm- or Pstr phenotypes of yersiniae. *Infect Immun*, 43, 895–900.

Chapter 11
Poxvirus Entry into Host Cells

Bernard Moss, Zain Bengali, Himani Bisht, Jason P. Laliberte, and P.S. Satheskumar

Abstract Vaccinia virus, the prototype member of the poxvirus family, assembles two related infectious forms: the mature virion (MV) and the enveloped virion (EV). The MV consists of a core surrounded by a lipoprotein membrane. The EV is essentially an MV with an additional antigenically distinct outer membrane that must be disrupted prior to or during entry. Entry of the MV occurs by fusion of the lipoprotein membrane with the plasma membrane or following endocytosis. Approximately 25 proteins are associated with the MV membrane. Of these, four have roles in cell attachment and twelve in entry and membrane fusion. The entry fusion proteins are conserved in all poxviruses and form a stable complex in the MV membrane.

Keywords Poxvirus · Membrane fusion · Endocytosis

11.1 Introduction

Poxviruses comprise a large family that infects many vertebrate and invertebrate species (Moss, 2007). Two members of grave medical importance are variola virus, the causative agent of smallpox, and monkeypox virus, which causes a smallpox-like disease in parts of Africa (Damon, 2007). These pathogens, as well as vaccinia virus (VACV), used as the vaccine to prevent smallpox, are closely related members of the orthopoxvirus genus. The more distantly related molluscum contagiosum virus, is a human specific virus that causes benign skin lesions in young children and more severe infections in immunocompromised individuals. Many other poxviruses cause infrequent zoonoses.

B. Moss (✉)
Laboratory of Viral Diseases, National Institute of Allergy and Infectious Diseases, National Institutes of Health, Bethesda, MD 20892, USA
e-mail: bmoss@nih.gov

A. Shafferman et al. (eds.), *The Challenge of Highly Pathogenic Microorganisms*,
DOI 10.1007/978-90-481-9054-6_11, © Springer Science+Business Media B.V. 2010

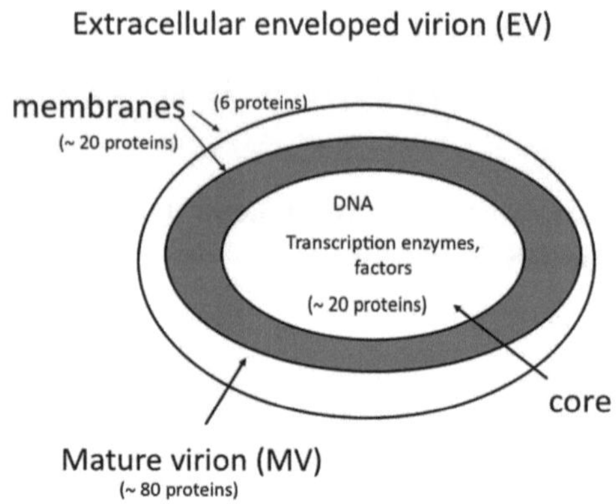

Fig. 11.1 Diagram of an EV

VACV is the best-characterized poxvirus and the prototype member of the family. There are two major infectious forms of VACV, the mature virion (MV) and the enveloped virion (EV). The MV consists of a nucleoprotein core containing a 200,000 bp double-stranded DNA genome, structural proteins and enzymes used for transcription of early genes surrounded by a lipoprotein membrane (Condit et al., 2006). The EV is an extracellular form of VACV that is formed from a subpopulation of MVs that become wrapped in modified trans-Golgi or endosomal membranes and are transported on microtubules through the cytoplasm and fuse with the plasma membrane (Smith et al., 2002). Thus, the EV has one additional membrane compared to the MV and is antigenically distinct (Fig. 11.1). Approximately 25 viral proteins are associated with the MV membrane and six others with the additional EV membrane. The spread of VACV is facilitated by EV attached to the tips of cellular projections called actin tails. However, the EV membrane is not fusogenic and is disrupted by interaction with glycosaminoglycans prior to entry of the MV (Law et al., 2006).

11.2 Entry of the VACV MV

11.2.1 Binding and Entry

Cell attachment is the first step in virus entry. Several proteins associated with the MV membrane are involved at this initial stage (Table 11.1). D8 binds to chondroitin sulfate [Hsiao et al., 1999 #7540], the A27 and H3 proteins to heparan sulfate (Chung et al., 1998; Hsiao et al., 1998; Lin et al., 2000; Vazquez and Esteban, 1999) and A26 to laminin (Chiu et al., 2007). Many enveloped viruses enter cells by endocytosis, while others enter directly through the plasma membrane (Sieczkarski and Whittaker, 2005). Until recently, MVs were thought to enter cells predominantly through the plasma membrane, based on electron micrographs (Armstrong et al.,

Table 11.1 VACV MV entry proteins

Protein	Description	Reference
Attachment		
A26	58 kDa; no TM[a]	Chiu et al., 2007
A27	13 kDa; no TM	Chung et al., 1998
D8	35 kDa; C-terminal TM	Hsiao et al., 1999
H3	37 kDa; C-terminal TM	Lin et al., 2000
Entry/fusion complex		
A16[b]	43 kDa; C-terminal TM; myr[c]	Ojeda et al., 2006b
A21	14 kDa; N-terminal TM	Townsley et al., 2005
A28	16 kDa; N-terminal TM	Senkevich et al., 2004
G3	13 kDa; N-terminal TM	Izmailyan et al., 2006
G9[b]	39 kDa; C-terminal TM; myr	Ojeda et al., 2006a
H2	22 kDa; N-terminal TM	Senkevich and Moss, 2005
I2[d]	8 kDa; C-terminal TM	Nichols et al., 2008
J5[b]	15 kDa; C-terminal TM	Senkevich et al., 2005
L5	15 kDa; N-terminal TM	Townsley et al., 2005
O3	4 kDa; N-terminal TM	Satheshkumar and Moss, 2009
EFC associated proteins		
F9	24 kDa; C-terminal	Brown et al., 2006
L1	27 kDa; C-terminal; myr	Bisht et al., 2008

[a]TM, Transmembrane domain
[b]sequence related
[c]myr, myristoylated
[d]not yet shown to be part of EFC

1973; Chang and Metz, 1976; Carter et al., 2005). However, other studies showed MVs enclosed within cytoplasmic vacuoles (Dales, 1963). The apparent disagreement was reconciled in a report by Townsley et al. (Townsley et al., 2006), where it was shown by electron microscopy and biochemical assays that MVs of the WR strain of VACV enter both by fusion with the plasma membrane and by low pH-mediated endocytosis (Fig. 11.2). The latter conclusion was supported by studies with a recombinant VACV that expresses firefly luciferase regulated by an early promoter. In one series of experiments, MVs were allowed to adsorb to cells at 4°C and then incubated with medium adjusted to a range of pH values for 2–3 min at 37°C. The pH was then neutralized and the incubation continued for 1 h. The pH 5 treatment enhanced luciferase activity 5 to10-fold compared to pH 7.4. Thus, mimicking the low pH of endosomes accelerated entry through the plasma membrane. In a second series of experiments, drugs that prevented the acidification of endosomes e.g. bafilomycin A1 and concanamycin, reduced luciferase expression unless the endosomal pathway were bypassed by lowering the pH of the medium. These studies indicated that the low pH endosomal pathway was the major route used by VACV (Townsley et al., 2006). Subsequent studies demonstrated that there are two distinct low pH steps (Townsley and Moss, 2007) and that the relative requirement for low pH varies with different strains of VACV (Bengali et al., 2009). Further work is needed to determine the preference of other poxviruses.

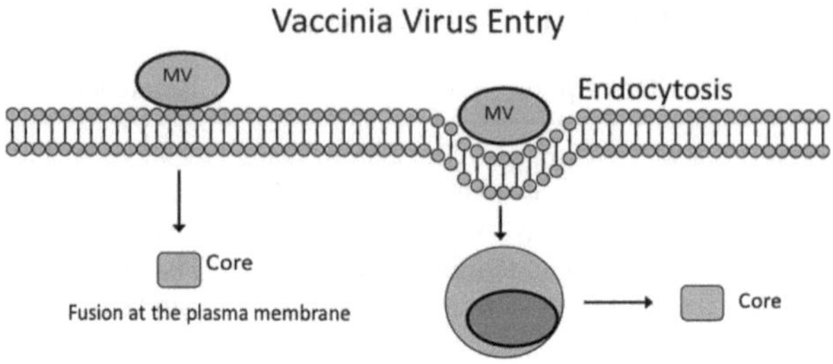

Fig. 11.2 Two pathways of MV entry into cells

A variety of endocytic mechanisms are employed by viruses (Sieczkarski and Whittaker, 2002). One such mechanism, employed by large viruses including VACV, is macropinocytosis (Mercer and Helenius, 2009, 2008). This is an actin-dependent process that results in the engulfment of particles. Mercer and Helenius (Mercer and Helenius, 2008) proposed that VACV MVs mimicked apoptotic bodies and that entry was dependent on a specific interaction of phosphatidylserine in the MV membrane with cell receptors. A subsequent study confirmed the importance of phospholipids in the MV membrane but showed that other phospholipids could replace phosphatidylserine indicating that the putative receptor was not specific for the latter lipid (Laliberte and Moss, 2009).

11.2.2 Membrane Fusion

The genomes of enveloped viruses enter cells following the fusion of viral and cellular membranes (White et al., 2008). This process is typically mediated by one or more viral fusion proteins, which bring the membranes in close apposition. Several years ago, our laboratory sought to identify the putative fusion proteins of VACV. Examination of the genome sequence revealed more than 40 potential transmembrane proteins, a subset of which was highly conserved with homologs in all poxviruses sequenced to date – even those with insect hosts. To determine the roles of these proteins, recombinant viruses in which the target gene was regulated by the *Escherichia coli lac* repressor were constructed. The first of these tested was the A28 protein, which has an N-terminal transmembrane domain (Senkevich et al., 2004). In the presence of inducer, the recombinant virus replicated like wild type virus. In the absence of inducer, virus particles were assembled but were non-infectious and the defect was shown to be at virus entry. In addition, the recombinant virus was unable to mediate syncytium formation indicating a membrane fusion defect. Subsequently, repression of genes encoding other transmembrane proteins was also found to produce a similar phenotype. In all, 12 proteins have now been identified that are involved in VACV entry (Table 11.1). Most of these proteins are assembled in a complex called the entry fusion complex (EFC) and others associate with the complex. The structure of the complex has not been elucidated, although interactions between A28 and H2 and between A16 and G9 have been determined (Nelson et al., 2008;

Wagenaar et al., 2008). A challenge for the future is to determine the individual roles of these proteins.

11.2.3 Regulation of MV Entry

Cells infected with VACV containing a mutation in either the A56 or the K2 gene undergo spontaneous fusion to form syncytia (Ichihashi and Dales, 1971; Law and Smith 1992; Turner and Moyer, 1992; Zhou et al., 1992). The A56 and K2 proteins form a dimer (or higher order oligomer) in the plasma membrane of the cell (Brum et al., 2003; Turner and Moyer, 2006). It was found that the A16 and G9 proteins of the EFC interact with A56-K2 and that this interaction prevents syncytium formation and also superinfection of cells by progeny VACV (Wagenaar and Moss, 2007, 2009; Wagenaar et al., 2008).

Acknowledgements This work was supported by the Division of Intramural Research, National Institute of Allergy and Infectious Diseases, National Institutes of Health.

References

Armstrong, J. A., Metz, D. H. & Young, M. R. (1973) The mode of entry of vaccinia virus into L cells. *J Gen Virol,* 21, 533–7.

Bengali, Z., Townsley, A. C. & Moss, B. (2009) Vaccinia virus strain differences in cell attachment and entry. *Virology,* 389, 132–40.

Bisht, H., Weisberg, A. S. & Moss, B. (2008) Vaccinia Virus L1 protein is required for cell entry and membrane fusion. *J Virol,* 82, 8687–94.

Brown, E., Senkevich, T. G. & Moss, B. (2006) Vaccinia virus F9 virion membrane protein is required for entry but not virus assembly, in contrast to the related 11 protein. *J Virol,* 80, 9455–64.

Brum, L. M., Turner, P. C., Devick, H., Baquero, M. T. & Moyer, R. W. (2003) Plasma membrane localization and fusion inhibitory activity of the cowpox virus serpin SPI-3 require a functional signal sequence and the virus encoded hemagglutinin. *Virology,* 306, 289–302.

Carter, G. C., Law, M., Hollinshead, M. & Smith, G. L. (2005) Entry of the vaccinia virus intracellular mature virion and its interactions with glycosaminoglycans. *J Gen Virol,* 86, 1279–90.

Chang, A. & Metz, D. H. (1976) Further investigations on the mode of entry of vaccinia virus into cells. *J Gen Virol,* 32, 275–82.

Chiu, W. L., Lin, C. L., Yang, M. H., Tzou, D. L. M. & Chang, W. (2007) Vaccinia virus 4c (A26L) protein on intracellular mature virus binds to the extracellular cellular matrix laminin. *J Virol,* 81, 2149–57.

Chung, C.-S., Hsiao, J.-C., Chang, Y.-S. & Chang, W. (1998) A27L protein mediates vaccinia virus interaction with cell surface heparin sulfate. *J Virol,* 72, 1577–85.

Condit, R. C., Moussatche, N. & Traktman, P. (2006) In a nutshell: structure and assembly of the vaccinia virion. *Adv Virus Res,* 66, 31–124.

Dales, S. (1963) The uptake and development of vaccinia virus in strain L cells followed with labeled viral deoxyribonucleic acid. *J Cell Biol,* 18, 51–72.

Damon, I. (2007) Poxviruses. In: Knipe, D. M. & Howley, P. M. (eds), Field's virology (pp. 2947–2975), Lippincott Williams & Wilkins, Philadelphia, PA.

Hsiao, J. C., Chung, C. S. & Chang, W. (1998) Cell surface proteoglycans are necessary for A27L protein- mediated cell fusion: Identification of the N-terminal region of A27L protein as the glycosaminoglycan-binding domain. *J Virol,* 72, 8374–9.

Hsiao, J. C., Chung, C. S. & Chang, W. (1999) Vaccinia virus envelope D8L protein binds to cell surface chondroitin sulfate and mediates the adsorption of intracellular mature virions to cells. *J Virol,* 73, 8750–61.

Ichihashi, Y. & Dales, S. (1971) Biogenesis of poxviruses: interrelationship between hemagglutinin production and polykaryocytosis. *Virology,* 46, 533–43.

Izmailyan, R. A., Huang, C. Y., Mohammad, S., Isaacs, S. N. & Chang, W. (2006) The envelope G3L protein is essential for entry of vaccinia virus into host cells. *J Virol,* 80, 8402–10.

Laliberte, J. P. & Moss, B. (2009) Appraising the apoptotic mimicry model and the role of phospholipids for poxvirus entry. *Proc Natl Acad Sci USA,* 106, 17517–21.

Law, K. M. & Smith, G. L. (1992) A vaccinia serine protease inhibitor which prevents virus-induced cell fusion. *J Gen Virol,* 73, 549–57.

Law, M., Carter, G. C., Roberts, K. L., Hollinshead, M. & Smith, G. L. (2006) Ligand-induced and non-fusogenic dissolution of a viral membrane. *Proc Natl Acad Sci USA,* 103, 5989–94.

Lin, C. L., Chung, C. S., Heine, H. G. & Chang, W. (2000) Vaccinia virus envelope H3L protein binds to cell surface heparan sulfate and is important for intracellular mature virion morphogenesis and virus infection in vitro and in vivo. *J Virol,* 74, 3353–65.

Mercer, J. & Helenius, A. (2008) Vaccinia virus uses macropinocytosis and apoptotic mimicry to enter host cells. *Science,* 320, 531–5.

Mercer, J. & Helenius, A. (2009) Virus entry by macropinocytosis. *Nat Cell Biol,* 11, 510–20.

Moss, B. (2007) Poxviridae: the viruses and their replication (pp. 2905–2945). In Knipe, D. M. & Howley, P. M. (Eds.) Fields virology. Lippincott Williams & Wilkins, Philadelphia, PA.

Nelson, G. E., Wagenaar, T. R. & Moss, B. (2008) A conserved sequence within the H2 subunit of the vaccinia virus entry/fusion complex is Important for interaction with the A28 subunit and infectivity. *J Virol,* 82, 6244–50.

Nichols, R. J., Stanitsa, E., Unger, B. & Traktman, P. (2008) The vaccinia I2L gene encodes a membrane protein with an essential role in virion entry. *J Virol,* 82, 10247–61.

Ojeda, S., Domi, A. & Moss, B. (2006a) Vaccinia virus G9 protein is an essential component of the poxvirus entry-fusion complex. *J Virol,* 80, 9822–30.

Ojeda, S., Senkevich, T. G. & Moss, B. (2006b) Entry of vaccinia virus and cell-cell fusion require a highly conserved cysteine-rich membrane protein encoded by the A16L gene. *J Virol,* 80, 51–61.

Satheshkumar, P. S. & Moss, B. (2009) Characterization of a newly identified 35 amino acid component of the vaccinia virus entry/fusion complex conserved in all chordopoxviruses. *J Virol,* 83, 12822–32.

Senkevich, T. G. & Moss, B. (2005) Vaccinia virus H2 protein is an essential component of a complex involved in virus entry and cell-cell fusion. *J Virol,* 79, 4744–54.

Senkevich, T. G., Ojeda, S., Townsley, A., Nelson, G. E. & Moss, B. (2005) Poxvirus multiprotein entry-fusion complex. *Proc Natl Acad Sci USA,* 102, 18572–77.

Senkevich, T. G., Ward, B. M. & Moss, B. (2004) Vaccinia virus entry into cells is dependent on a virion surface protein encoded by the A28L gene. *J Virol,* 78, 2357–66.

Sieczkarski, S. B. & Whittaker, G. R. (2002) Dissecting virus entry via endocytosis. *J Gen Virol,* 83, 1535–45.

Sieczkarski, S. B. & Whittaker, G. R. (2005) Viral entry. *Curr Top Microbiol Immunol,* 285, 1–23.

Smith, G. L., Vanderplasschen, A. & Law, M. (2002) The formation and function of extracellular enveloped vaccinia virus. *J Gen Virol,* 83, 2915–31.

Townsley, A., Senkevich, T. G. & Moss, B. (2005) Vaccinia virus A21 virion membrane protein is required for cell entry and fusion. *J Virol,* 79, 9458–69.

Townsley, A. C. & Moss, B. (2007) Two distinct low-pH steps promote entry of vaccinia virus. *J Virol,* 81, 8613–20.

Townsley, A. C., Weisberg, A. S., Wagenaar, T. R. & Moss, B. (2006) Vaccinia virus entry into cells via a low pH-dependent-endosomal pathway. *J Virol,* 80, 8899–8908.

Turner, P. C. & Moyer, R. W. (1992) An orthopoxvirus serpin-like gene controls the ability of infected cells to fuse. *J Virol,* 66, 2076–85.

Turner, P. C. & Moyer, R. W. (2006) The cowpox virus fusion regulator proteins SPI-3 and hemagglutinin interact in infected and uninfected cells. *Virology,* 347, 88–99.
Vazquez, M. I. & Esteban, M. (1999) Identification of functional domains in the 14-kilodalton envelope protein (A27L) of vaccinia virus. *J Virol,* 73, 9098–109.
Wagenaar, T. R. & Moss, B. (2007) Association of vaccinia virus fusion regulatory proteins with the multicomponent entry/fusion complex. *J Virol,* 81, 6286–93.
Wagenaar, T. R. & Moss, B. (2009) Expression of the A56 and K2 proteins is sufficient to inhibit vaccinia virus entry and cell fusion. *J Virol,* 83, 1546–54.
Wagenaar, T. R., Ojeda, S. & Moss, B. (2008) Vaccinia virus A56/K2 fusion regulatory protein interacts with the A16 and G9 subunits of the entry fusion complex. *J Virol,* 82, 5153–60.
White, J. M., Delos, S. E., Brecher, M. & Schornberg, K. (2008) Structures and mechanisms of viral membrane fusion proteins: multiple variations on a common theme. *Crit Rev Biochem Mol Biol,* 43, 189–219.
Zhou, J., Sun, X. Y., Fernando, G. J. P. & Frazer, I. H. (1992) The vaccinia virus K2L gene encodes a serine protease inhibitor which inhibits cell-cell fusion. *Virology,* 189, 678–86.

[illegible] (2006) The complex [illegible] [illegible] [illegible] [illegible] [illegible]

Vazquez, M. I. & Esteban, M. (1999) Identification of functional domains in the 14-kilodalton envelope protein (A27L) of vaccinia virus. *J Virol*, 73, 9098–109.

Wagenaar, T. R. & Moss, B. (2007) Association of vaccinia virus fusion regulatory proteins with the [illegible] entry/fusion complex. *J Virol*, 81, 6286–93.

[illegible]

[illegible]

White, J. M., Delos, S. E., Brecher, M. & Schornberg, K. [illegible] Structures and mechanisms of viral membrane fusion proteins: [illegible] variations on a common theme. [illegible]

[illegible]

Chapter 12
Role of A33R Amino-Acid 118L in the Interactions of Vaccinia Virus with the Host

Nir Paran, Shlomo Lustig, Anat Zvi, Tomer Israely, Sharon Melamed, Noam Erez, Boaz Politi, David Ben-Nathan, Paula Schneider, Batel Lachmi, Ofir Israeli, Dana Stein, Reuven Levin, and Udy Olshevsky

Abstract Immunization of BALB/c mice with vaccinia virus protein A33 ($A33^{VACV}$) protects mice from intranasal challenge with the WR strain of vaccinia virus or with ectromelia virus making A33 an important candidate to be included in experimental smallpox subunit vaccines. Single vaccination with a recombinant Sindbis virus expressing $A33^{VACV}$ protect mice against lethal VACV-WR and ectromelia virus (ECTV) but not against the closely related cowpox virus (CPXV). Furthermore, even recombinant Sindbis virus expressing the cowpox virus A33 ortholog ($A33^{CPXV}$) failed to protect either against cowpox or against VACV-WR challenge. Our attempts to map the regions which may account for this differential behavior were directed against a region of difference between the two orthologs. A stretch of 7 amino acids in A33 was mapped as important for protection which contain the following changes in $A33^{CPXV}$: L112F, Q117K and L118S. This region maps to a single putative prevalent 9-mer CTL epitope with L112 as an essential anchoring residue, and a major target epitope for neutralizing antibodies encompassing L118. Vaccination with A33 harboring these individual substitutions highlighted the crucial role of L118 in induction of protective immunity.

Keywords Smallpox · A33 · Cowpox · Vaccinia · Alphavirus

12.1 Introduction

Various vaccinia virus (VACV) strains served as smallpox vaccines in the world wide campaign for smallpox eradication (Fenner et al., 1988). The high sequence similarity (>97%) within the orthopox genus, allows for Vaccinia virus to efficiently cross protect from variola virus, the causative agent of smallpox. In animal models, VACV protects mice from ectromelia (ECTV) and cowpox (CPXV) challenge,

N. Paran (✉)
Department of Infectious Diseases, Israel Institute for Biological Research, P.O. Box 19, Ness-Ziona, 74100, Israel
e-mail: nirp@iibr.gov.il

Nir Paran and Shlomo Lustig equally contributed to the study.

A. Shafferman et al. (eds.), *The Challenge of Highly Pathogenic Microorganisms*,
DOI 10.1007/978-90-481-9054-6_12, © Springer Science+Business Media B.V. 2010

rabbits from rabbitpox (RPXV) and monkeys from monkeypox (MPXV) virus (Damon, 2007).

Although VACV is an efficient vaccine, severe side effects associated with the vaccine caused the discontinuation of mass vaccination after smallpox has been eradicated. However, smallpox still remains a potential hazard. In view of the risk of adverse reactions, highly attenuated vaccine strains – e.g. MVA and LC16m8 are being extensively evaluated to assure their potency as safer vaccines. In addition, the approach of using protective vaccinia virus antigens is pursued by several labs, aiming at generating an efficacious and safe subunit vaccine. Several viral antigens were tested throughout the last few years as protective antigens including A33, L1, B5, A27, D8, H3. In this regard, A33 antigen is one of the most studied and promising candidates. Several labs demonstrated its contribution to protection against VACV, ECTV and MXPV induced diseases using A33-based protein or DNA vaccine (Fang et al., 2006; Fogg et al., 2007, 2004; Heraud et al., 2006; Hooper et al., 2000, 2003, 2009, 2007, 2004; Sakhatskyy et al., 2006, 2008; Xiao et al., 2007). Moreover, antibodies to A33 were shown to have a major contribution to protection against VACV challenge (Chen et al., 2007; Lustig et al., 2005).

Several forms of subunit vaccines can be used to evaluate the protective efficacy of a given antigen/antigens, including purified protein, DNA expression vectors and recombinant viruses or replicons. The use of a replication competent Sindbis virus vector (Hahn et al., 1992) as a vaccine platform for the evaluation of the protective potential of single VACV proteins allows for expression of the viral antigen in the infected cells avoiding the need for protein expression and purification and allows for mammalian post translational modifications to take place. In addition, the use of a Sindbis virus based vaccine allows for robust protein expression due to the unique mechanism of RNA replication encoded by the virus. This recombinant Sindbis vector encoding A33, allows in a single vaccination regimen to fully protect BALB/c mice from VACV-WR intranasal (i.n.) and intravenous (i.v.) challenge and from ECTV foot pad (f.p.) and i.n. challenges. However, the same Sindbis A33 vector failed to protect BALB/c mice against i.n. challenge with CPXV. The present study maps the protective epitope/s in A33 based on the variations in $A33^{CPXV}$ sequence in comparison with $A33^{VACV}$.

12.2 Results

12.2.1 Protection Against VACV, ECTV and CPXV Challenge with Sindbis A33

To evaluate the protective efficacy of Sindbis $A33^{VACV}$, BALB/c mice were vaccinated by single injection with Sindbis A33 or Sindbis pTE as negative control. Two weeks post vaccination mice were challenged with VACV-WR (200 intranasal (i.n.) LD_{50}), ECTV-Moscow (10 i.n. LD_{50} or 3 f.p. LD_{50}) or CPXV (3 i.n. LD_{50}). All $A33^{VACV}$ vaccinated mice survived the challenge in contrast to control vaccinated

Table 12.1 Protection against Orthopoxvirus infection by Sindbis A33

Sindbis A33 based vaccine	Challenge virus	Challenge dose	Infection route	Survival (%)[a]
$A33^{VACV}$	VACV-WR	200 LD_{50} (5×10^6 pfu)	Intranasal (i.n.)	100
	ECTV (Moscow)	10 LD_{50} (10 pfu)	i.n.	100
		3 LD_{50} (10 pfu)	Foot-pad (f.p.)	90
$A33^{VACV}$	CPXV	3 LD_{50}	i.n.	0
$A33^{CPXV}$	(Brighton-red)	(3×10^5 pfu)		0
$A33^{CPXV}$	VACV-WR	20 LD_{50} (5×10^5 pfu)		60

[a]All unvaccinated and control vaccinated mice (Sindbis pTE empty vector) did not survive the challenge

mice or non vaccinated mice that did not survive the challenge (Table 12.1). In spite of the high sequence similarity between $A33^{VACV}$ and $A33^{CPXV}$, $A33^{VACV}$ vaccinated mice were not protected against lethal challenge with CPXV (3 i.n. LD_{50}) (Table 12.1). Furthermore, even vaccination with the homologous $A33^{CPXV}$ vaccine did not protect against CPXV challenge (3 i.n. LD_{50}) and only partially protected against VACV-WR challenge (60%) to which $A33^{VACV}$ conferred full protection (Table 12.1). These results indicated to the poor protective ability of $A33^{CPXV}$ as an antigen, and furthermore, suggested that CPXV evades $A33^{VACV}$ induced immunity.

12.2.2 Mapping the Protective Region in A33

Alignment of the amino acid (aa) sequences of $A33^{VACV}$ and $A33^{CPXV}$ orthologs revealed two main regions of differences spanning aa S82-S89 and L112-L118 which we denote CP-I and CP-II, respectively (Fig. 12.1).

To elaborate on the role of these substitutions to the ability of A33 to confer protection, two chimeric molecules were constructed by replacement of the $A33^{VACV}$ CP-I and CP-II regions with the corresponding CPXV sequences. Mice were vaccinated with the recombinant Sindbis virus encoding $A33^{VACV}$, $A33^{CPXV}$ or the chimeric A33 proteins, followed by lethal intranasal challenge with either VACV-WR (20LD_{50}) or CPXV (3LD_{50}). Vaccination with Sindbis vector encoding A33 harboring the CP-I replacement did not affect the inherent protective ability of $A33^{VACV}$. However, A33 containing the CPXV CP-II region, failed to protect BALB/c mice (data not shown). None of the tested constructs provided protection against CPXV challenge even though the challenge dose was much lower than with VACV-WR.

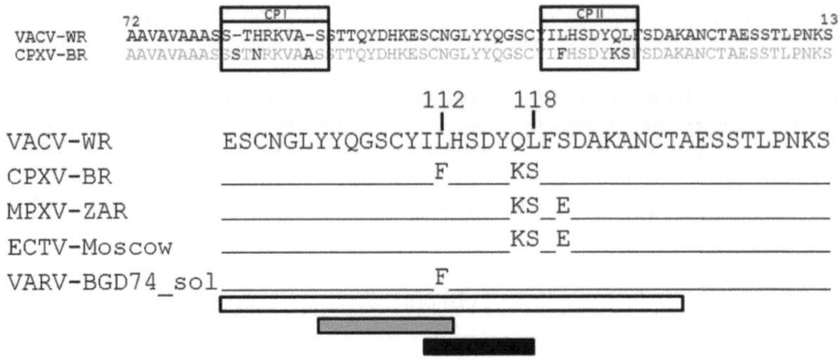

Fig. 12.1 Protective epitope mapping in VACV A33. Alignment of A33[VACV] from VACV-WR (aa 72–137) with the corresponding region of its orthologs: cowpox virus Brighton-red (CPXV-BR) highlighting CP I and CP II regions. The *lower panel* shows multiple sequence alignment of A33[VACV] from VACV-WR (aa 98–137) and CPXV-BR, monkeypox virus strain Zair 76 (MPXV-ZAR), ectromelia virus strain Moscow (ECTV-Moscow) and variola virus major strain Bangladesh 74(VARV-BGD74_sol). Mapped regions and epitopes are marked by boxes as follows: Immunogenic epitope (Heraud et al., 2006) (*white filled box*), predicted CTL epitope (*grey filled box*) and protective CP-II region (*black filled box*)

12.2.3 In-Silico Analysis of A33 Potential CTL Epitopes

In order to evaluate the role of CTL-mediated immune response in the protection conferred by A33, we analyzed the sequence of A33 *in-silico* for prediction of potential CTL epitopes, utilizing 3 independent programs: NetMHC, SYFPEITHI and BIMAS (Nielsen et al., 2003; Parker et al., 1994; Rammensee et al., 1999). The analysis was set to predict H-2Kd epitopes since evaluation of the protective ability of A33 was performed in inbred BALB/c mice. By all 3 programs we independently identified a major 9mer epitope spanning aa Y104 to L112 in A33. In the prediction, the L112 residue which is part of the CP-II region is an anchoring residue in the predicted CTL epitope. Furthermore, it appears that the L112F substitution (corresponding to the CPXV orthologous peptide), abolishes the prediction of this 9mer to serve as a CTL epitope. This variation in sequence is also found in A33[VARV] (Fig. 12.1). In view of the possibility that this substitution might be crucial for eliciting a CTL-based immunoprotection, we tested the effect of L112F and additional A33[VACV]/A33[CPXV] variations within the CP-II region to the protective ability of A33.

12.2.4 Mapping of Specific Amino Acids which are Crucial for Protection

The CP-II region in A33[CPXV] includes the following amino acid substitutions: L112F, Q117K and L118S. Corresponding single aa replacements were introduced into the A33[VACV] gene in a Sindbis vector and the resulting constructs were

Table 12.2 Mapping of the protective epitope of VACV A33

Sindbis A33 based vaccine	% survival[a]
A33^{VACV}	100
A33^{CPXV}	60
A33^{VACV} L112F	100
A33^{VACV} Q117K	100
A33^{VACV} L118S	0
A33^{VACV} QL117-8KS	40
pTE (control)	0
Unvaccinated	0

[a]Following challenge with VACV-WR (20 LD^{50} intranasal)

used to vaccinate BALB/c mice. In addition, one double mutant was generated (Q117K-L118S). All A33 derivatives induced anti-A33 antibodies with no significant difference in antibody titer between groups (data not shown).Similar results were also obtained by comet inhibition assay (data not shown). Vaccinated mice were challenged by intranasal instillation with VACV-WR (20LD$_{50}$) (Table 12.2).

The results revealed that both L112F and Q117K substitutions protected mice similarly to A33^{VACV}. In contrast, L118S substitution resulted in a detrimental effect leading to complete loss of protection. Interestingly, incorporation of the double replacement Q117K – L118S partially restored the protection conferred by A33^{VACV} against VACV-WR. Morbidity (Fig. 12.2) correlated with mortality

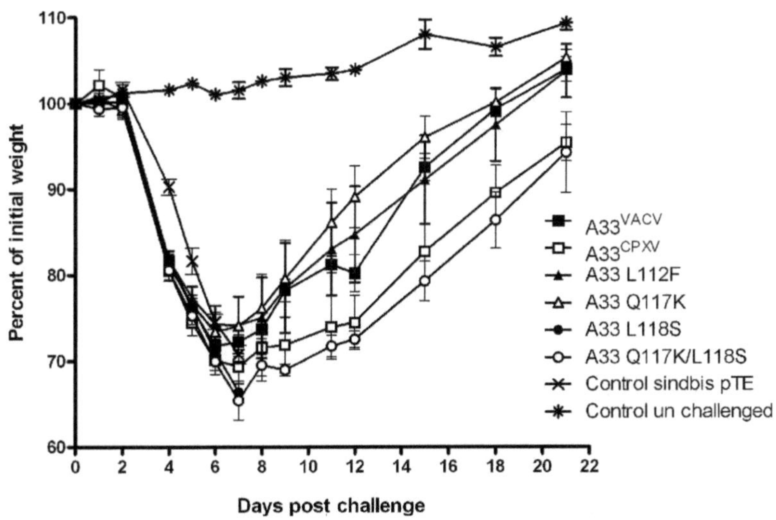

Fig. 12.2 Morbidity of mice vaccinated with various A33 constructs. BALB/c mice were vaccinated with Sindbis virus expressing various A33 proteins or with a control Sindbis pTE as indicated. Two weeks after vaccination mice were challenged with VACV-WR by intranasal instillation (20 LD$_{50}$). Control non-infected and back-titration groups were also included (not shown). ($n = 5$). Morbidity is demonstrated as weight loss from initial weight. Error bars represent standard errors of the mean (SEM)

rates: Control (Sindbis pTE without A33) and A33-L118S vaccinated mice lost about 30–35% of their initial weight and did not survive the challenge (MTTD of 6.6 and 7.6, respectively), whereas mice vaccinated with A33^{VACV}, A33-L112F or A33-Q117K survived, lost about 27% of their initial weight and recovered significantly faster (maximal morbidity at day 6). Only 60% of the mice vaccinated with A33^{CPXV} or 40% with the double substitution A33-Q117K L118S were protected but survivors had a significantly longer and severe disease compared to A33^{VACV} vaccinated mice ($p<0.015$ and $p<0.003$ respectively) (Fig. 12.2).

12.3 Discussion

During the last few years several vaccinia virus antigens were evaluated as potential subunit vaccines, with A33 being a strong candidate. Several reports demonstrated efficient protection conferred by A33 against VACV and ECTV in mice and also showed that A33 as part of a multi-subunit vaccine, contributes to protection of non-human primates (NHPs) against MPXV (Fang et al., 2006; Fogg et al., 2007, 2004; Heraud et al., 2006; Hooper et al., 2000, 2003, 2009, 2004; Sakhatskyy et al., 2006, 2008; Xiao et al., 2007). Using the Sindbis virus recombinant system, A33-based vaccine efficiently protected mice against lethal VACV (200 intranasal LD_{50}) and ECTV (10 intranasal and 3 dermal LD_{50}) in a single vaccination regimen. However, the inability of A33R^{VACV} – based vaccine to protect mice against challenge with the closely related CPXV and the inability of A33^{CPXV} to protect against CPXV or to only partially protect against VACV-WR, indicate that A33R^{CPXV} might be an immune-escape, non-protective version of the VACV A33. Using a domain swapping approach, we mapped the region in A33^{VACV} important for protection to a short 7-mer (LHSDYQL, residues 112–118 in A33R^{VACV}) denoted CP-II. The homologous region in CPXV contains 3 aa substitutions L112F, Q117K and L118S.

Some of the sequence variations found in A33^{CPXV}, are also present in MPXV and VARV (Fig. 12.1). Since A33 is considered as an important contributor to any future subunit vaccine against smallpox, and since A33 might also be indicated against other orthopoxvirus infections (e.g. MPXV, CPXV and VACV) unraveling the basis for this protective ability at the single aa-level is of a major importance.

Our in-silico analysis of potential CTL epitopes in A33, identified a single prevalent 9-mer spanning from Y104 to L112, in which L112 is as essential anchoring residue. The analysis revealed that the L112F substitution "abolishes" the predicted binding of the peptide to the H-2Kd molecule, impairing the ability of A33 to putatively elicit a cellular immune response. Nevertheless, our finding that A33 L112F retained its ability to induce protective immunity indicates that at least in the BALB/c mouse model, this putative CTL epitope contributes very little if at all, to the protection acquired by A33 vaccination and/or that protection afforded by A33 is mainly acquired by the humoral rather than cellular arm. In that respect, we can assume that in VARV (having an F112 residue in its A33^{VARV}), the L112F substitution by itself will presumably not render the virus resistant to A33-based vaccination.

Our data demonstrate that out of the few changes found in $A33^{CPXV}$, the L118S substitution markedly influence the protection ability of $A33^{VACV}$, and that addition of Q117K mutation to L118S, partially restored protective ability. Interestingly, in a recent study, the binding epitope of the protective/neutralizing monoclonal antibody 1G10 was determined (Golden and Hooper, 2008). In that study, the L118S change in MPXV was shown to be detrimental for the 1G10 binding and the addition of Q117K partially restored this binding (Golden and Hooper, 2008). In our hands, sera from A33 L118S vaccinated mice retained the ability to bind A33 and to inhibit comet, suggesting that in addition to 1G10 type protective antibodies, antibodies to additional epitopes bind A33 and inhibit comet. The inability of A33 to protect BALB/c mice against CPXV challenge, results most likely from the inability of the $A33^{VACV}$-induced protective antibodies to recognize $A33^{CPXV}$. On the other hand, the inability of $A33^{CPXV}$ to protect mice from VACV-WR challenge would mean that $A33^{CPXV}$ is incapable of mounting 1G10-type protective immune response. It appears that $A33^{CPXV}$ seems to lack both the immunogenicity required to induce protective immunity, and the antigenicity required to be recognized by protective antibodies, attributing both properties to the L118 residue. This residue and the CP-II region which we mapped in this study as central for protection, overlap a region previously mapped in primates as immunogenic (Chen et al., 2007; Heraud et al., 2006; Sirven et al., 2009) further emphasizing the importance of this region for future vaccines (Fig. 12.1).

The loss of protection resulting from the L118S substitution and the partial restoration of protection by the double mutant Q117K-L118S, correlate with partial restoration of the binding of 1G10 antibodies to A33 (Golden and Hooper, 2008) suggesting that other weak epitopes, masked by the major L118 residue, might contribute to the protection. Indeed, $A33^{VACV}$ protects BALB/c against ECTV challenge. $A33^{ECTV}$ bears the same Q117K-L118S substitutions, but similarly to MPXV has an additional S120E substitution. This substitution was shown to improve binding of the 1G10 neutralizing antibody (Golden and Hooper, 2008), pointing toward E120 as a compensatory residue. It is possible that a paradigm which includes boosting immunizations, induces an immune response through additional epitopes that support protection in low challenge doses (Thornburg et al., 2007), and contribute to protection in multi-subunits vaccines and in other hosts (Fang et al., 2006; Fogg et al., 2007, 2004; Heraud et al., 2006; Hooper et al., 2000, 2003, 2009, 2007, 2004; Sakhatskyy et al., 2006, 2008; Xiao et al., 2007).

As the immune protection conferred by VACV against VARV and other Orthopox viruses is based on cross-protection, sequence variation between the species in the protective antigens, as demonstrated here by mapping the basis for the protective ability of A33 to a single amino acid change, should be carefully assessed.

Acknowledgments We would like to thank C. Rice from the Rockefeler University for providing us with the pTE32J plasmids, and T. Waner, Y. Shlomovich, B. Sharabi and S. Abraham for assistance with animal husbandry.

References

Chen, Z., Earl, P. L., Americo, J. L., Damon, I., Smith, S. K., Fujuan, Y., Sebrell, A., Emerson, S., Cohen, G. H., Eisenberg, R. J., Gorshkova, I., Schuck, P., Satterfield, W., Moss, B. & Purcell, R. (2007) Characterization of Chimpanzee/Human monoclonal antibodies to vaccinia virus A33 glycoprotein and its variola virus homolog in vitro and in a vaccinia virus mouse protection model. *J Virol,* 81, 8989–95.

Damon, I. K. (2007) Poxviruses. In: Knipe, D. M. & Howley, P. M. (eds), Fields virology. Wolters Kluwer Lippincott Williams & Wilkins, Philadelphia, PA.

Fang, M., Cheng, H., Dai, Z., Bu, Z. & Sigal, L. J. (2006) Immunization with single extracellular enveloped virus protein produced in bacteria provides partial protection from lethal orthopoxvirus infection in a natural host. *Virology,* 345, 231–43.

Fenner, F., Henderson, D. A., Arita, I., Jezek, Z. & Ladnyi, I. D. (1988) Smallpox and its eradication, World Health Organization, Geneva.

Fogg, C., Americo, J. L., Lustig, S., Huggins, J. W., Smith, S. K., Damon, I., Resch, W., Earl, P. L., Klinman, D. M. & Moss, B. (2007) Adjuvant-enhanced antibody resposes to recombinant proteins correlates with protection of mice and monkeys to orthopoxvirus challenges. *Vaccine,* 25, 2787–99.

Fogg, C., Lustig, S., Whitbeck, J. C., Eisenberg, R. J., Cohen, G. H. & Moss, B. (2004) Protective immunity to vaccinia virus induced by vaccination with multiple recombinant outer membrane proteines of intracellular and extracellular virions. *J Virol,* 78, 10230–7.

Golden, J. W. & Hooper, J. W. (2008) Heterogeneity in the A33R protein impacts the cross-protective efficacy of a candidate amallpox DNA vaccine. *Virology,* 20; 377(1), 19–29.

Hahn, C. S., Hahn, Y. S., Braciale, T. J. & Rice, C. M. (1992) Infectious Sindbis virus transient expression vectors for studying antigen precessing and presentation. *Proc Natl Acad Sci USA* 89, 2679–83.

Heraud, J. M., Edghill-Smith, Y., Ayala, V., Kalisz, I., Parrino, J., Kalyanaraman, V. S., Manischewitz, J., King, L. R., Hryniewicz, A., Trindade, C. J., Hasset, M., Tsai, W. P., Venzon, D., Nalca, A., Vaccari, M., Silvera, P., Bray, M., Graham, B. S., Golding, H., Hooper, J. W. & Franchini, G. (2006) Subunit recombinant vaccine protects against monkeypox. *J Immunol,* 177, 2552–64.

Hooper, J. W., Custer, D. M., Schmaljohn, C. J. & Schmaljohn, A. L. (2000) DNA vaccination with vaccinia virus L1R and A33R genes protects mice against a lethal poxvirus challenge. *Virology,* 266, 329–39.

Hooper, J. W., Custer, D. M. & Thompson, E. (2003) Four-gene-combination DNA vaccine protects mice against a lethal vaccinia virus challenge and ellicits appropriate antibody response in nonhuman primates. *Virology,* 306, 181–95.

Hooper, J. W., Ferro, A. M., Golden, J. W., Silvera, P., Dudek, J., Alterson, K., Custer, D. M., Rivers, B., Morris, J., Owens, G., Smith, J. F. & Kamrud, K. I. (2009) Molecular smallpox vaccine delivered by alfavirus replicons elicits protctive immunity in mice and non-human primates. *Vaccine,* 11; 28(2), 494–511.

Hooper, J. W., Golden, J. W., Ferro, A. M. & King, A. D. (2007) Smallpox DNA vaccine deliverd by novel skin electroporation device protects mice against intranasal poxvirus challenge. *Vaccine,* 25, 1814–23.

Hooper, J. W., Thompson, E., Wilhelmsen, C., Zimmerman, M., Ichou, M. A., Steffen, S. E., Schmaljohn, C. J., Schmaljohn, A. L. & Jahrling, P. (2004) Smallpox DNA vaccine protects nonhuman primates against lethal monkeypox. *J Virol,* 78, 4433–43.

Lustig, S., Fogg, C., Whitbeck, J. C., Eisenberg, R. J., Cohen, G. H. & Moss, B. (2005) Combination of pokyclonal or monoclonal antibodies to proteins of the outer membranes of the two infectious forms of vaccinia virus protect mice against a lethal respiratory challenge. *J Virol,* 79, 13454–62.

Nielsen, M., Lundegaard, C., Worning, P., Lauemoller, S. L., Lamberth, K., Buus, S., Brunak, S. & Lund, O. (2003) Reliable prediction of T-cell epitopes using neural networks with novel sequence representation. *Protein Sci,* 12, 1007–17.

Parker, K. C., Bednarek, M. A. & Coligan, J. E. (1994) Scheme for ranking potential HLA-A2 binding peptides based on independent binding of individual peptide side chains. *J Immunol,* 152, 163–75.

Rammensee, H. G., Bachmann, J., Emmerich, N. K., Bachor, O. A. & Stevanovic, S. (1999) SYFPEITHI: Database for MHC ligands and peptide motifs. *Immunogenetics,* 50, 213–9.

Sakhatskyy, P., Wang, S., Chou, T. H. & Lu, S. (2006) Immunogenicity and protection efficacy of monovalant and polyvalant poxvirus vaccines that include the D8 antigen. *Virology,* 355, 89–107.

Sakhatskyy, P., Wang, S., Zhang, C., Chou, T. H., Kishko, M. & Lu, S. (2008) Immunogenicity and protection efficacy of subunit-based smallpox vaccines using variola major antigens. *Virology,* 376, 89–107.

Sirven, P., Castelli, F. A., Probst, A., Szely, N. & Maillere, B. (2009) In vitro human CD4+ T cell response to the vaccinia protective antigens B5R and A33R. *Mol Immunol*, 46(7), 1481–7.

Thornburg, N. J., Ray, C. A., Collier, M. L., Liao, H. X., Pickup, D. J. & Johnston, R. E. (2007) Vaccination with Venezuelan equine encephlitis replicons encoding cowpox virus structural proteins protects mice from itranasal cowpox virus\challenge. *Virology,* 362, 441–52.

Xiao, Y., Aldaz-Carroll, L., Oritz, A. M., Whitbeck, J. C., Alexander, E., Lou, H., Davis, H. L., Bracial, T. J., Eisenberg, R. J., Cohen, G. H. & Isaacs, S. N. (2007) A protein-based smallpox vaccine protects mice from vaccinia and ectromelia virus challenges when given as a prime and single boost. *Vaccine,* 25, 1214–24.

Chapter 13
Inhibitors of Innate Immunity from Vaccinia Virus

Geoffrey L. Smith

Abstract Vaccinia virus (VACV) is an enigma, for it is the only vaccine to have eradicated a human disease, smallpox, and yet its origin and natural host remain unknown. After the eradication of smallpox, VACV has continued to be studied intensively because of the potential use of recombinant VACVs as vaccines against other infectious diseases, and because it is an excellent model for studying virus-host interactions. This short article considers some of the strategies used by VACV to suppress the host innate immune response to infection. These strategies include proteins that are secreted from infected cells to bind complement factors, cytokines, chemokines or interferons, and intracellular proteins that can synthesize steroid hormones, or block apoptosis or innate signalling pathways leading to production of inflammatory mediators.

13.1 Introduction

Vaccinia virus (VACV) is the best studied member of the *Orthopoxvirus* (OPV) genus of the *Poxviridae* (Moss, 2007). VACV is well known as the live vaccine that was used in the World Health Organisation (WHO) campaign to eradicate smallpox, a disease caused by *Variola virus* (VARV), a related OPV (Fenner et al., 1988). However, despite its fame, the origin and natural host of VACV remain unknown. It is believed that Edward Jenner used *Cowpox virus* (CPXV), another OPV, when he introduced vaccination to the world in 1796 (Jenner, 1798), but in 1939 Alan Downie showed that all the smallpox vaccine strains being used in the 20th century were not CPXV but were a distinct OPV species that became known as VACV, after *vacca*, Latin for cow (Downie, 1939). Thus sometime between 1796 and 1939 VACV replaced CPXV as the smallpox vaccine. The exact time and place this happened are lost in history, but it probably occurred early in the 19th century.

G.L. Smith (✉)
Department of Virology, Faculty of Medicine, Imperial College London, St. Mary's Campus, Norfolk Place, London W2 1PG, England
e-mail: geoffrey.l.smith@imperial.ac.uk

A. Shafferman et al. (eds.), *The Challenge of Highly Pathogenic Microorganisms*,
DOI 10.1007/978-90-481-9054-6_13, © Springer Science+Business Media B.V. 2010

Shortly after smallpox was declared eradicated by the WHO in 1980, two research groups working in USA developed techniques that enabled the construction of genetically engineered (recombinant) VACV strains that expressed genes derived from other organisms (Mackett et al., 1982; Panicali and Paoletti, 1982). These genes were expressed during the normal replication cycle of VACV and so when the recombinant VACV was used as a vaccine, the foreign antigen was simultaneously synthesized and presented to the immune system (Panicali et al., 1983; Smith et al., 1983a, b). Thus VACV recombinants could immunise against not only smallpox but also the pathogen from which the foreign gene was derived. Further, because the coding capacity of VACV for foreign DNA was large (Smith and Moss, 1983), the virus could be engineered to express several foreign antigens simultaneously and so immunise against multiple pathogens with a single vaccination (Perkus et al., 1985). Additional benefits of recombinant VACV vaccines were the ease with which VACV could be grown and formulated for vaccination, the stability of the freeze-dried vaccine and its ability to induce both antibody and T cell immunity. On the other hand, VACV had an imperfect safety record as a vaccine against smallpox and could produce serious complications in vaccinees with immunological deficiencies or eczema (Lane et al., 1969). Therefore, it was recognized that if recombinant VACVs were to be used for widespread human vaccination the safety of the vaccine needed improvement.

To obtain safer VACV strains several approaches were adopted. The first was to passage VACV extensively in cell culture and at the end of this process identify strain(s) that were attenuated. This was the classical way that live attenuated vaccines were developed for yellow fever, polio, measles mumps and rubella, for instance. This process led to the isolation in Germany of modified virus Ankara (MVA) from the parental Ankara strain (Stickl and Hochstein-Mintzel, 1971), and the isolation in Japan of LC16m8 derived from the Lister strain (Hashizume et al., 1985). These strains were definitely attenuated in man, but their effectiveness as vaccines against smallpox were never established because by this time smallpox was no longer endemic in these countries and so those vaccinated were not exposed to VARV. The biological properties and applications of MVA are described elsewhere in this volume by Professor Gerd Sutter.

A second approach was to identify VACV genes that contributed to virulence and then delete these from the virus. If the gene(s) deleted encoded proteins that suppressed the host response to infection, the resultant virus might be attenuated and have increased immunogenicity. This strategy was adopted in my laboratory. The first objective was to sequence regions of the VACV genome that were likely to encode virulence genes. Comparison of the genomes of many OPVs had demonstrated that the central region of the large double stranded (ds)DNA genomes of these viruses were highly conserved, whereas the terminal regions were more variable (Mackett and Archard, 1979). Indeed, replication competent viruses with large deletions adjacent to either terminus had been described and it was proposed that these would be attenuated in vivo (Panicali et al., 1981; Moss et al., 1981). Accordingly, we sequenced 42 kb of VACV strain Western Reserve (WR) adjacent to the right inverted terminal repeat (ITR) and identified a wide array of virulence genes including many that were shown subsequently to inhibit innate immunity (Smith et al., 1991). Sequencing of the left end of the VACV WR genome (Kotwal and Moss, 1988) and the complete genome of VACV strain Copenhagen (Goebel et al., 1990) revealed many other non-essential virulence genes.

13.2 Inhibitors of Innate Immunity

Nearly 20 years after these sequence data were generated we know a great deal about the proteins encoded by these genes and at least one function of many of them has been elucidated. There are literally scores of VACV immunomodulatory proteins (Smith, 1999; Seet et al., 2003) and these can be grouped according to their site of action. The first group comprises proteins that are secreted from the infected cell and function in the extracellular environment to capture host cytokines, chemokines, interferons (IFNs) or complement factors. By binding these host proteins, they are prevented for engaging their natural binding partners and so cannot activate complement or co-ordinate the inflammatory response to infection. A second group includes proteins that function inside the infected cell to block innate immune signalling pathways, inhibit apoptosis, or interfere with the anti-viral activity of IFN-induced proteins. There is even an enzyme (3β-hydroxysteroid dehydrogenase 3β-HSD) that synthesizes steroid hormones which suppress the host inflammatory response to infection and is considered further below. The third group are proteins that are present on the cell surface where they can bind host chemokines or IFNs, or influence how the infected cell is recognised by immune cells. In this article, examples of the first and second group are described.

13.3 A Soluble Interleukin-1β Receptor

Sequencing the right end of the VACV WR genome revealed a gene called *B15R* that was predicted to encode a soluble glycoprotein containing three immunoglobulin (Ig)-like domains (Smith and Chan, 1991). Of particular interest was the observation that these Ig domains showed convincing similarity to the human and mouse interleukin (IL)-1 receptor (IL-1R). Unlike the type I IL-1R expressed on the surface of mammalian cells, the virus protein lacked a hydrophobic transmembrane anchor sequence and an intracellular signalling domain, and so was likely to be secreted. This led to the hypothesis that the VACV WR B15 protein was a soluble IL-1R that would bind IL-1 in solution and prevent it reaching its natural receptor on cells.

This hypothesis was confirmed and B15 was shown to be a secreted glycoprotein made late during infection that bound human and mouse IL-β (Alcami and Smith, 1992). The affinity constant for the interaction of B15 with human IL-1β was K_d 234 ± 49 pM, at least as high as the affinity of IL-1β for the IL-1R on cells. In addition, the virus protein was abundant with about 100,000 copies being produced from each infected cell in 24 h. The abundance of B15 and its high affinity for IL-1β enabled the virus protein to effectively block IL-1β binding to IL-1Rs on cells. Consequently, IL-1-induced signal transduction and the subsequent inflammatory response were inhibited. Interestingly, B15 bound only IL-1β and not IL-1α or the natural regulator of IL-1 action, the IL-1 receptor antagonist (IL-1ra). The failure to bind IL-1ra was logical if the purpose of the virus protein was to block IL-1 mediated signal transduction, but the failure to bind IL-1α was unexpected and indicated that the inhibition of IL-1β was more important for the virus.

The contribution of B15 to virus virulence was investigated by the construction of deletion mutants lacking the *B15R* gene. Using a mouse intracranial infection model Spriggs et al., (1992) reported that the virus deletion mutant caused less serious illness when compared to wild type virus (Spriggs et al., 1992). A similar observation was made using the mouse intradermal model (Tscharke et al., 2002). Conversely, Alcami and Smith (1992) reported that a deletion mutant, vΔB15R, caused enhanced disease compared to wild type and revertant controls when mice were infected by the intranasal route (Alcami and Smith, 1992). These different outcomes reflected the infection models used. The intranasal infection model is characterised by a systemic infection and loss of significant body weight and the production of systemic levels of circulating IL-1β that contribute to immune pathology. So after infection with the wild type virus, the B15 protein prevented IL-1β reaching its receptors and consequently such a systemic reaction was prevented. In contrast, infection with vΔB15R led to production of IL-1β that was able to bind IL-1Rs and trigger a systemic response.

Further analysis of the outcome of infection with vΔB15R via the intranasal model showed that this virus induced a fever that was sustained for several days, whereas animals infected with the wild type and revertant viruses did not. This result was novel for virus pathogenesis, in that it demonstrated that VACV had evolved a mechanism to regulate the body temperature of the mammalian host it infects. But the result was also informative about mammalian physiology, for because B15 only bound IL-1β and not IL-1α, IL-6, tumour necrosis factor (TNF) or IFNs, it showed that IL-1β was the principal endogenous pyrogen (Alcami and Smith, 1996).

Two other features of the B15 protein are noteworthy. First, although the encoding gene was conserved in several VACV strains and some other OPVs, the gene was fragmented by multiple frameshift and nonsense mutations in VARV strain Harvey (Aguado et al., 1992). Further sequencing confirmed this was also the case in every VARV strain analysed (> 45 strains) (Esposito et al., 2006). The lack of an IL-1β-binding protein made by VARV, the causative agent of smallpox, was consistent with the induction of high fever in patients with smallpox (Fenner et al., 1988). This was also consistent with observations made with VACV strains in mice. Using VACV strains WR and Copenhagen, viruses engineered to express the B15 protein prevented fever in infected animals, whereas infection with viruses not expressing the protein induced fever (Alcami and Smith, 1996). The sequencing of VARV strains also showed that during VARV evolution the gene encoding the IL-1β-binding protein had become non-functional although the gene fragments remain. So in this case VARV evolution was by loss of function rather than gain of function.

The second notable feature of B15 was that it was expressed by MVA, despite many other immunomodulatory proteins having being lost from this VACV strain (Blanchard et al., 1998). Moreover, deletion of the gene from VACV strain WR or MVA caused an increase in immunogenicity (Staib et al., 2005). This demonstration confirmed that improvement to VACV immunogenicity could be achieved by deletion of immunomodulators and a similar conclusion was reached following deletion of gene *A41L*, encoding a chemokine binding protein, from VACV WR or MVA (Clark et al., 2006).

13.4 3β-Hydroxysteroid Dehydrogenase

VACV strains WR and Copenhagen contain a gene, *A44L*, predicted to encode a protein with 31% amino acid identity to the human 3β-hydroxysteroid dehydrogenase (3β-HSD) (Goebel et al., 1990; Smith et al., 1991), an enzyme required for the synthesis of all classes of steroid hormones. The presence of such an enzyme in a virus (or other microbial pathogen) was without precedent and led to the hypothesis that VACV synthesized steroid hormones at the site of infection to diminish the host inflammatory response to infection and thereby enable virus escape from the immune system.

A characterization of the VACV WR A44 protein confirmed this hypothesis (Moore and Smith, 1992). The *A44L* gene was expressed early during infection and produced an intracellular 38-kDa protein that converted pregnenolone to progesterone, a steroid hormone. The *A44L* gene was non-essential for virus replication in cell culture but a deletion mutant lacking the gene, vΔA44L, had reduced virulence in vivo (Moore and Smith, 1992). Further analysis of the response to infection with vΔA44L, compared to wild type and revertant controls, showed there was a reduced level of local and plasma corticosterone early (days 1–3) after infection (Reading et al., 2003). Thereafter, there was a strong local inflammatory response with enhanced levels of IFN-γ, rapid recruitment of CD4+ and CD8+ lymphocytes and increased CTL activity. These observations demonstrated that the VACV 3β-HSD contributed to virus virulence by inhibiting the inflammatory response to infection.

There are two interesting parallels between the VACV 3β-HSD and the VACV IL-1β-binding protein. First, VACV strain MVA retains a gene equivalent to *A44L* that encodes a 3β-HSD (Sroller et al., 1998) and it follows that deletion of this gene from MVA might increase immunogenicity. Second, an analysis of VARV genomes, showed that VARV strains retain an *A44L*-like gene, but this is disrupted by frameshift and non-sense mutations so that it is non-functional (Aguado et al., 1992; Esposito et al., 2006). This was particularly surprising given that 3β-HSD increases the virulence of VACV, the vaccine used to eradicate smallpox, and so it might have been expected that VARV, a highly virulent virus in man, would have retained this virulence factor.

13.5 A Family of Bcl-2 Proteins Encoded by VACV

Bioinformatic analysis of the VACV genome identified several families of related genes (Smith et al., 1991) and one family includes proteins that are now called N1, B14, A52, K7 and F1. These proteins share a low level of amino acid identity (<20%) indicating a common origin but diversification long ago. The structures of these proteins have been determined and show they are all B cell lymphoma (Bcl)-2 family members (Aoyagi et al., 2007; Cooray et al., 2007; Graham et al., 2008; Kvansakul et al., 2008; Kalverda et al., 2009). The function of some of these proteins is considered with focus on N1 and B14.

Protein N1. N1 was described originally as a non-essential, secreted protein that contributed to virulence (Kotwal et al., 1989) but subsequent analysis showed it was an intracellular, non-covalent homodimer that is expressed early during infection (Bartlett et al., 2002). Functionally, N1 was reported to target the I-kappa B kinase (IKK) complex, inhibiting signalling to nuclear factor kappa B (NF-κB) from TNF receptor, and inhibiting NF-κB and IFN regulatory factor (IRF)3 signalling by toll-like receptors (DiPerna et al., 2004). However, a re-examination of N1 function demonstrated that it did not interact with the IKK complex (Chen et al., 2008) or block TNF-induced NF-κB activation, but blocked NF-κB activation downstream of IL-1R signalling (Graham et al., 2008). The crystal structure of N1 was solved by two groups and revealed that N1 has a Bcl-2 fold (Aoyagi et al., 2007; Cooray et al., 2007) despite lacking amino acid similarity to Bcl-2 family proteins. On its surface N1 contains a groove, similar to the groove on the surface of anti-apoptotic Bcl-2 proteins, and consistent with this, N1 inhibited staurosporine-induced apoptosis and bound several pro-apoptotic Bcl-2 proteins (Cooray et al., 2007). Therefore N1 is an intracellular virulence factor that inhibits both innate immune signalling and apoptosis.

Protein B14. The VACV strain *B14R* gene was studied because it is conserved in most VACV strains and OPVs despite being encoded in the right variable region of the genome, and this conservation implied an important function. The gene is expressed early during infection and encodes a cytoplasmic protein that is non-essential for virus replication in cell culture but affects virulence in vivo (Chen et al., 2006). After infection by the deletion mutant, vΔB14R, there was an enhanced infiltration of cells into the infected lesion and reduced virus titres. Investigation of the mechanism of action of B14 showed that it inhibited NF-κB activation in response to TNFα, IL-1β and the phorbol ester PMA (Chen et al., 2008). The fact that B14 inhibited activation of NF-κB by signalling pathways from all these stimuli suggested that B14 functioned where these pathways converge or downstream of it. This was confirmed when B14 was shown to bind to the IKK complex via IKKβ. In the presence of B14, the phosphorylation of IKKβ was prevented. Consequently, the IKK complex did not phosphorylate the inhibitor of NK-κB (IκBα) which remained bound to NF-κB in the cytoplasm.

The crystal structure of B14 was determined and showed that, like N1, it is a Bcl-2 family member. However, unlike N1 it lacks a surface groove for binding

BH3 peptides from pro-apoptotic Bcl-2 family members and consequently is not anti-apoptotic (Graham et al., 2008). So whereas N1 is able to block both innate immune signalling and the induction of apoptosis, B14 only blocks innate immune signalling. However, the block in innate immune signalling imposed by B14 is more profound and it can block NF-κB activation from multiple pathways (Chen et al., 2008), whereas N1 only blocks downstream of IL-1/TRAF6 (Graham et al., 2008).

The structure of protein A52 was also solved by X-ray crystallography and revealed a Bcl-2 protein that was more closely related to B14 than N1 (Graham et al., 2008). Like B14, A52 lacked a surface groove and was not able to inhibit apoptosis. Both A52 and B14 show evidence of an ancestral groove that has become occluded by 2 types of modification. First, the alpha helix 2 is longer in B14 and A52 so that it protrudes up into and fills the groove. Second, there are bulky aromatic or hydrophobic amino acid side chains that protrude into the groove from the sides and also partially fill it. In consequence, A52 is not able to inhibit apoptosis by binding BH3 peptides from pro-apoptotic Bcl-2 family members.

Bioinformatic analysis predicts that there are 9 Bcl-2 proteins encoded by the VACV genome. The structures of 5 of these have been determined and the others await functional and structural characterisation. One hypothesis to explain this remarkable collection of Bcl-2 proteins encoded by a single virus is that an ancestral virus acquired a gene encoding an anti-apoptotic Bcl-2 protein from its host. Subsequently, this gene was duplicated and its function was diversified by mutation. The myxoma virus M11 protein or the VACV F1 protein might represent proteins similar to that acquired from the host, for these proteins have a surface groove to bind BH3 peptides and are anti-apoptotic, but are unable to modulate innate immune signalling pathways, at least those that have been analysed. As gene duplication and diversification progressed, the stable Bcl-2 protein fold was adapted by VACV to acquire binding surfaces for interaction with molecules regulating signalling pathways. The N1 protein represents such a molecule, in that it retains an open surface groove for binding pro-apoptotic Bcl-2 family proteins, but it also binds host protein(s) that regulate IL-1-induced activation of NF-κB. As the diversification progressed further, the ability to regulate apoptosis was lost by mutations that filled in the groove, and these proteins became solely modulators of innate immune signalling. Proteins B14 and A52 represent this group of molecules.

In conclusion, VACV is a remarkable virus that has evolved a wide range of strategies to inhibit the innate immune response to infection. The study of these pathways is increasing our understanding of virus pathogenesis and how our innate immune system functions, and is also providing knowledge that can be applied to the development of safer and more immunogenic VACV based vaccines.

Acknowledgements The author is most grateful to the conference organisers for the kind invitation to visit Israel and attend the 46th Oholo Conference. The work in the author's laboratory has been supported by grants from the Wellcome Trust, the Medical Research Council of UK. GLS is a Wellcome Principal Research Fellow.

References

Aguado, B., Selmes, I. P. & Smith, G. L. (1992) Nucleotide sequence of 21.8 kbp of variola major virus strain Harvey and comparison with vaccinia virus. *J Gen Virol,* 73, 2887–902.

Alcami, A. & Smith, G. L. (1992) A soluble receptor for interleukin-1 beta encoded by vaccinia virus: a novel mechanism of virus modulation of the host response to infection. *Cell,* 71, 153–67.

Alcami, A. & Smith, G. L. (1996) A mechanism for the inhibition of fever by a virus. *Proc Natl Acad Sci USA*, 93, 11029–34.

Aoyagi, M., Zhai, D., Jin, C., Aleshin, A. E., Stec, B., Reed, J. C. & Liddington, R. C. (2007) Vaccinia virus N1L protein resembles a B cell lymphoma-2 (Bcl-2) family protein. *Protein Sci,* 16, 118–24.

Bartlett, N., Symons, J. A., Tscharke, D. C. & Smith, G. L. (2002) The vaccinia virus N1L protein is an intracellular homodimer that promotes virulence. *J Gen Virol,* 83, 1965–76.

Blanchard, T. J., Alcami, A., Andrea, P. & Smith, G. L. (1998) Modified vaccinia virus Ankara undergoes limited replication in human cells and lacks several immunomodulatory proteins: implications for use as a human vaccine. *J Gen Virol*, 79, 1159–67.

Chen, R. A., Jacobs, N. & Smith, G. L. (2006) Vaccinia virus strain Western Reserve protein B14 is an intracellular virulence factor. *J Gen Virol*, 87, 1451–8.

Chen, R. A., Ryzhakov, G., Cooray, S., Randow, F. & Smith, G. L. (2008) Inhibition of IkappaB kinase by vaccinia virus virulence factor B14. *PLoS Pathog*, 4, e22.

Clark, R. H., Kenyon, J. C., Bartlett, N. W., Tscharke, D. C. & Smith, G. L. (2006) Deletion of gene A41L enhances vaccinia virus immunogenicity and vaccine efficacy. *J Gen Virol*, 87, 29–38.

Cooray, S., Bahar, M. W., Abrescia, N. G., Mcvey, C. E., Bartlett, N. W., Chen, R. A., Stuart, D. I., Grimes, J. M. & Smith, G. L. (2007) Functional and structural studies of the vaccinia virus virulence factor N1 reveal a Bcl-2-like anti-apoptotic protein. *J Gen Virol*, 88, 1656–66.

Diperna, G., Stack, J., Bowie, A. G., Boyd, A., Kotwal, G., Zhang, Z., Arvikar, S., Latz, E., Fitzgerald, K. A. & Marshall, W. L. (2004) Poxvirus protein N1L targets the I-kappaB kinase complex, inhibits signaling to NF-kappaB by the tumor necrosis factor superfamily of receptors, and inhibits NF-kappaB and IRF3 signaling by toll-like receptors. *J Biol Chem*, 279, 36570–8.

Downie, A. W. (1939) Immunological relationship of the virus of spontaneous cowpox to vaccinia virus. *Br. J. Exp. Pathol.*, 20, 158–76.

Esposito, J. J., Sammons, S. A., Frace, A. M., Osborne, J. D., Olsen-Rasmussen, M., Zhang, M., Govil, D., Damon, I. K., Kline, R., Laker, M., Li, Y., Smith, G. L., Meyer, H., Leduc, J. W. & Wohlhueter, R. M. (2006) Genome sequence diversity and clues to the evolution of variola (smallpox) virus. *Science,* 313, 807–12.

Fenner, F., Anderson, D. A., Arita, I., Jezek, Z. & Ladnyi, I. D. (1988) Smallpox and Its Eradication, World Health Organisation, Geneva.

Goebel, S. J., Johnson, G. P., Perkus, M. E., Davis, S. W., Winslow, J. P. & Paoletti, E. (1990) The complete DNA sequence of vaccinia virus. *Virology*, 179, 247–66, 517–63.

Graham, S. C., Bahar, M. W., Cooray, S., Chen, R. A. J., Whalen, D. M., Abrescia, N. G. A., Alderton, D., Owens, R. J., Stuart, D. I., Smith, G. L. & Grimes, J. M. (2008) Vaccinia virus proteins A52 and B14 share a Bcl-2-like fold but have evolved to inhibit NF-kappa B rather than apoptosis. *Plos Pathogens*, 4, e100128.

Hashizume, S., Yoshizawa, H., Morita, M. & Suzuki, K. (1985) Properties of attenuated mutant of vaccinia virus, LC16m8, derived from Lister strain. In Quinnan, G. V. (ed.), Vaccinia viruses as vectors for vaccine antigens. Elsevier, New York.

Jenner, E. (1798) An Enquiry into the Causes and Effects of Variolae Vaccinae, a Disease Discovered in some Western Countries of England, particularly Gloucestershire, and known by the Name of Cow Pox., London, Reprinted by Cassell, 1896.

Kalverda, A. P., Thompson, G. S., Vogel, A., Schroder, M., Bowie, A. G., Khan, A. R. & Homans, S. W. (2009) Poxvirus K7 protein adopts a Bcl-2 fold: biochemical mapping of its interactions with human DEAD box RNA helicase DDX3. *J Mol Biol*, 385, 843–53.

Kotwal, G. J., Hugin, A. W. & Moss, B. (1989) Mapping and insertional mutagenesis of a vaccinia virus gene encoding a 13,800-Da secreted protein. *Virology*, 171, 579–87.

Kotwal, G. J. & Moss, B. (1988) Analysis of a large cluster of nonessential genes deleted from a vaccinia virus terminal transposition mutant. *Virology*, 167, 524–37.

Kvansakul, M., Yang, H., Fairlie, W. D., Czabotar, P. E., Fischer, S. F., Perugini, M. A., Huang, D. C. & Colman, P. M. (2008) Vaccinia virus anti-apoptotic F1L is a novel Bcl-2-like domain-swapped dimer that binds a highly selective subset of BH3-containing death ligands. *Cell Death Differ*, 15, 1564–71.

Lane, J. M., Ruben, F. L., Neff, J. M. & Millar, J. D. (1969) Complications of smallpox vaccination, 1968. National surveillance in the United States. *N Engl J Med*, 281, 1201–8.

Mackett, M. & Archard, L. C. (1979) Conservation and variation in Orthopoxvirus genome structure. *J Gen Virol*, 45, 683–701.

Mackett, M., Smith, G. L. & Moss, B. (1982) Vaccinia virus: a selectable eukaryotic cloning and expression vector. *Proc Natl Acad Sci USA*, 79, 7415–9.

Moore, J. B. & Smith, G. L. (1992) Steroid hormone synthesis by a vaccinia enzyme: a new type of virus virulence factor. *EMBO J*, 11, 3490.

Moss, B. (2007) Poxviridae: the viruses and their replicaton. In Knipe, D. M. (ed.), Fields virology. 5th ed., Lippincott Williams & Wilkins, Philadelphia, PA.

Moss, B., Winters, E. & Cooper, J. A. (1981) Deletion of a 9,000-base-pair segment of the vaccinia virus genome that encodes nonessential polypeptides. *J Virol*, 40, 387–95.

Panicali, D., Davis, S. W., Mercer, S. R. & Paoletti, E. (1981) Two major DNA variants present in serially propagated stocks of the WR strain of vaccinia virus. *J Virol*, 37, 1000–10.

Panicali, D., Davis, S. W., Weinberg, R. L. & Paoletti, E. (1983) Construction of live vaccines by using genetically engineered poxviruses: biological activity of recombinant vaccinia virus expressing influenza virus hemagglutinin. *Proc Natl Acad Sci USA*, 80, 5364–8.

Panicali, D. & Paoletti, E. (1982) Construction of poxviruses as cloning vectors: insertion of the thymidine kinase gene from herpes simplex virus into the DNA of infectious vaccinia virus. *Proc Natl Acad Sci USA*, 79, 4927–31.

Perkus, M. E., Piccini, A., lipinskas, B. R. & Paoletti, E. (1985) Recombinant vaccinia virus: immunization against multiple pathogens. *Science*, 229, 981–4.

Reading, P. C., Moore, J. B. & Smith, G. L. (2003) Steroid hormone synthesis by vaccinia virus suppresses the inflammatory response to infection. *J. Exp. Med.*, 197, 1269–78.

Seet, B. T., Johnston, J. B., Brunetti, C. R., Barrett, J. W., Everett, H., Cameron, C., Sypula, J., Nazarian, S. H., Lucas, A. & Mcfadden, G. (2003) Poxviruses and immune evasion. *Annu Rev Immunol*, 21, 377–423.

Smith, G. L. (1999) Vaccinia virus immune evasion. *Immunol. Lett*, 65, 55–62.

Smith, G. L. & Chan, Y. S. (1991) Two vaccinia virus proteins structurally related to the interleukin-1 receptor and the immunoglobulin superfamily. *J Gen Virol*, 72, 511–8.

Smith, G. L., Chan, Y. S. & Howard, S. T. (1991) Nucleotide sequence of 42 kbp of vaccinia virus strain WR from near the right inverted terminal repeat. *J Gen Virol*, 72, 1349–76.

Smith, G. L., Mackett, M. & Moss, B. (1983a) Infectious vaccinia virus recombinants that express hepatitis B virus surface antigen. *Nature*, 302, 490–5.

Smith, G. L. & Moss, B. (1983) Infectious poxvirus vectors have capacity for at least 25 000 base pairs of foreign DNA. *Gene*, 25, 21–8.

Smith, G. L., Murphy, B. R. & Moss, B. (1983b) Construction and characterization of an infectious vaccinia virus recombinant that expresses the influenza hemagglutinin gene and induces resistance to influenza virus infection in hamsters. *Proc Natl Acad Sci USA*, 80, 7155–9.

Spriggs, M. K., Hruby, D. E., Maliszewski, C. R., Pickup, D. J., Sims, J. E., Buller, R. M. & Vanslyke, J. (1992) Vaccinia and cowpox viruses encode a novel secreted interleukin-1-binding protein. *Cell*, 71, 145–52.

Sroller, V., Kutinova, L., Nemeckova, S., Simonova, V. & Vonka, V. (1998) Effect of 3-beta-hydroxysteroid dehydrogenase gene deletion on virulence and immunogenicity of different vaccinia viruses and their recombinants. *Arch Virol*, 143, 1311–20.

Staib, C., Kisling, S., Erfle, V. & Sutter, G. (2005) Inactivation of the viral interleukin 1beta receptor improves CD8+ T-cell memory responses elicited upon immunization with modified vaccinia virus Ankara. *J Gen Virol*, 86, 1997–2006.

Stickl, H. & Hochstein-Mintzel, V. (1971) [Intracutaneous smallpox vaccination with a weak pathogenic vaccinia virus ("MVA virus")]. *Munch Med Wochenschr*, 113, 1149–53.

Tscharke, D. C., Reading, P. C. & Smith, G. L. (2002) Dermal infection with vaccinia virus reveals roles for virus proteins not seen using other inoculation routes. *J Gen Virol*, 83, 1977–86.

Chapter 14
Brucella Species Synchronize Their Life Cycle to the Gestation Cycle of Their Ruminant Hosts

Svetlana Bardenstein and Menachem Banai

Abstract Brucellosis is a zoonostic disease caused by Gram negative *Brucella* species bacteria. *B. melitensis*, *B. abortus* and *B. suis* cause third trimester abortions in sheep and goats, cattle and swine, respectively. Serological diagnosis, in the past century, was based upon measurements of agglutinins and complement fixing antibodies while current state of art techniques include ELISA and fluorescence polarization. Our data show that following abortion or normal parturition, *Brucella* are secreted in the milk of lactating animals prior to development of a detectable humoral response, thereby reducing the reliability of serological diagnosis as the cornerstone of eradication programs. Accordingly, we hypothesized that infection starts as a clandestine invasion of the host by *Brucella* where, by some as yet unrecognized mechanism, the organisms manage to evade the immune response. Pregnancy then exerts physiological changes on the brucellae organisms that activate their virulence properties, allowing invasion and colonization of the trophoblasts in the third trimester, causing abortion and spread of the organisms to the external environment. Changes in the virulence properties of the bacteria within the blood stream also trigger their invasion of and propagation in the mammary glands, sustaining their secretion in the milk. These dual characteristics of *Brucella*'s stealth invasion of the host on one hand, and escape via abortive or secretion pathways on the other, demonstrate a unique virulence pattern specifically attainable due to the synchronization of their life cycle to the animal's gestation cycle.

Keywords *Brucella* · Macrophages · Replicative compartment · Trophoblasts

14.1 Introduction

Brucellosis is a zoonotic disease caused by Gram negative bacteria of genus *Brucella*. The hallmark of the disease is the strong affiliation between *Brucella* species and their respective natural hosts, where in domesticated farm animals,

M. Banai (✉)
Department of Bacteriology, Kimron Veterinary Institute, P.O. Box 12, Bet Dagan 50250, Israel
e-mail: menachemba@moag.gov.il

A. Shafferman et al. (eds.), *The Challenge of Highly Pathogenic Microorganisms*,
DOI 10.1007/978-90-481-9054-6_14, © Springer Science+Business Media B.V. 2010

B. melitensis infects small ruminants, *B. abortus* infects bovine and *B. suis* infects swine. *Brucella* causes abortion storms in their natural hosts in the last trimester of pregnancy and orchitis and epididymis in males. Abortion constitutes the major mechanism of spread and transmission of the organism in the field, as the fetus and membranes are heavily contaminated by these organisms (Alexander et al., 1981). In addition, females excrete the bacteria in the milk. Humans contract the disease by direct contact with infected materials and consumption of milk and dairy products (Corbel, 1997). Human brucellosis is most often presented as undulant fever but it may progress to a chronic stage, and complications such as splenomegaly, hepatomegaly and lymph node enlargement develop frequently. Several other complications such as osteomyelitis, endocarditis, meningitis and local manifestations of inflammatory responses are commonly seen. In rare cases the disease may be fatal (Madkour, 1989). Major therapy includes combined treatment with antibiotics with intracellular penetration such as tetracyclines, rifampicin, aminoglycosides and new generation antibiotics such as co-trimoxazole, fluroquinolones and macrolides (Lang et al., 1995).

Although recognized for over a century, the disease remains intriguing in its taxonomic position regarding newly emerging species such as sea mammals (Foster et al., 2007; Hernandez-Mora et al., 2008), rodents (Scholz et al., 2008) and an environmental strain recently isolated from a breast transplant in a woman from which the animal source has not yet been identified (Whatmore, 2009). The deciphering of the complete genome sequence of at least 10 *Brucella* strains including reference, vaccine and field strains has added controversy regarding the place of molecular tools in establishing accurate taxonomy (Chain et al., 2005; Foster et al., 2009; Rajashekara et al., 2004; Wattam et al., 2009; Whatmore et al., 2007; Whatmore, 2009).

Serological diagnosis of the disease was developed as early in time as the recognition of the Brucellosis problem (Wright and Smith, 1897). Improved diagnosis was achieved a few decades later following establishment of an international serum standard assigned with 1000 international agglutination and CFT units. During this period, a *B. abortus* S19 vaccine was developed, affording effective protection against bovine brucellosis (Buck, 1930) and later *B. melitensis* Rev.1 as a sheep and goat vaccine (Herzberg and Elberg, 1953, 1955; Herzberg et al., 1953). It was then understood that smooth vaccines elicit cross reactive antibodies with field infections which would hamper serological surveillance (Stevens, 1994). In order to circumvent this problem reducing the vaccine dose combined with change of the inoculation route was proposed (Alton and Corner, 1981; Fensterbank et al., 1985). In addition, the serological tests have been modified in order to afford differential diagnosis of infections amongst vaccinated populations (Letesson et al., 1997). In this regard, it was shown that reduced vaccine doses did not confer sufficient protection (Fensterbank et al., 1985) and therefore, this approach was almost completely abandoned in the last decades.

The new ELISAs and FPA assays were reported superior to CFT in diagnosing brucellosis (Nielsen and Ewalt, 2008). For local purposes, however, our laboratory continued to implement CFT serology which is calibrated according to the international serum standard (Alton et al., 1988). One of the major advantages of CFT diagnosis was that this calibration has enabled the test to distinguish field infection from vaccine responses due to an increased sensitivity to antibodies elicited by field strains. In Israel, S19 vaccination of female calves is mandatory and adopting CFT has helped in monitoring the herds. The epidemiology of brucellosis in Israel has, however, involved *B. melitensis* infection in the small ruminant population and transmission of the disease to cattle herds (see results). Thus, a major outcome of the higher prevalence of the disease with increased numbers of contaminated areas has proven that the vaccine, per se, did not confer sufficient protection, neither by *B. melitensis* Rev.1 in the small ruminant population nor by *B. abortus* S19 in cattle.

14.2 Results

14.2.1 Transmission of B. melitensis to Cattle

Small ruminant brucellosis was endemic in Israel during the years of the early 1990s which constituted the source of transmission of *B. melitensis* to the cattle population (Banai, 2002). The implementation of S19 vaccination of the young female calf progeny did not fully protect the herds against *B. melitensis* as was evidenced in several cases where calves contracted the disease in close proximity to vaccination. In the period 1991–1995 nine dairy herds and 23 beef herds were infected. Interestingly, in six dairy herds and all beef cattle infections, the disease was limited to only a few cows per a herd whereas in the other three dairy herds the disease affected over 10% of the herd population and eradication was usually a prolonged procedure till the successful completion of cleansing of the herds was confirmed (Data not shown). This difference in the level of infection and time required to complete eradication raised questions pertaining to the mechanism that governed each of the two scenarios. Similar data were revealed during the eradication of *B. melitensis* in intensively managed small ruminant flocks that were severely infected, despite their thorough *B. melitensis* Rev. 1 vaccine coverage (Banai, 2002). It was initially considered that the false negative diagnosis of true reactors contributed to the perpetuation of the chronicity of the disease at the farm level due to lack of sensitivity of the CFT method in comparison to ELISA and FPA (Gall and Nielsen, 2004).

14.2.2 B. melitensis Infection in Israeli Cattle, a Case Report

In 2006, we again encountered the reemergence of *B. melitensis* transmission to dairy cattle herds (data not shown). In one case, a large dairy herd that included more than 800 milking cows was suspected to be infected following identification of a CFT positive cow that had parturated prematurely. At this point of time, because no clinical signs had been observed, we assumed that this was the onset of the outbreak, allowing us to pursue the question regarding the performance of the CFT in

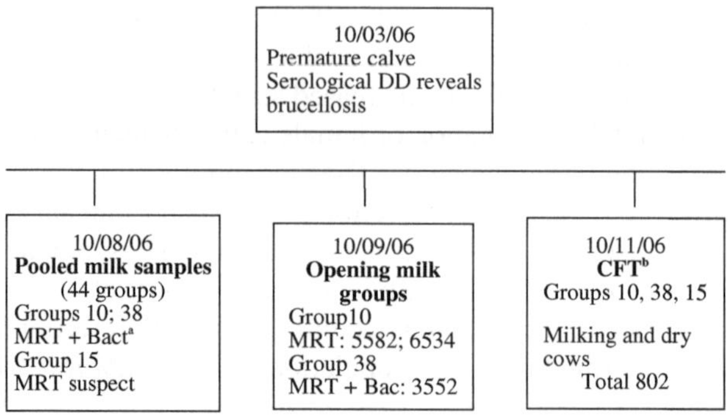

[a]Bacteriology yielded *B. melitensis* bv. 1 after 4 days.
[b]Cow 3552 was positive by all serological tests (CFT, i-ELISA, c-ELISA and FPA), cow 6534 was positive by c-ELISA (marginally) and CFT and cow 5582 was CFT only positive in the first serological round. In addition 14 cows were CFT only positive. All together 17 serologically positive cows were immediately slaughtered

Fig. 14.1 Initial diagnosis carried out in the farm following identification of a serology positive cow that gave birth to a premature calve

diagnosing brucellosis in a real time scenario. As can be seen in Fig. 14.1, immediate steps were taken in order to obtain early information regarding the health status of the herd. The milking animals were divided into groups of twenty, from which milk was pooled for MRT and bacteriological culture. As can be seen, two groups, 10 and 38, reacted positively by the MRT test and a third group, # 10 produced suspect results. *B. melitensis* biovar 1 was isolated 4 days later from the two positive milk pools. Immediately after this, the cows comprising the two reactive pools were tested individually, identifying cow no. 3552 as positive by MRT and isolation of *B. melitensis* biovar 1 strain. Serological analysis of 802 dry and in milk cows in addition to individuals from each of the three milk pools resulted in 17 cows in total that were positive by one or more tests. All 17 cows were immediately taken from the herd and slaughtered. In order to prevent the risk to human health during the slaughter process, cow 3552 was dried off and euthanized.

A second round of serological testing of the cows revealed 4 additional cows that yielded *B. melitensis* from milk samples and were positive by CFT. These cows were immediately removed and euthanized. Subsequently, the serum samples from the first and second round of CFT tests were sent to Klaus Nielsen in Canada for further comparison of their serological responses by i-ELISA, c-ELISA and FPA, respectively. Interestingly, these tests corroborated our CFT results regarding cow 3552 and three out of 4 cows that had tested positive by CFT in the second round of investigation, but failed to diagnose cow 6221 that was included amongst this group. The tests identified 3 more CFT positive cows from the first round but were negative in 13 more cows from this group. There were, however, 9 more reactors by one or more of the 3 tests that had been found negative by CFT. Because CFT was used as the sole criterion for culling, these cows remained on the farm till the end of the eradication program, showing no clinical signs of disease or positive bacteriological

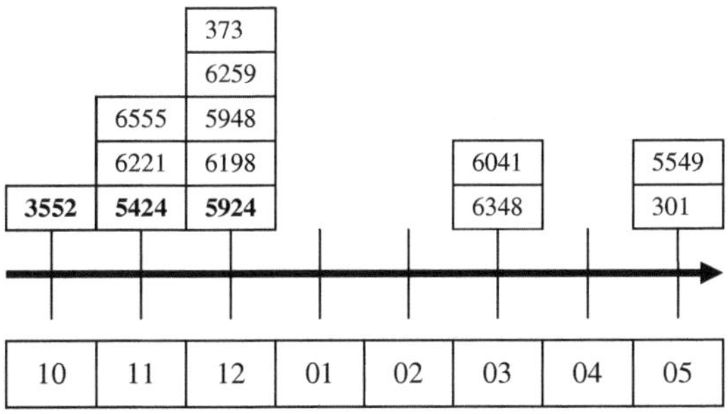

Fig. 14.2 Chronological representation of milk culture surveillance of cows that yielded *Brucella melitensis* biovar 1 isolates. Cows were monthly tested by CFT, i-ELISA, c-ELISA and FPA and milk samples were cultured for *Brucella* species isolation. The cows (ID numbers) and time of *Brucella* isolation are indicated in a monthly time-scale. Cows 3552 was serologically positive by all four tests in the first round of testing whereas cows 5424 and 5924 seroconverted by these tests in the second round of tests. Cow 6221 was negative by ELISAs and FPA but positive by CFT. The other cows were serologically negative in the first two rounds of tests but seroconverted by CFT at time of *Brucella* isolation

cultures from the milk. We therefore attributed the positive serological results to the previous S19 vaccination background. In support of this conclusion was the continuous identification of FPA positive reactors by testing performed in the Israeli laboratory in animals that had been serologically negative by all other tests. These specific cows were maintained in the milking group without further complications.

As can be seen in Fig. 14.2, retrospective analysis of the results shows that additional cows eventually developed brucellosis as evidenced by isolation of *B. melitensis* biovar 1 strain from the milk. Eight cows that had been negative by CFT, the ELISAs and FPA, were found positive by CFT during the following months and *Brucella* was isolated from their milk. This feature, where the cows were negative by all possible serological tests during the period of incubation of the disease, strongly suggests that no anti-smooth *Brucella* antibodies existed in these animals at the time of testing.

Interestingly, as can be seen in Table 14.1, most of the cows that eventually yielded *Brucella* in their milk were part of the same group that included the index case which was the cow that had calved prematurely and was therefore removed from the dairy prior to the onset of the eradication campaign. It can thus be surmised that these cows were likely to have been infected simultaneously by the same initial source of the outbreak, but each individually developed acute brucellosis accompanied by a humoral response only after parturition. In contrast, two cows that were the first to be identified as *B. melitensis* secretors, 3552 and 5424, were months away from parturition, the former being 6 months pregnant and the latter had been in the milking group for the previous 7 months. It could thus be argued that the disease was introduced to the farm around 7–8 months prior to the first diagnosis in October. If this assumption is correct, the cows from the October parturition group were, at the time of infection, only in the early stages of pregnancy, which is the period of highest risk to contract brucellosis. Moreover, as treatment in the farm was not pursued

Table 14.1 Correlation between dates of Brucella isolation and parturition of the cows

Cow ID	Birth date	Milking cycle	Date of Brucella isolation	Last parturition
3552	17/10/2000	4	09/10/06	Gest 188d
6221	21/10/2003	2	06/11/06	Parturition 21.10.06
5424	18/12/2000	4	06/11/06	Empty, last parturition 11/03/06
6555	11/11/2004	1	06/11/06	Parturition 22.10.06
373	01/01/2001	4	28/12/06	Parturition 16.10.06
5924	05/10/2002	3	28/12/06	Parturition 16.10.06
5948	29/10/2002	3	28/12/06	Parturition 25.12.06
6198	02/10/2003	2	28/12/06	Parturition 03.10.06
6259	01/12/2003	2	28/12/06	Parturition 30.10.06
6041	01/02/2003	2	15/03/07	Gest 205d
6348	01/03/2004	2	15/03/07	Parturition 18.02.07
5549	02/07/2001	4	20/05/07	Parturition 15.10.06
301	01/09/2000	5	31/05/07	Parturition 24.12.06

throughout these months, the disease was possibly transmitted to several other cows which could explain several instances that were observed in the past where high prevalence rates of the disease were seen (Data not shown). Alternatively, it could be speculated that only clonal lineages of virulent pathovars could play a determinative role in the pathogenesis of the disease suggesting unknown virulence mechanisms in *Brucella*.

14.3 Discussion

Data shown here (Figs. 14.1 and 14.2) strongly suggest that antibody responses to *Brucella* infection are delayed till after parturition (Nicoletti, 1980; Saegerman et al., 2004). This is surmised because highly sensitive techniques, such as i-ELISA, c-ELISA and FPA failed to detect antibodies in sera from pregnant cows that secreted the organisms in the milk (Fig. 14.2, Table 14.1). We are, therefore, the first to show that the failure to diagnose certain infected cows may not be due to the limitations in sensitivity of the serological assays (Nielsen et al., 2005; Gall and Nielsen, 2004), rather that this is a major characteristic of the pathogenesis of the disease. This phase, as observed by field data, is in agreement with previous publications that emerged from experimental brucellosis in animal models and tissue cultures regarding stealth *Brucella* pathogenicity (Barqueiro-Calvo et al., 2007, 2009; Cardoso et al., 2006; Goldstein et al., 1992; Gorvel, 2008; Neta et al., 2008; Pei et al., 2008; Tsolis et al., 2008).

Besed on the literature, we propose that *Brucellae* maintain a life cycle that alternates between three different replicative compartments in the host (Roop II et al., 2009; Fig. 14.3). Inside monocyte-derived macrophages of the reticulo-endothelial system (Pappas et al., 2005; Roop II et al., 2009) brucellae cells remain enclosed in *Brucella* containing vacuoles (BCV) labeled with ER specific markers that confer permissive bacterial replication by aid of the *virB* secretion apparatus (Seleem et al., 2008; Celli and Gorvel, 2004). Because most likely the BCVs remain intact, the replication of brucellae cells inside BCV is significantly restricted due to shortage of space and nutrients.

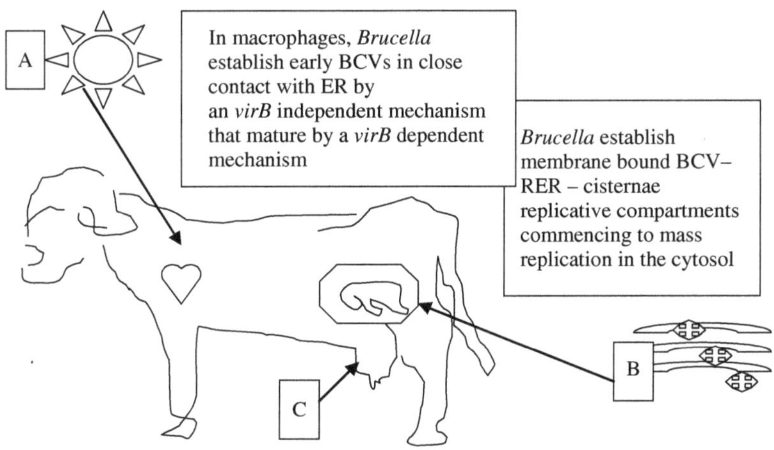

Fig. 14.3 Macrophage, trophoblast and supra-mammary gland possibly represent three *Brucella* replicative compartments. (**a**) In a mouse-macrophage model (Rambow-Larsen et al., 2009), *Brucella* establish an early BCV replicative compartment following a limited fusion event with ER that is segregated from the endocytic pathway in a non-autophagic dependent process, and subsequently mature to late BCV by the involvement of the *virB* type IV secretion apparatus (Celli and Gorvel, 2004). In this model, *Brucella* organisms remain contained in the intact BCV thereby being limited in space and nutrition for mass proliferation in the host. (**b**) In a bovine placentitis model, *Brucella* initially infect erythro-phagocytic trophoblasts and spread to chorioallantoic trophoblasts. In the trophoblasts, *Brucella* establish membrane bound BCV-RER-cistenae replicative compartments (Meador and Deyoe, 1989). In necrotic trophoblasts, cisternae are fragmented and *Brucella* loosen to the cytoplasm where they are free to replicate to numerous numbers of organisms in a single cell. This would eventually lead to abortion and mass bacterial spread in the fetus and membranes. (**c**) In supra-mammary gland infection, similar to B, *Brucella* are secreted to the milk in mass amounts. The process may involve non-professional phagocyte infection, however, possibly by an autophagosome bypass of late, but not early endosomal compartments, establishing a replicative compartment in contact with ER (Pizarro-Cerdá et al., 1998)

In the uterus brucellae cells initially encounter erythro-phagocytic trophoblasts from which they spread to chorioallantoic trophoblasts (Samartino and Enright, 1993). Inside the trophoblasts, brucellae cells establish membrane bound BCV-RER-cisternae replicative compartments. Infected cisternae respond, however, by development of a voluminous structure that may disrupt (Meador and Deyoe, 1989) thus releasing *Brucella* organisms into the cytosol. In favor of this hypothesis, a putative hemolysin and corresponding secretary proteins have been demonstrated in the *Brucella* chromosome (DelVecchio et al., 2002). Erythritol has been shown to be a significant catabolite for *Brucella* species and its production by trophoblasts in the gravid uterus has been inferred as a significant facilitator of *Brucella* growth (Smith et al., 1965). Here we propose that erythritol catabolism may not be directly involved with *Brucella* virulence. Instead, it is very likely that *Brucella* exploit this substrate (Sperry and Robertson, 1975) that is available in the uterus for their rapid proliferation in the cytosol achieving mass cultural growth in the exudates of the gravid uterus to mass amounts of organisms.

Brucella secretion into the milk has been documented as a major source of human infection following consumption of infected dairy products (Pappas et al., 2005). We propose that in the event of udder infection *Brucella* encounter non-professional phagocytic cells in which, yet, a third mechanism of BCV maturation involves autophagosome-dependent development of *Brucella* replicative compartments (Pizarro-Cerdá et al., 1998). It is speculated that *Brucella* may eventually access the cytosol of infected cells by the hemolysin cascade in order to gain free replication in the supra-mammary glands thereby sustaining their secretion in the milk.

The difference between *Brucella* proliferation in macrophage BCV-ER compartments to trophoblast BCV-ER transformation into cytoplasmic replicative compartments could explain evolutionary adaptation of *Brucella* to its preferred host. Cytoplasmic replication exposes the pathogen to cytosolic immune surveillance pathways involving cytoplasmic pathogen recognition receptors (PRR) and proinflammatory cytokine production such as type I IFNs and activation of an inflammasome complex. Inflammasome mediates cell death as a central immune defense mechanism against intracellular pathogens (Kumar and Valdivia, 2009). Thus, *Brucella* containment in intact BCV prevents macrophage death and perpetuates a chronic survival of the organisms in the host. In contrast, during trophoblast infection this process may induce cellular necrosis and placentitis leading to abortion and the release of *Brucella* organisms into the environment in massive amounts (Alexander et al., 1981; Samartino and Enright, 1993). Recently a light, oxygen and voltage (LOV-HK) blue-light virulence mechanism has been described in *B. abortus* (Swartz et al., 2007). As abortion results in changing the environmental dark to light exposure for *Brucella* in the uterus it is intriguing to hypothesize that an off-on switch exists in *Brucella* that could synchronize the life cycle of this pathogen. Taken the stealth pathogenesis of *Brucella* this mechanism may exert a biological function in preparing *Brucella* to infect new naïve hosts.

This model may explain our observation that *Brucella* secretion in the milk precedes the humoral response. If not diagnosed, pregnant animals would very likely abort and spread the organisms in the farm explaining Nicoletti's observation (1980) that eradication of *B. abortus* is almost impossible to achieve without including repeated vaccination as a major control tool.

Therefore, pregnant animals that have successfully parturated or following abortion should be targeted as a prioritized group in serological monitoring of the herds. By focusing on this group, eradication of the disease should be feasible based on bacteriological culturing of milking cows and serological removal of positive responders by MRT and blood samples. Latent brucellosis would then be the reason for unsuccessful removal of all infected animals in the first round of tests necessitating repeated surveillance of the whole herd until confirmed clean.

Our data further show that despite recent reemergence of *B. melitensis* in the small ruminant population and humans, dairy cattle brucellosis remained under reasonable control in Israel within the last decade (data not shown). As in past years S19 vaccination failed to fully protect the herds against occasional infections by *B. melitensis*, the continued mandatory vaccination of the young replacement females at ages of between 3 and 7 months did not support the supposition that vaccination in itself was sufficient in controlling the disease. In contrast, since 2000, changes in management policies have been implemented including the merging of small herds to form larger units as well as strengthening bio-security in the farms. This may emphasize the importance of bio-security as a major factor in preventing additional new cases. To this effect, it may be possible to control, and possibly eradicate, small ruminant brucellosis in Israel by *B. melitensis* Rev.1 vaccination of the young replacement females, control of animal movement and registration, increasing bio-security measures particularly amongst nomadic flocks and eventually eradicating the disease by a test and slaughter program of serologically positive reactors rather than repeating adult vaccination.

Acknowledgements We are indebted to Klaus Nielsen and his team for conducting i-ELISA, c-ELISA and FPA on the serum samples. Work in the author's laboratory was partly supported by Binational Agricultural Research and Development Grant US-3829-06 R.

References

Alexander, B., Schnurrenberger P. R. & Brown R. R. (1981) Numbers of *Brucella abortus* in the placenta, umbilicus and fetal fluid of two naturally infected cows. *Vet Rec*, 108, 500.

Alton, G. G. & Corner, L. A. (1981) Vaccination of heifers with a reduced dose of *Brucella abortus* strain 19 vaccine before first mating. *Aust Vet J*, 57, 548–50.

Alton, G. G., Jones, L. M., Angus, R. D. & Verger, J. M. (1988) Techniques for the brucellosis laboratory. Institut National De La Recherche Agronomique, Paris.

Banai, M. (2002) Control of small ruminant brucellosis by use of *Brucella melitensis* Rev.1 vaccine: laboratory aspects and field applications. *Vet Micobiol*, 90, 497–519.

Barqueiro-Calvo, E., Chaves-Olarte, E., Weiss, D. S., Guzmán-Verri, C., Chacon-Diaz, A., Rucavado, A., Moriyon, I. & Moreno, E. (2007) *Brucella abortus* uses a stealthy strategy to avoid activation of the innate immune system during the onset of infection. *PLoS ONE* 2(7), e631.

Barqueiro-Calvo, E., Conde-Alvarez, R., Chac6n-Diaz, C., Quesada-Lobo, L., Martirosyan, A., Guzmán-Verri, C., Iriarte, M., Mancek-Keber, M., Jeral, R., Gorvel, J. -P., Moriy6n, I,

Moreno, E. & Chaves-Olarte, E. (2009) The differential interaction of *Brucella* and *Ochrobactrum* with innate immunity reveals traits related to the evolution of stealthy pathogens. *PLoS ONE* 4(6), e5893.
Buck, J. M. (1930) Studies of vaccination during calfhood to prevent bovine infectious abortion. *J Agricul Res*, 41, 667-89.
Cardoso, P. G., Macedo, G. C., Azevedo, V. & Oliveira, S. C. (2006) *Brucella* spp. noncanonical LPS: structure, biosynthesis, and interaction with host immune system. *Microbial Cell Fact*, 5, 13.
Celli, J. & Gorvel, J. -P. (2004) Organelle roberry: *Brucella* interactions with the endoplasmic reticulum. *Curr Opin Microbiol.*, 7, 93–7.
Chain, P. S. G., Comerci, D. J., Tolmasky, M. E., Larimer, F. W., Malfatti, S. A., Vergez, L. M., Aguero, F., Land, M. L., Ugalde, R. A. & Garcia, E. (2005) Whole-genome analyses of speciation events in pathogenic *Brucellae Infect Immun*, 73, 8353–61.
Corbel, M. J. (1997) Brucellosis: an overview. *Emerg Infect Dis*, 3, 213–21.
Delvecchio, V. G., Kapatral, V., Redkar, R. J., Patra, G., Mujer, C., Los, T., Ivanova, N., Anderson, I., Bhattacharyya, A., Lykidis, A., Reznik, G., Jablonski, L., Larsen, N., D'souza, M., Bernal, A., Mazur, M., Goltsman, E., Selkov, E., Elzer, P. H., Hagius, S., O'callaghan, D., Letesson, J. J., Haselkorn, R., Kyrpides, N. & Overbeek, R. (2002) The genome sequence of the facultative intracellular pathogen *Brucella melitensis*. *Proc Natl Acad Sci* USA, 99, 443–8.
Fensterbank, R., Pardon, P. & Marly, J. (1985) Vaccination of ewes by a single conjuctival administration of *Brucella melitensis* Rev. 1 vaccine. *Ann Rech Vet*, 16, 351–6.
Foster, J. T., Beckstrom-Sternberg, S. M., Pearson, T., Beckstrom-Sternberg, J. S., Chain, P. S. G., Roberto, F. F., Hnath, J., Brettin, T. & Keim, P. (2009) Whole-genome-based phylogeny and divergence of the genus *Brucella J Bacteriol*, 191, 2864–70.
Foster, G., Osterman, B. J., Godfroid, J., Jacques, I. & Cloeckaert, A. (2007) *Brucella ceti* sp. nov. and *Brucella pinnipedialis* sp. nov. for *Brucella* strains with cetaceans and seals as their preferred hosts. *Intl J Sys Evol Microbiol*, 57, 2688–93.
Gall, D. & Nielsen, K. (2004) Serological diagnosis of bovine brucellosis: a review of test performance and cost comparison. *Revue Scientifique et Technique* (The Office International des Epizooties, Paris) 23, 989–1002.
Goldstein, J., Hoffman, T., Frasch, C., Lizzio, E. F., Beining, P. R., Hochstein, D., Lee, Y. L., Angus, R. D. & Golding, B. (1992) Lipopolysaccharide (LPS) from *Brucella abortus* is less toxic than that from *Escherichia coli*, suggesting the possible use of *B. abortus* or LPS from *B. abortus* as carrier in vaccines. *Infect Immun*, 60, 1385–9.
Gorvel, J. -P. (2008) *Brucella*: a Mr "Hide" converted into Dr Jekyll. *Microbes Infect*, 10, 1010–13.
Hernandez-Mora, G., Gonzales-Barrientos, R., Morales, J. A., Chaves-Olarte, E., Guzman-Verri, C., Baqueiro-Calvo, E., De-Miguel, M. J., Marin, C. M., Blasco, J. M., & Moreno, E. (2008) Neurobrucellosis in stranded dolphins, Costa Rica. *Emerg Infect Dis*, 14, 1430–3.
Herzberg, M. & Elberg, S. (1953) Immunization against *Brucella* infection 1. Isolation and characterization of a streptomycin-dependent mutant. *J Bacteriol*, 66, 585–99.
Herzberg, M. & Elberg, S. S. (1955) Immunization against *Brucella* infection. III. Response of mice and guinea pigs to injection of viable and nonviable suspensions of a streptomycin-dependent mutant of *Brucella melitensis*. *J Bacteriol*, 69, 432–5.
Herzberg, M., Elberg, S. S. & Meyer, K. F. (1953) Immunization against *Brucella* infection. II. Effectiveness of a streptomycin-dependent strain of *Brucella*. *J Bacteriol*, 66, 600–5.
Kumar, Y. & Valdivia, R. H. (2009) Leading a sheltered life: intracellular pathogens and maintenance of vacuolar compartments. *Cell Host Microbe*, 5, 593–1001.
Lang, R., Banai, M., Lishner, M. & Rubinstein, E. (1995) Brucellosis. *Intl J Antimicrob Agents*, 5, 203–8.
Letesson, J. J., Tibor, A., Van Yynde, G., Wansard, V., Weynants, V., Denoel, P. & Saman, E. (1997) Humoral immune responses of *Brucella*-infected cattle, sheep, and goats to eight purified recombinant *Brucella* proteins in an indirect enzyme-linked immunosorbent assay. *Clin Diagn Lab Immun*, 4, 556–64.

Madkour, M. M. (1989) Brucellosis. Butterworths, London, 294.
Meador, V. P. & Deyoe, B. L. (1989) Intracellular localization of *Brucella abortus* in bovine placenta. *Vet Pathol*, 26, 513–15.
Neta, A. V. C., Stynen, A. P. R., Paixäo, T. A., Miranda, K. L., Silva, F. L., Rouz, C. M., Tsolis, R. M., Everts, R. E., Lewin, H. A., Carvalho, A. F., Lage, A. P. & Santo, R. L. (2008) Modulation of the bovine trophoblastic innate immune response by *Brucella abortus*. *Infect Immun*, 76, 1897–907.
Nicoletti, P. (1980) The epidemiology of bovine brucellosis. *Adv Vet Sci Comp Med*, 24, 69–98.
Nielsen, K. & Ewalt, D. R. (2008) Bovine brucellosis. Manual of diagnostic tests and vaccines for terrestrial animals, 6th ed. Vol. 2, The Office International des Epizooties, Paris, pp. 624–59,
Nielsen, K., Smith, P., Yu, W., Nicoletti, P., Elzer, P., Robles, C., Bermudez, R., Renteria, T., Moreno, F., Ruiz, A., Massengill, C., Muenks, Q., Jurgersen, G., Tollersrud, T., Samartino, L., Conde, S., Forbes, L., Gall, D., Perez, B., Rojas, X. & Minas, A. (2005) Towards single screening tests for brucellosis. *Rev Sci Tec* (The Office International des Epizooties, Paris) 24, 1027–38.
Pappas, G., Akritidis, N., Bosilkovski, M. & Tsianos, E. (2005) Brucellosis. *New Eng J Med*, 352, 2325–36.
Pei, J., Turse, J. E. & Ficht, T. A. (2008) Evidence of *Brucella abortus* OPS dictating uptake and restricting NF-κB activation in murine macrophages. *Microbes Infect*, 10, 582–90.
Pizarro-Cerdá, J., Méresse, S., Parton, R. G., Van Der Goot, G., Sola-Landa, A., Lopez-Goñi, I., Moreno, E. & Gorvel, J. -P. (1998) *Brucella abortus* transits through the autophagic pathway and replicates in the endoplasmic reticulum of nonprofessional phagocytes. *Infect Immun*, 66, 5711–24.
Rajashekara, G., Glasner, J. D., Glover, D. A. & Splitter, G. A. (2004) Comparative whole-genome hybridization reveals genomic islands in *Brucella* species. *J Bacteriol*, 186, 5040–51.
Rambow-Larsen, A. A., Peterson, E. M., Gourley, C. R. & Splitter, G. A. (2009) *Brucella* regulators: self-control in a hostile environment. *Trends Microbiol*, 17, 371–77.
Roop II R. M., Gains, J. M., Anderson, E. S., Caswell, C. C. & Martin, D. W. (2009) Survival of the fittest: how *Brucella* strains adapt to their intracellular niche in the host. *Med Microbiol Immun*, 198, 221–38.
Saegerman, C., De Waele, L., Gilson, D., Godfroid, J., Thiange, P., Michel, P., Limbourg, B., Vo T. K. -O., Limet, J., Letesson, J. -J. & Berkvens, D. (2004) Evaluation of three serum i-ELISAs using monoclonal antibodies and protein G as peroxidase conjugate for the diagnosis of bovine brucellosis. *Vet Microbiol*, 100, 91–105.
Samartino, L. E. & Enright, F. M. (1993) Pathogenesis of abortion of bovine brucellosis. *Comp Immunol Microbio. Infect Dis*, 16, 95–101.
Scholz, H. C., Hubalek, Z., Sedláček, I., Vergnaud, G., Tomaso, H., Dahouk, S., Melzer, F., Kämpfer, P., Neubauer, H., Cloeckaert, A., Maquart, M., Zygmunt, M. S., Whatmore, A. M., Falsen, E., Bahn, P., Göllner, C., Pfeffer, M., Huber, B., Busse, H. -J. & Nöckler, K. (2008) *Brucella microti* sp. nov., isolated from the common vole *Microtus arvalis*. *Intl J Sys Evol Microbiol*, 58, 375–82.
Seleem, M. N., Boyle, S. M. & Sriranganathan, N. (2008) *Brucella*: a pathogen without classic virulence genes. *Vet Microbiol*, 129, 1–14.
Smith, H., Williams, A. E., Pearce, J. H., Keppie, J., Harris-Smith, P. W., Fitz-George, R. B. & Witt, K. (1965) Foetal erythritol: a cause of the localization of *Brucella abortus* in bovine contagious abortion. *Nature*, 193, 47–9.
Sperry, J. F. & Robertson, D. C. (1975) Erythritol catabolism by *Brucella abortus*. *J Bacteriol*, 121, 619–30.
Stevens, M. G., Hennager, S. G., Olsen, S. C. & Cheville, N. F. (1994) Serologic responses in diagnostic tests for brucellosis in cattle vaccinated with *Brucella abortus* 19 or RB51. *J Clin Microbiol*, 32, 1065–6.
Swartz, T. E., Tseng, T. -S., Frederickson, M. A., Paris, G., Comerci, D. J., Rajashekara, G., Kim, J. -G., Mudgett, M. B., Splitter, G. A.., Ugalde, R. A., Goldbaum, F. A., Briggs, W. R. &

Bogomolni, R. A. (2007) Blue-light-activated histidine kinases: two-component sensors in bacteria. *Science*, 317, 1090–93.
Tsolis, R. M., Young, G. M., Solnic, J. V. & Bäumler, A. J. (2008) From bench to bedside: stealth of enteroinvasive pathogens. *Nat Rev*, 6, 883–92.
Wattam, A. R., Williams, K. P., Snyder, E. E., Almeida, N. F. Jr., Shukla, M., Dickerman, A. W., Crasta, O. R., Kenyon, R., Lu, J., Shallom, J. M., Yoo, H., Ficht, T. A., Tsolis, R. M., Munk, C., Tapia, R., Han, C. S., Detter, J. C., Bruce, D., Brettin, T. S., Sobral, B. W., Boyle, S. M. & Setubal, J. C. (2009) Analysis of ten *Brucella* genomes reveals evidence for horizontal gene transfer despite a preferred intracellular lifestyle. *JBacteriol*, 191, 3569–79.
Whatmore, A. M. (2009) Current understanding of the genetic diversity of *Brucella*, an expanding genus of zoonotic pathogens. *Infect Genet Evol* (doi: 10.1016/j.meegid. 2009.07.001).
Whatmore, A. M., Perrett, L. L. & Mackmillan, A. P. (2007) Characterization of the genetic diversity of *Brucella* by multilocus sequencing. *BMC Microbiol*, 7, 34.
Wright, A. E. & Smith, F. (1897) On the application of the serum test to the differential diagnosis of typhoid and Malta fever. *Lancet*, 1, 656.

Chapter 15
MMP-9 as a Marker for Vaccinia Virus Related Encephalitis

Tomer Israely, Sharon Melamed, Noam Erez, Boaz Politi, Udy Olshevsky, Reuven Levin, Shlomo Lustig, and Nir Paran

Abstract Smallpox vaccination might be associated with adverse reactions, ranging in severity from benign to lethal. One of the most serious complications is postvaccinal encephalitis. The aim of this study was to identify early markers for vaccinia virus (VACV) induced encephalitis. For this purpose we infected mice intracranially with VACV-Lister or VACV-WR. Histopathological analysis showed that following infection with VACV-WR tissue damage in the brain spatially and temporally correlated with virus replication, infiltration of white blood cells and apoptosis. None of the above markers was observed upon infection with the vaccine strain Lister.

MMP-9 is a serum factor known for its correlation to BBB integrity and encephalitis in humans. We found that in sera of VACV-Lister infected animals, MMP-9 levels did not change throughout the infection. However, during VACV-WR infection, in the first 2 days levels of MMP-9 were significantly low than the controls and subsequently rose to levels which were significantly higher than the controls. Elevated MMP-9 was associated with damage to the brain and to BBB integrity.

In conclusion, efficient virus replication in the brain causes significant brain damage followed by BBB brake-down and release of MMP-9 to the serum, markers which were not observed when the attenuated strain was examined. Thus, we suggest MMP-9 as a possible non-invasive serum indicator for encephalitis caused by VACV virus.

Keywords Blood-brain-barrier · Encephalitis · Matrix metalloproteinase-9 · Smallpox · Vaccinia

---

T. Israely (✉)
Department of Infectious Diseases, Israel Institute for Biological Research, P.O. Box 19, Ness-Ziona 74100, Israel
e-mail: tomeri@iibr.gov.il

Equal contribution by Tomer Israely and Sharon Melamed.

A. Shafferman et al. (eds.), *The Challenge of Highly Pathogenic Microorganisms*,
DOI 10.1007/978-90-481-9054-6_15, © Springer Science+Business Media B.V. 2010

15.1 Introduction

Smallpox, a disease caused by Variola virus (VARV) was considered in the past as one of the most fatal and dangerous diseases to human kind. Following intensive worldwide vaccination campaign, the WHO declared in 1980 that smallpox has been eradicated (Fenner et al., 1988; WHO, 1980) and from that point vaccination gradually ceased. Although the disease was eradicated there is a concern nowadays that Smallpox can reemerge accidentally or intentionally as an agent of bioterrorism. Due to the fact that most of the world population is unvaccinated, this malicious intention becomes to be a major concern. In the United States for example, vaccination was discontinued at the 80′s and reinstituted among military personnel in 2002 (CDC, 2004) and select civilian public health care workers a year later (Sejvar et al., 2005).

Smallpox vaccination is associated with several side effects as eczema vaccinatum, generalized vaccinia and progressive vaccinia, dilated cardiomyopathy and myocarditis heart problems and postvaccinal encephalitis (PVE) which refer to central nervous system (CNS) inflammation (Casey et al., 2005). In 1922, 11 fatalities due to PVE were reported in Great Britain, and this complication was recognized from this time onward (Scott, 1967). Although rare, PVE is a severe complication of the vaccination. The rate and severity of PVE vary significantly among studies and between vaccine strains ranging from 2.9 cases per million in Unites States in 1968 (Henderson and Moss, 1999) up to 449 cases per million in Germany in the 1940s and 1950s (Sejvar et al., 2005). The case-fatality rate for PVE has been reported at 25% (Lane et al., 1970).

PVE signs usually began abruptly 10–12 days after onset of the rash. Patients initially suffer from headache, fever, dizziness, vomiting, disorientation and pains in the neck and back. Then, 1–2 days later, some subjects develop high fever and neurological signs. The acute phase last approximately one week and death often occurred 2–3 weeks after the onset of encephalitis. In those that survived, recovery was often complete and occurred over weeks to months (Booss and Davis, 2003). As PVE can be diagnosed only after the outset of the symptoms and no predisposing indications are known, the choice of treatment is supportive.

In search for PVE markers that would allow early diagnosis and discrimination from other morbidity symptoms, matrix metalloproteinases (MMPs) appeared to be potential markers. MMPs constitute a family of zinc and calcium dependent endopeptidases that function in the breakdown of extracellular matrix (ECM). They play an important role in many normal physiological processes such as organogenesis, neovascularization, or wound healing (Birkedal-Hansen, 1993; Canete Soler et al., 1995; Del Bigio and Jacque, 1995; Romanic and Madri, 1994). They also participate in many pathological processes such as arthritis, cancer and cardiovascular disease (Overall and Lopez-Otin, 2002).

Matrix metalloproteinase-9 (MMP-9; gelatinase B) degrades components of the ECM with high specific activity for denatured collagens (gelatin). MMP-9 has been shown to degrade collagen type IV in the basement membranes on endothelial walls, which may lead to disruption of the blood-brain barrier (BBB) after both cerebral ischemia and inflammation, allowing entry of leukocytes or cytokines into the brain (Rosenberg et al., 1994). It has been shown that changes in the MMP-9 level were correlated in humans with several brain diseases as multiple sclerosis (Bever and Rosenberg, 1999; Boz et al., 2006) (Waubant et al., 1999; Yong et al., 2007), subacute sclerosing panencephalitis (Ichiyama et al., 2007b), acute disseminated encephalomyelitis (Ichiyama et al., 2006) and acute encephalopathy following prolonged febrile seizures (Suenaga et al., 2008).

In addition, impaired MMP-9 levels correlate with various pathogens infection such as Japanese encephalitis virus (Mishra et al., 2009), Semliki Forest virus (Keogh et al., 2003), West Nile virus (Wang et al., 2008) Herpes simplex virus encephalitis (Martinez-Torres et al., 2004), Influenza-associated encephalopathy (Ichiyama et al., 2007a) and *Angiostogylus cantonesis* (Chen et al., 2005). Recent works showed that MMP-9 gene silencing or knock out maintain the integrity of the BBB and protects the brain after ischemic damage, respectively (Bonoiu et al., 2009; Svedin et al., 2007).

The aims of this study were to characterize poxvirus induced neurovirulence in a mouse model by direct inoculation of VACV-WR and VACV-Lister, neurovirulent adapted and vaccine strains, respectively. Having an established model, we further characterized MMP-9 as a PVE tracer.

15.2 Results

15.2.1 Characterization of VACV-WR and VACV-Lister Following Intracranial Infection in Mice

In order to compare the neurovirulent strain VACV-WR to the vaccine strain VACV-Lister, mice were infected intracranialy (i.c.) with 1×10^2 and 1×10^6 plaque forming units (pfu) respectively. Mice infected with VACV-Lister at the same doses didn't show any sign of disease (data not shown). Mice were monitored for morbidity and mortality and brain viral titers were determined at different time points post exposure. In both cases, during the first 4–5 days animals lost weight in a comparable manner – about 20% of their initial weight (Fig. 15.1a). However, mice infected with VACV-WR did not survive the infection while mice infected with VACV-Lister regained weight starting at day 7 onward and recovered completely within 11–12 days. From the 4th day post infection animals infected with VACV-WR suffered from loss of balance and orientation which implies on involvement/damage to the brain. On the other hand, animals infected with VACV-Lister didn't show these phenomena at any time (data not shown). Brains of VACV-Lister and VACV-WR infected animals were evaluated for viral titer and were histopathological examined throughout the period of the weight loss. In VACV-Lister infected mice, the viral titer remained constant and even dropped slightly. On the other hand, VACV-WR replicated efficiently reaching 1×10^4 times the given dose within 2 days and

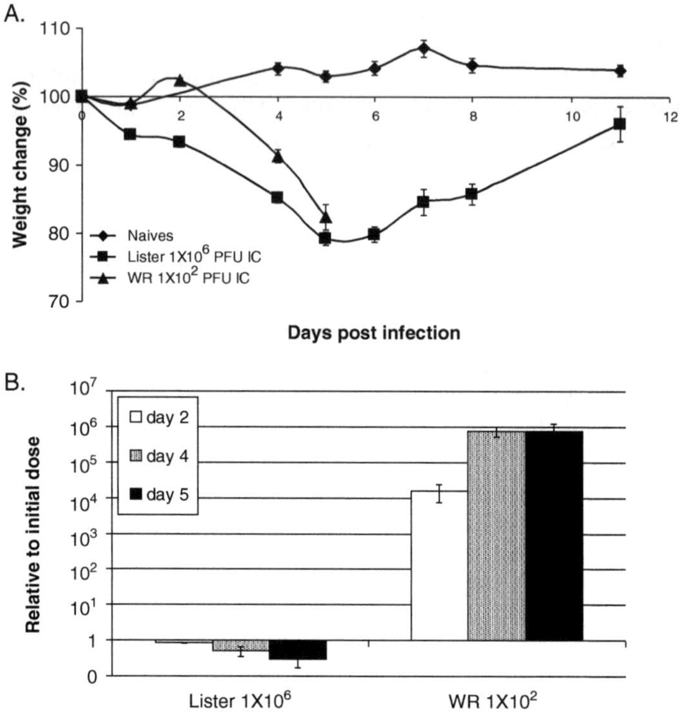

Fig. 15.1 Morbidity and brain viral titer following intracranial infection. Animals were inoculated with VACV-Lister (1×10^6) or VACV-WR (1×10^2). (**a**) Weight change following infection. (**b**) Change in brain viral titers relative to the given dose of infection on days 2, 4, and 5 post infection

1×10^6 times within 4 days post infection. On that day and on the 5th day (1 day prior to death), total viral load in the brain reached 1×10^8 pfu (Fig. 15.1b).

15.2.2 Brain Examination Following VACV-WR Infection

In order to define more closely the influence of VACV-WR infection on the brain tissue, brains were fixed and serially sliced. Four days post exposure, a local damage was mainly observed (Fig. 15.2a–b). The damaged appeared as gradual loss of tissue which in some parts of the brain was quite excessive. At the edges of the damaged area, immune brain system cells were detected adjacent to cells containing particles resembling viral factories (Fig. 15.2c). At the 5th day post infection, the damage expanded and was detected in larger parts of the brain. The damage was accompanied with hematomas (Fig. 15.2d). At the damaged area the virus was detected immunohistochemically using specific anti-vaccinia antibodies (Fig. 15.2e). Within the damaged area, apoptotic cells were detected using TUNEL staining (Fig. 15.2f). The above markers were not observed neither in control mock injected nor in

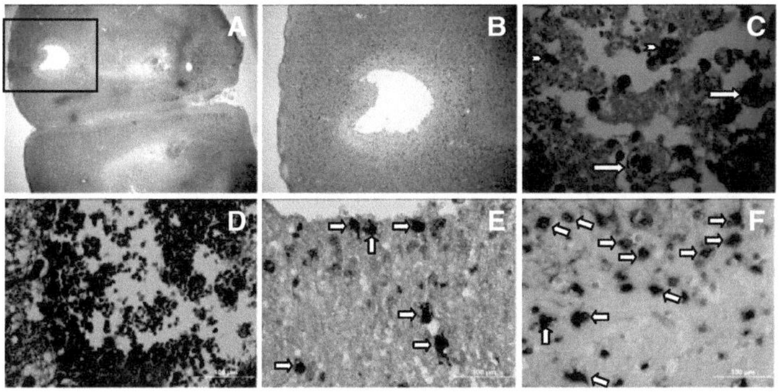

Fig. 15.2 Histopathological brain sections of VACV-WR infected animals. Mice were infected with 1×10^2 pfu and examined 4 days (**a–c**) and 5 days (**d–f**) post infection. (**a–b**) Local tissue damage was observed 4 days post infection (**a**) box which is magnified in (**b**). (**c**) The edges of the damage area contained immune cells (*arrow*) and viral factories (*arrow heads*). (**d**) Hematoma in a damage area 5 days post infection. (**e**) Positive viral stain at the damaged area (*arrows*). (**f**) Positive apoptotic cells detected by TUNEL at the damaged area (*arrows*)

VACV-Lister injected brains, suggesting an active damage induced by VACV-WR replication rather than by the injection process.

15.2.3 Determination of MMP-9 in Mice Serum

In order to track systemic markers as possible markers for encephalitis, sera were examined for total MMP-9 concentration at different time points post inoculation by ELISA (R&D Systems). The levels of serum MMP-9 in VACV-Lister infected animals remained as in the control of uninfected animals, indicating that neither the injection process nor the introduction of a live virus resulted in changes in serum MMP-9. In contrast, at the first 2 days following VACV-WR infection, a significant decrease in serum MMP-9 was detected (Fig. 15.3). Then, on day 3 post infection, a significant increase in the serum MMP-9 of these animals was detected compared to day 2. Along with disease progression, at day 5, one day before death, the concentration of the serum MMP-9 peaked and was significantly higher compared to either naïve or VACV-Lister infected groups ($P < 0.05$) (Fig. 15.3). It is important to emphasize that both infected animals exhibited similar weight lost and yet between these two groups, serum MMP-9 levels differed significantly (Fig. 15.3). These results imply that VACV-WR cause a possible damage to BBB integrity.

15.2.4 Examination of BBB Integrity by Evans Blue

In order to examine the BBB integrity, Evans blue (EB) dye was injected systemically for 1 h and brains were examined following systemic perfusion with PBS (Fig. 15.4). In agreement with the MMP-9 levels the BBB remained intact during

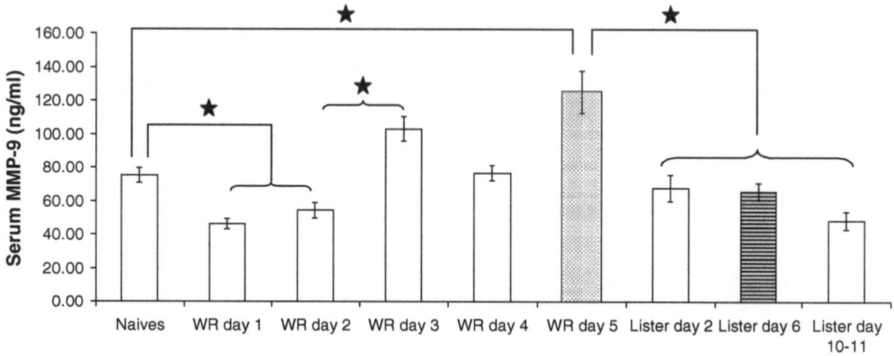

Fig. 15.3 Levels of MMP-9 in serum samples of mice infected i.c. with either VACV-WR or VACV-Lister. Quantitative determination of mouse MMP-9 in serum samples were done by ELISA (R&D Systems). Each bar represents at-least 10 different animal serum. Doted and striped bars represent the time of peak morbidity in the VACV-WR and VACV-Lister infected animals, respectively. Asterisk denote for $P < 0.05$

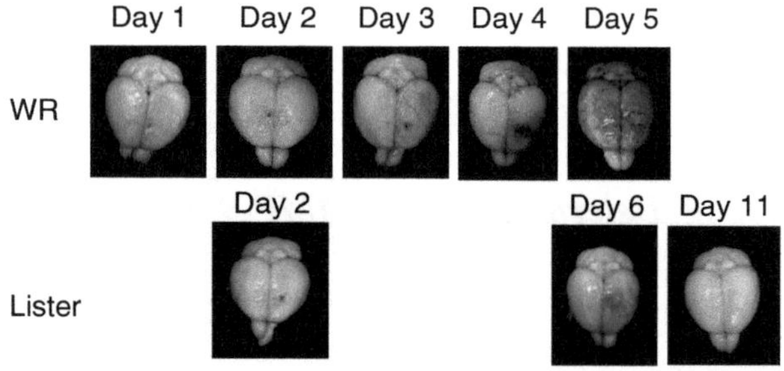

Fig. 15.4 BBB integrity examined by Evans Blue (EB) systemic injection. Mice infected with VACV-WR (*top panel*) or VACV-Lister (*bottom panel*) were IP injected with EB at various days post infections. After an hour brains were perfused with PBS, dissected and examined following fixation

the first 2 days following VACV-WR IC inoculation. Already 3 days post inoculation some signs of EB staining could be detected at the site of inoculation and the positive staining EB area expanded at the following days. Five days post inoculation a dramatic increase of EB staining was detected throughout the brain implying on major damage to the BBB. In comparison, the brains of the VACV-Lister infected animals did not show any signs of EB staining up until day 6 in which light staining was observed. Eleven days post infection, when mice resumed their weight, EB staining was no longer observed (Fig. 15.4).

15.3 Discussion

The currently approved smallpox vaccines are associated with rare but severe adverse reactions, and post vaccinal encephalitis is one of them. Accordingly, vaccine lots are tested for their encephalitis potential to ensure their safety (EMEA).

In the current work we aimed to identify markers for poxvirus induced encephalitis in mice after i.c. inoculation. We demonstrate that elevated serum MMP-9 is associated with elevated viral load in the brain, increased BBB permeability and brain tissue damage. These markers were obtained following intracranial infection of mice with the VACV-WR strain but not with the vaccine strain VACV-Lister. VACV-WR, a widely used challenge strain was originated from the vaccine strain NYCBH and was mouse brain adapted to yield a neurovirulent strain. The use of VACV-WR injected directly into the brain allowed for induction of defined and consistent encephalitis, associated with consistent morbidity and mortality. In contrast, VACV-Lister even when introduced at much higher levels poorly replicated in the brain, if at all, and mice developed mild disease and survived the infection. The use of direct inoculation of viruses into the brain allowed for the discrimination between neurovirulence and non-neurovirulence viral infection of the brain and to exclude effects of virus spread and involvement of other tissues in the early stages of infection allowing for the screening of differential serum encephalitic markers. MMP-9 raise in sera is well documented following tissue damage and increased permeability of the BBB, both markers were observed following injection of VACV-WR but not VACV-Lister even though mice lost weight and looked morbid. Yet, morbidity associated with death and neurological defect behavioral symptoms were only observed upon infection with VACV-WR (data not shown). Preliminary results indicated that serum MMP-9 levels are not elevated in other poxvirus induced non-encephalitic diseases allowing for non-invasive monitoring of neurovirulence in infected mice.

The genetic background rendering people as having high risk to develop post vaccinal encephalitis is unknown, yet the higher incidence of encephalitis in small children and the elderly population calls for weak immune response as a possible indication allowing for virus spread from the infection site to the brain. In our work we have been using immune competent mice (BALB/c) to which the Lister vaccine strain is attenuated even if the virus is directly introduced to the brain (sporadic deaths can be observed only when injected doses are 1×10^6 pfu and higher, data not shown). Interestingly, in VACV-Lister infected mice, light and transient Evans blue staining was observed 6 days post infection, which was followed by recovery. Mice not only survived brain infection with VACV-Lister, but these mice developed rapid protective humoral immune response (data not shown), indicating for a role of the immune system in virus clearance from the brain and prevention of encephalitis. Whether the transient EB staining indicating for transient increased BBB permeability allows for the efficient protective immunity in those mice remains elusive. Overall, it seems that while in the VACV-Lister the immune system is under control and neutralizes the virus, in the VACV-WR the virus is in command. At the first days a drop in the MMP-9 can be attributed to a viral strategy in which the virus creating a sterile area for propagation and at the peak of the disease, in correlation with the high viral titer and MMP-9 the virus is spreading out using destruction of the BBB. In agreement to that, VACV-WR in the blood of these animals was detected only from day four (data not shown). Currently, there aren't any established models for encephalitis in mice following dermal vaccination with the vaccine strains. Whether immune deficient mice will be more susceptible to development of encephalitis with the vaccine strain and whether serum MMP-9 will be elevated similarly to immune competent mice is yet to be discovered.

Acknowledgments We wish to thank Dr. Shira Chapman, Dr. Tamar Kadar and Dr. Trevor Waner for their valuable contribution to this study.

References

Bever, C. T., Jr. & Rosenberg, G. A. (1999) Matrix metalloproteinases in multiple sclerosis: targets of therapy or markers of injury? *Neurology,* 53, 1380–1.

Birkedal-Hansen, H. (1993) Role of matrix metalloproteinases in human periodontal diseases. *J Periodontol,* 64, 474–84.

Bonoiu, A., Mahajan, S. D., Ye, L., Kumar, R., Ding, H., Yong, K. T., Roy, I., Aalinkeel, R., Nair, B., Reynolds, J. L., Sykes, D. E., Imperiale, M. A., Bergey, E. J., Schwartz, S. A. & Prasad, P. N. (2009) MMP-9 gene silencing by a quantum dot-siRNA nanoplex delivery to maintain the integrity of the blood brain barrier. *Brain Res,* 1282, 142–55.

Booss, J. & Davis, L. E. (2003) Smallpox and smallpox vaccination: neurological implications. *Neurology,* 60, 1241–5.

Boz, C., Ozmenoglu, M., Velioglu, S., Kilinc, K., Orem, A., Alioglu, Z. & Altunayoglu, V. (2006) Matrix metalloproteinase-9 (MMP-9) and tissue inhibitor of matrix metalloproteinase (TIMP-1) in patients with relapsing-remitting multiple sclerosis treated with interferon beta. *Clin Neurol Neurosurg,* 108, 124–8.

Canete Soler, R., Gui, Y. H., Linask, K. K. & Muschel, R. J. (1995) MMP-9 (gelatinase B) mRNA is expressed during mouse neurogenesis and may be associated with vascularization. *Brain Res Dev Brain Res,* 88, 37–52.

Casey, C. G., Iskander, J. K., Roper, M. H., Mast, E. E., Wen, X. J., Torok, T. J., Chapman, L. E., Swerdlow, D. L., Morgan, J., Heffelfinger, J. D., Vitek, C., Reef, S. E., Hasbrouck, L. M., Damon, I., Neff, L., Vellozzi, C., Mccauley, M., Strikas, R. A. & Mootrey, G. (2005) Adverse events associated with smallpox vaccination in the United States, January-October 2003. *Jama,* 294, 2734–43.

CDC (2004) Centers for Disease Control and Prevention (CDC). Update: adverse events following civilian smallpox vaccination-United States, 2003. *Morb Mortal Wkly Rep,* 53, 106–7.

Chen, K. M., Lee, H. H., Chou, H. L., Liu, J. Y., Tsai, B. C. & Lai, S. C. (2005) Upregulation of MMP-9/TIMP-1 enzymatic system in eosinophilic meningitis caused by Angiostrongylus cantonensis. *Int J Exp Pathol,* 86, 81–9.

Del Bigio, M. R. & Jacque, C. M. (1995) Localization of proteinase expression in the developing rabbit brain. *Brain Res Dev Brain Res,* 86, 345–7.

Fenner, F., Henderson, D. A., Arita, I., Jezek, Z. & Ladnyi, I. D. (1988) Smallpox and its Eradication. World Health Organization, Geneva, Switzerland.

Henderson, D. A. & Moss, B. (1999) Smallpox and vaccinia. In Plotkin, S. A. & Orenstein, W. A. (eds.), Vaccines, W.B Saunders Company, Philadelphia, PA.

Ichiyama, T., Kajimoto, M., Suenaga, N., Maeba, S., Matsubara, T. & Furukawa, S. (2006) Serum levels of matrix metalloproteinase-9 and its tissue inhibitor (TIMP-1) in acute disseminated encephalomyelitis. *J Neuroimmunol,* 172, 182–6.

Ichiyama, T., Morishima, T., Kajimoto, M., Matsushige, T., Matsubara, T. & Furukawa, S. (2007a) Matrix metalloproteinase-9 and tissue inhibitors of metalloproteinases 1 in influenza-associated encephalopathy. *Pediatr Infect Dis J,* 26, 542–4.

Ichiyama, T., Siba, P., Suarkia, D., Takasu, T., Miki, K., Kira, R., Kusuhara, K., Hara, T., Toyama, J. & Furukawa, S. (2007b) Serum levels of matrix metalloproteinase-9 and tissue inhibitors of metalloproteinases 1 in subacute sclerosing panencephalitis. *J Neurol Sci,* 252, 45–8.

Keogh, B., Sheahan, B. J., Atkins, G. J. & Mills, K. H. (2003) Inhibition of matrix metalloproteinases ameliorates blood-brain barrier disruption and neuropathological lesions caused by avirulent Semliki Forest virus infection. *Vet Immunol Immunopathol,* 94, 185–90.

Lane, J. M., Ruben, F. L., Abrutyn, E. & Millar, J. D. (1970) Deaths attributable to smallpox vaccination, 1959 to 1966, and 1968. *Jama,* 212, 441–4.

Martinez-Torres, F. J., Wagner, S., Haas, J., Kehm, R., Sellner, J., Hacke, W., Meyding-Lamade, U. (2004) Increased presence of matrix metalloproteinases 2 and 9 in short- and long-term experimental herpes simplex virus encephalitis, *Neuroscience letters*, 368, 274–278.

Mishra, M. K., Dutta, K., Saheb, S. K. & Basu, A. (2009) Understanding the molecular mechanism of blood-brain barrier damage in an experimental model of Japanese encephalitis: correlation with minocycline administration as a therapeutic agent. *Neurochem Int,* 55, 717–23.

Overall, C. M. & Lopez-Otin, C. (2002) Strategies for MMP inhibition in cancer: innovations for the post-trial era. *Nat Rev Cancer,* 2, 657–72.

Romanic, A. M. & Madri, J. A. (1994) Extracellular matrix-degrading proteinases in the nervous system. *Brain Pathol,* 4, 145–56.

Rosenberg, G. A., Dencoff, J. E., Mcguire, P. G., Liotta, L. A. & Stetler-Stevenson, W. G. (1994) Injury-induced 92-kilodalton gelatinase and urokinase expression in rat brain. *Lab Invest,* 71, 417–22.

Scott, T. F. (1967) Postinfectious and vaccinal encephalitis. *Med Clin North Am,* 51, 701–17.

Sejvar, J. J., Labutta, R. J., Chapman, L. E., Grabenstein, J. D., Iskander, J. & Lane, J. M. (2005) Neurologic adverse events associated with smallpox vaccination in the United States, 2002–2004. *Jama,* 294, 2744–50.

Suenaga, N., Ichiyama, T., Kubota, M., Isumi, H., Tohyama, J. & Furukawa, S. (2008) Roles of matrix metalloproteinase-9 and tissue inhibitors of metalloproteinases 1 in acute encephalopathy following prolonged febrile seizures. *J Neurol Sci,* 266, 126–30.

Svedin, P., Hagberg, H., Savman, K., Zhu, C. & Mallard, C. (2007) Matrix metalloproteinase-9 gene knock-out protects the immature brain after cerebral hypoxia-ischemia. *J Neurosci,* 27, 1511–8.

Wang, P., Dai, J., Bai, F., Kong, K. F., Wong, S. J., Montgomery, R. R., Madri, J. A. & Fikrig, E. (2008) Matrix metalloproteinase 9 facilitates West Nile virus entry into the brain. *J Virol,* 82, 8978–85.

Waubant, E., Goodkin, D. E., Gee, L., Bacchetti, P., Sloan, R., Stewart, T., Andersson, P. B., Stabler, G. & Miller, K. (1999) Serum MMP-9 and TIMP-1 levels are related to MRI activity in relapsing multiple sclerosis. *Neurology,* 53, 1397–401.

WHO (1980) World Health Organization. The global eradication of smallpox: final report of the global commission for the certification of smallpox eradication. History of International Public Health No. 4. World Health Organization, Geneva.

Yong, V. W., Zabad, R. K., Agrawal, S., Goncalves Dasilva, A. & Metz, L. M. (2007) Elevation of matrix metalloproteinases (MMPs) in multiple sclerosis and impact of immunomodulators. *J Neurol Sci,* 259, 79–84.

Chapter 16
Epidemic Cowpox Virus Infections in Germany

Andreas Kurth and Andreas Nitsche

Abstract Several zoonotic infections by orthopoxviruses represent a potential threat to humans today. While monkeypox is endemic in certain areas of Africa, infections with vaccinia-like viruses occur naturally in Brazil and India (Buffalopox). Cowpox (caused by CPXV) used to be enzootic in cattle in Europe. However, no infections in cattle were diagnosed in Germany over the last decades. Instead, individual cases of CPXV infections are increasingly detected in cats and their owners. Both cats and humans commonly present local exanthema on arms and legs or in the face. Although cowpox is generally regarded as a self-limiting disease, immunosuppressed patients can develop lethal systemic disease resembling variola virus infection. Here we report the cumulated incidence of cowpox transmitted from rats to humans, with a focus on pet rats.

Keywords Poxvirus · Cowpox virus · Zoonoses · Pet rats · Rats

16.1 Poxviruses

The earliest classification of the family *Poxviridae* was based on symptoms characterized by pocks on the skin of vertebrates (Fields et al., 1996; Fenner, 2006). Of the eight genera within the subfamily *Chordopoxvirinae*, only four contain human-pathogenic poxviruses, namely *Orthopoxvirus*, *Parapoxvirus*, *Molluscipoxvirus* and *Yatapoxvirus* (see Table 16.1). The most dangerous ones were infections with the orthopoxvirus variola (Moore et al., 2006) which was eradicated by vaccination in the last century (Fenner et al., 1988). Currently, infections with monkeypox, vaccinia or cowpox virus can also be fatal, depending on the immunological state of the infected individual (Lewis-Jones, 2004). Infections with parapoxviruses, molluscipoxviruses and yatapoxviruses are usually self-limited and mostly remain localized (Fields et al., 1996). While in the past vaccination and administration of

---

A. Kurth (✉)
Centre for Biological Safety 1, German Consultant Laboratory for Poxviruses, Robert Koch Institute, Nordufer 20, 13353 Berlin, Germany
e-mail: kurtha@rki.de

A. Shafferman et al. (eds.), *The Challenge of Highly Pathogenic Microorganisms*,
DOI 10.1007/978-90-481-9054-6_16, © Springer Science+Business Media B.V. 2010

Table 16.1 Human-pathogenic poxviruses

Genus	Species	Abbreviation	Non-human hosts
Orthopoxviruses	Variola virus	VARV	
	Monkeypox virus	MPXV	Primates
	Cowpox virus	CPXV	Cats, cattle, exotic animals, rodents
	Vaccinia virus	VACV	Rabbits, cattle, water buffalos
Parapoxvirus	Orf virus	ORFV	Sheep, goats, ruminants, artiodactyls
	Pseudocowpox virus	PCPV	Cattle
	Bovine papular stomatitis virus	BPSV	Cattle
Molluscipoxvirus	Molluscum contagiosum virus	MOCV	
Yatapoxvirus	Tanapox virus	TANV	Rodents
	Yaba monkey tumor virus	YMTV	Primates

Modified from ICTVdb

vaccinia immunoglobulins were the only options for prevention and therapy of variola virus infections, today the panel of available antipoxviral drugs is expanded by either the FDA-approved drug cidofovir or the new small compound ST-246 (Yang et al., 2005) which was shown to be a potent treatment option in a recent vaccinia virus transmission from a vaccinee to an infant (Vora et al., 2008). However, to enable a reliable risk assessment of suspected cases and, if necessary, a timely treatment of the patient, the diagnosis of the causative agent of poxvirus-like lesions and its clear differentiation from variola virus are the most important tasks in poxvirus diagnostics (Kurth and Nitsche, 2007).

16.2 Orthopoxvirus (OPV)

Variola virus (VARV, smallpox) is nowadays the most relevant virus in the context of possible biological crime and bioterrorism (Beeching et al., 2002). Although it has been eradicated from nature by vaccination with vaccinia virus, the virus is still stored in two selected WHO depositories, and illegal possession of VARV is discussed (Halloran et al., 2002). A deliberate release of VARV could have catastrophic consequences in today's increasingly unvaccinated population (Madeley, 2003; Halloran et al., 2002; Whitley, 2003). Smallpox was transmitted naturally by direct contact with a diseased individual or contaminated material. So far, no natural reservoir for VARV is known, a fact that rendered the eradication by immunization of the human population possible. Skin eruption appeared after 3–4 days, developing through several stages of macule, papule, vesicle and pustule within 5 or 6 days with a characteristic distribution involving the face and extremities. In the last century mortality was estimated to be 30–40%.

Human infections by monkeypox virus (MPXV) have been found in the rainforest areas of west and central Africa (Di Giulio and Eckburg, 2004), although one outbreak with a zoonotic transmission involving 72 patients occurred in the USA in 2003 (Centers for Disease Control and Prevention, 2003). Rodents are discussed to be a natural reservoir for MPXV. The clinical features of human monkeypox are similar to those of smallpox and therefore difficult to differentiate. Hence, after the eradication of smallpox, MPXV is considered the most severe human infection with poxviruses.

Vaccinia virus (VACV) infection-related complications in humans have been reported since the beginning of vaccination (Fulginiti et al., 2003). Although, like for smallpox, no natural reservoir for VACV is known so far, rare but possibly serious complications following vaccination or contact with a recent vaccinee are documented. These complications include erythema multiforme, congenital vaccinia, generalized vaccinia, progressive vaccinia, postvaccinal encephalitis and eczema vaccinatum. Recently, in Brazil natural human infections with VACV were observed (Nagasse-Sugahara et al., 2004). These viruses were closely related to VACV strains formerly used for vaccination and must have survived over decades in unknown animal reservoirs.

Cowpox virus (CPXV) received its name as a result of its association with pustular lesions on the teats of cows and the hands of milkers, although infections in cows are not common (Chantrey et al., 1999). CPXV is supposedly the virus used by Edward Jenner in his first vaccinations against smallpox. In contrast to smallpox, infections with CPXV display a broad host range and are restricted to the Old World with wild rodents as their natural reservoir. Cowpox has been observed in several animal species like cats, rodents, dogs and exotic animals like elephants. Increased numbers of CPXV infections, also in humans, have been detected during the last decade. Possible reasons are an increased susceptibility to OPV infections after smallpox vaccination was abandoned, but also an increased awareness of physicians of this rare infectious disease. Today, human cases of CPXV infections are commonly described after transmission of CPXV from domestic cats (Coras et al., 2005) and, in a few rare cases, directly from rats. Infections in humans usually remain self-limiting and localized on fingers, hands and the face but can become fatal after generalization, particularly in immunosuppressed patients (Eis-Hubinger et al., 1990; Bennett and Baxby, 1996).

CPXV have the largest genome of all OPV and share the highest number of genes found in OPV. Wide regions of the CPXV genome have been shown to be highly homologous to VARV but, in addition, CPXV are genetically very diverse (Meyer et al., 2005). Due to these facts, the pathogenic potential of CPXV is not entirely understood.

Interestingly, CPXV isolated from humans or animals have generally been shown to be case-specific and genetically diverse. According to the HA gene sequence, each individual cowpox case is caused by a virus strain at least slightly varying when compared to other cases even from the same geographical area.

16.3 Diagnostics

Besides virus propagation in cell culture or on the CAM of embryonated chicken eggs, electron microscopy used to be the first-line diagnostic tool during the smallpox eradication era in the last century. While propagation of poxviruses is laborious and time consuming, electron microscopy is rapid, but lacks sensitivity (Hazelton and Gelderblom, 2003). Therefore, over the past 10 years, these powerful methods have been complemented or replaced by PCR- and real-time PCR-based approaches, which display a tremendous sensitivity and are fast and easy to establish in a diagnostic laboratory (Mackay et al., 2002). Several real-time PCR-based assays have been published that identify and type OPV in less than two hours after the specimen's arrival in the lab, either by specific amplification or specific detection of the virus species, or by fluorescence melting curve analysis following the PCR reaction (Olson et al., 2004; Nitsche et al., 2004). Recently, pyrosequencing-based techniques have found their way into rapid viral typing. However, the best PCR-based approach established to identify OPV species is by sequencing the open reading frame of the hemagglutinin gene and comparison of the obtained sequences to the 193 sequences that have been published in Genbank so far (Damaso et al., 2007). It is important to note that all these PCR-based approaches require an exact knowledge of the target sequences used for identification and typing of virus species. Since CPXV are highly diverse, no CPXV-specific assays have been published so far that recognise all CPXV isolates and discriminate them from all other OPV.

16.4 CPXV Infections Transmitted from Rats

As described occasionally in the literature, it was assumed that rodents transmit CPXV directly to humans and other vertebrates by direct contact, like biting or scratching (Honlinger et al., 2005; Postma et al., 1991; Wolfs et al., 2002). These cases have been attributed to incidental contacts of humans to rodents. Over the last two years we observed several cowpox cases that were verifiably transmitted either by wild rats (elephant infection), pet rats (human infection) and feeder rats (animal and human infection).

As reported previously, in the spring of 2007 a circus elephant was infected with CPXV in northern Germany (Kurth et al., 2008). After generalisation of the infection the elephant had to be euthanized. The source of infection was not clear at the time, but the animal keepers reported the sighting of numerous rats in the circus area over the past weeks, requiring a professional exterminator. Four rats could be captured and investigated by PCR and serology. All rats were found to be serologically positive for OPV with three out of the four rats being also PCR-positive for the same CPXV variant as the elephant. This underlines that the infection was probably transmitted to the elephant by wild rats. The infection of one animal keeper who was in charge of the elephant's care during the disease was probably transmitted from the elephant.

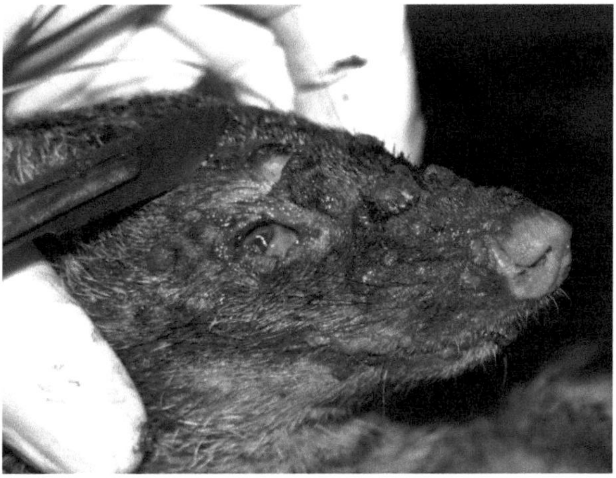

Fig. 16.1 Acute lesions on the head of a mongoose with a generalized infection

In February 2008 we observed an outbreak of fatal CPXV infections (Fig. 16.1) in banded mongooses (*Mungos mungo*) and jaguarundis (*Herpailurus yagouaroundi*) at a zoological garden in Krefeld (Kurth et al., 2009). Again, according to the HA sequence, this CPXV was a new variant. Interestingly, the same virus variant was identified later in two time-delayed human CPXV infections occurring in two different geographical areas of Germany. One human case was observed seven months later in September 2008 in an employee (Fig. 16.2) at a private reptile zoo more than 300 km away from the zoo in Krefeld. The patient developed a lesion in the face which was not immediately identified as cowpox. Since the patient worked in the office the infection could not be traced back to direct contact of the patient with diseased animals. The second human case caused by the same CPXV variant was found 15 months after the mongoose infection in Krefeld in a girl working in falconry, also about 300 km away from Krefeld. She used to feed frozen rodents to falcons and developed a typical CPXV lesion on her hand. Eventually, we were able to link all three cowpox cases to one supplier providing the zoos and the falconry with feeder rats.

Starting in March 2008 further human cowpox cases (Fig. 16.3) were diagnosed also in the area of Krefeld (Becker et al., 2009). All patients had a case history revealing close contact to pet rats. The rats were either symptomatically or asymptomatically infected (Fig. 16.4). In total, crust material, swabs and serum from 16 patients presenting skin lesions typical of cowpox were subjected to either real-time PCR assays specific for human-pathogenic poxviruses or to serology testing by immunofluorescence staining of CPXV-infected cells with the patient's serum. OPV-positive specimens were typed by sequencing of the complete hemagglutinin ORF. All patients were PCR positive for OPV only and the virus could be identified as an identical CPXV strain, although no direct connection could be established between the patients. However, in comparison to recent CPXV infections found in Germany, including the elephant and the mongoose cases, this virus represents a

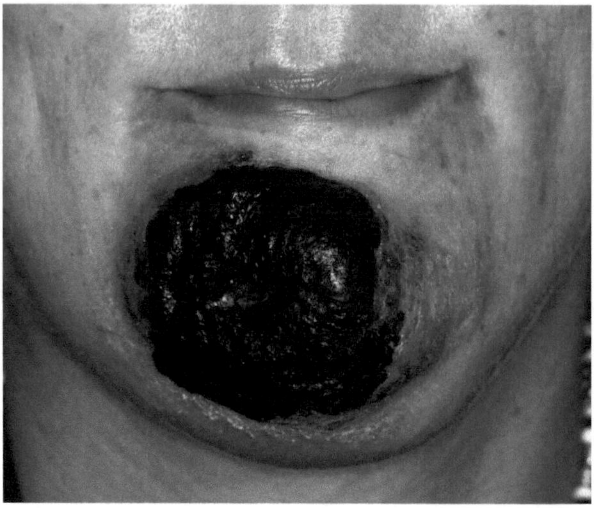

Fig. 16.2 Severe cowpox lesion on the patient's chin caused by the identical virus strain that was isolated from deceased mongooses and jaguarundis

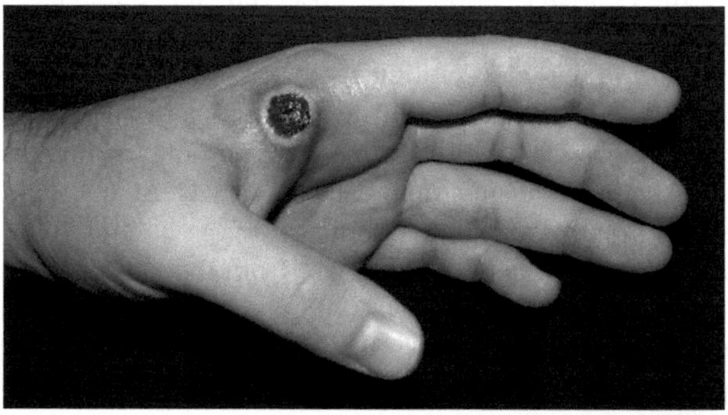

Fig. 16.3 Typical cowpox infection in a human. Subacute stage on the back of the hand

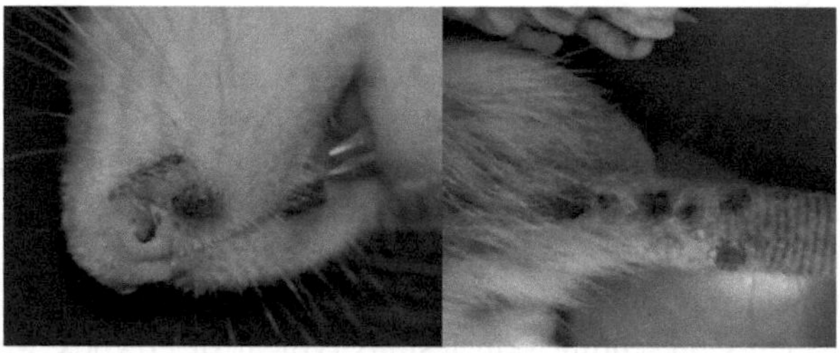

Fig.16.4 Cowpox infection in pet rats. Characteristic circumscribed pox lesions on nose and tail base

new strain. In addition, all patients showed significant OPV-specific antibody titers. As shown later, the same CPXV variant caused pet rat-related CPXV infections in southern Germany and France (Ninove et al., 2009; Campe et al., 2009). Our findings point to a common source of infection of the pet rats which has not been identified so far.

16.5 Discussion

In summary, we could show that the same CPXV strain was transmitted from rats to animals or humans over a time period of six months for pet rats and more than one year for frozen feeder rats. Moreover, in the area of Krefeld we observed the co-circulation of different CPXV strains at the same time. The infection route into the pet rat and feeder rat populations is not known.

The pathogenic potential of this CPXV strain can not be predicted yet. Due to the genomic homologies to VARV and the fact that CPXV are co-circulating in various animals, the emergence of a highly pathogenic CPXV strain and potentially human-to-human transmission can not be excluded. The recent findings have clearly shown that the pathogenicity of CPXV is dependent on the interaction with the host. While certain CPXV strains are highly virulent in one species, they may not be in another one. Sequence data of whole genomes of CPXV isolates may contribute to a better understanding of this virus–host interaction.

While contact to wild rats is usually rare and can be controlled, feeder rats represent a risk to other animals and to humans working with these animals. In this context, animal suppliers should be discussed as potential sources for pan-European transmission and potential multiplication of CPXV by distributing infected feeder animals. Finally, the increasing popularity of pet rats implicates the risk of CPXV transmission to humans. The screening of pet rats for poxviruses should be discussed in the future.

References

Becker, C., Kurth, A., Hessler, F., Kramp, H., Gokel, M., Hoffmann, R., Kuczka, A. & Nitsche, A. (2009) Cowpox virus infection in pet rat owners: not always immediately recognized. *Dtsch Arztebl Int*, 106, 329–34.

Beeching, N. J., Dance, D. A., Miller, A. R. N. & Spencer, R. C. (2002) Biological warfare and bioterrorism. *BMJ*, 324, 336–9.

Bennett, M. & Baxby, D. (1996) Cowpox. *J.Med Microbiol*, 45, 157–8.

Campe, H., Zimmermann, P., Glos, K., Bayer, M., Bergemann, H., Dreweck, C., Graf, P., Weber, B. K., Meyer, H., Buttner, M., Busch, U. & Sing, A. (2009) Cowpox virus transmission from pet rats to humans, Germany. *Emerg Infect Dis*, 15, 777–80.

Centers for Disease Control and Prevention. (2003) Multistate outbreak of monkeypox-Illinois, Indiana, and Wisconsin, 2003. *MMWR Morb Mortal Wkly Rep*, 52, 537–40.

Chantrey, J., Meyer, H., Baxby, D., Begon, M., Bown, K. J., Hazel, S. M., Jones, T., Montgomery, W. I. & Bennett, M. (1999) Cowpox: reservoir hosts and geographic range. *Epidemiol Infect*, 122, 455–60.

Coras, B., Essbauer, S., Pfeffer, M., Meyer, H., Schroder, J., Stolz, W., Landthaler, M. & Vogt, T. (2005) Cowpox and a cat. *Lancet*, 365, 446.

Damaso, C. R., Reis, S. A., Jesus, D. M., Lima, P. S. & Moussatche, N. (2007) A PCR-based assay for detection of emerging vaccinia-like viruses isolated in Brazil. *Diagn Microbiol Infect Dis*, 57, 39–46.

Di Giulio, D. B. & Eckburg, P. B. (2004) Human monkeypox: an emerging zoonosis. *Lancet Infect Dis*, 4, 15–25.

Eis-Hubinger, A. M., Gerritzen, A., Schneweis, K. E., Pfeiff, B., Pullmann, H., Mayr, A. & Czerny, C. P. (1990) Fatal cowpox-like virus infection transmitted by cat. *Lancet*, 336, 880.

Fenner, F., Henderson, D. A., Arita, I., Jezek, Z., & Ladnyi, I. D. (1988) Smallpox and its eradication, World Health Organization, Geneva.
Fenner, F. (2006) Medical virology Chapter 21 Poxviridae, Academic, New York.
Fields, B. N., Knipe, D. M. & Howley, P. M. (1996) Virology, Lippincott-Raven, Philadelphia, PA.
Fulginiti, V. A., Papier, A., Lane, J. M., Neff, J. M. & Henderson, D. A. (2003) Smallpox vaccination: a review, part II. Adverse events. *Clin Infect Dis*, 37, 251–71.
Halloran, M. E., Longini, I. M., Jr., Nizam, A. & Yang, Y. (2002) Containing bioterrorist smallpox. *Science.*, 298, 1428–32.
Hazelton, P. R. & Gelderblom, H. R. (2003) Electron microscopy for rapid diagnosis of infectious agents in emergent situations. *Emerg Infect Dis*, 9, 294–303.
Honlinger, B., Huemer, H. P., Romani, N., Czerny, C. P., Eisendle, K. & Hopfl, R. (2005) Generalized cowpox infection probably transmitted from a rat. *Br J Dermatol*, 153, 451–3.
Kurth, A. & Nitsche, A. (2007) Fast and reliable diagnostic methods for the detection of human poxvirus infections. *Future Virology*, 2, 467–79.
Kurth, A., Straube, M., Kuczka, A., Dunsche, A. J., Meyer, H. & Nitsche, A. (2009) Cowpox virus outbreak in banded mongooses (Mungos mungo) and jaguarundis (Herpailurus yagouaroundi) with a time-delayed infection to humans. *PLoS ONE*, 4, e6883.
Kurth, A., Wibbelt, G., Gerber, H.P., Petschaelis, A., Pauli, G. & Nitsche, A. (2008) Rat-to-Elephant-to-Human Transmission of Cowpox Virus. *Emerg InfectDis*, 14, 670–71.
Lewis-Jones, S. (2004) Zoonotic poxvirus infections in humans. *Curr Opin Infect Dis*, 17, 81–9.
Mackay, I. M., Arden, K. E. & Nitsche, A. (2002) Real-time PCR in virology. *Nucleic Acids Res*, 30, 1292–305.
Madeley, C. R. (2003) Diagnosing smallpox in possible bioterrorist attack. *Lancet*, 361, 97–8.
Meyer, H., Totmenin, A., Gavrilova, E. & Shchelkunov, S. (2005) Variola and camelpox virus-specific sequences are part of a single large open reading frame identified in two German cowpox virus strains. *Virus Res*, 108, 39–43.
Moore, Z. S., Seward, J. F. & Lane, J. M. (2006) Smallpox. *Lancet*, 367, 425–35.
Nagasse-Sugahara, T. K., Kisielius, J. J., Ueda-Ito, M., Curti, S. P., Figueiredo, C. A., Cruz, A. S., Silva, M. M., Ramos, C. H., Silva, M. C., Sakurai, T. & Salles-Gomes, L. F. (2004) Human vaccinia-like virus outbreaks in Sao Paulo and Goias States, Brazil: virus detection, isolation and identification. *Rev Inst Med Trop Sao Paulo.*, 46, 315–22.
Ninove, L., Domart, Y., Vervel, C., Voinot, C., Salez, N., Raoult, D., Meyer, H., Capek, I., Zandotti, C. & Charrel, R. N. (2009) Cowpox virus transmission from pet rats to humans, France. *Emerg Infect Dis*, 15, 781–4.
Nitsche, A., Ellerbrok, H. & Pauli, G. (2004) Detection of orthopoxvirus DNA by real-time PCR and identification of variola virus DNA by melting analysis. *J Clin Microbiol*, 42, 1207–13.
Olson, V. A., Laue, T., Laker, M. T., Babkin, I. V., Drosten, C., Shchelkunov, S. N., Niedrig, M., Damon, I. K. & Meyer, H. (2004) Real-time PCR system for detection of orthopoxviruses and simultaneous identification of smallpox virus. *J Clin Microbiol*, 42, 1940–6.
Postma, B. H., Diepersloot, R. J., Niessen, G. J. & Droog, R. P. (1991) Cowpox-virus-like infection associated with rat bite. *Lancet*, 337, 733–4.
Vora, S., Damon, I., Fulginiti, V., Weber, S. G., Kahana, M., Stein, S. L., Gerber, S. I., Garcia-Houchins, S., Lederman, E., Hruby, D., Collins, L., Scott, D., Thompson, K., Barson, J. V., Regnery, R., Hughes, C., Daum, R. S., Li, Y., Zhao, H., Smith, S., Braden, Z., Karem, K., Olson, V., Davidson, W., Trindade, G., Bolken, T., Jordan, R., Tien, D. & Marcinak, J. (2008) Severe eczema vaccinatum in a household contact of a smallpox vaccinee. *Clin Infect Dis*, 46, 1555–61.
Whitley, R. J. (2003) Smallpox: a potential agent of bioterrorism. *Antiviral Res*, 57, 7–12.
Wolfs, T. F., Wagenaar, J. A., Niesters, H. G. & Osterhaus, A. D. (2002) Rat-to-human transmission of Cowpox infection. *Emerg Infect Dis*, 8, 1495–6.
Yang, G., Pevear, D. C., Davies, M. H., Collett, M. S., Bailey, T., Rippen, S., Barone, L., Burns, C., Rhodes, G., Tohan, S., Huggins, J. W., Baker, R. O., Buller, R. L., Touchette, E., Waller, K., Schriewer, J., Neyts, J., Declercq, E., Jones, K., Hruby, D. & Jordan, R. (2005) An orally bioavailable antipoxvirus compound (ST-246) inhibits extracellular virus formation and protects mice from lethal orthopoxvirus Challenge. *J Virol*, 79, 13139–49.

Chapter 17
The Two Partner Secretion Transporters of *Yersinia pestis:* Cloning, Immunogenicity and In Vivo Expression Following Airway Infection

Yinon Levy, Anat Zvi, Haim Grosfeld, Avital Tidhar, David Gur, Yehuda Flashner, Eitan Elhanany, Emanuelle Mamroud, Sara Cohen, and Avigdor Shafferman

Abstract A bioinformatic screen of *Yersinia pestis* genome identified 11 TPS components, carrying the POlypeptide TRansport-Associated (POTRA) domain, found in other pathogens to be involved in immunogenicity and pathogenicity. All rTpsBs (transporters) cloned and expressed in *Escherichia coli* induced a humoral responses in immunized mice. Mice surviving intranasal infection with fully virulent *Y. pestis* Kimberley53 strain generated sera cross reacting with some recombinants TpsBs. Moreover, at least 3 TpsB polypeptides were directly detected in bacteria isolated following intranasal *Y. pestis* infection. All in vivo produced TpsBs appear to be truncated, missing about 75–85 amino acids. MALDI-TOF-MS analysis permitted to identify the truncation at a region overlapping a domain of a closely related *Bordetella pertussis* TpsB (FhaC), recently proposed to undergo a major conformational change that allows effector translocation.

Keywords *Yersinia pestis* · Bioinformatics · Airway infection · Two partner secretion · POTRA

17.1 Introduction

Yersinia pestis, the causative agent of plague, is classified by the CDC as agent of group A (Inglesby et al., 2000; Riedel, 2005), and a continuous research and efforts are conducted to develop new treatment modalities against the pathogen. To date, no efficient and safe vaccine is available. Vaccines based on the highly immunogenic antigens F1 and V, as well as a genetic fusion product, are currently under study and clinical trials (Anderson et al., 1998; Heath et al., 1998; Williamson

Y. Levy (✉)
Department of Biochemistry and Molecular Genetics, Israel Institute for Biological Research, P.O. Box 19, Ness-Ziona 74100, Israel
e-mail: levy-ynon@iibr.gov.il; yinonl@iibr.gov.il

Y. Levy and A. Zvi contributed equally to this work.

A. Shafferman et al. (eds.), *The Challenge of Highly Pathogenic Microorganisms*,
DOI 10.1007/978-90-481-9054-6_17, © Springer Science+Business Media B.V. 2010

et al., 2005). Attempts to identify virulence-related and/or surface exposed antigens as novel vaccine candidates were recently reported (Andrews et al., 1999; Benner et al., 1999; Leary et al., 1999; Garbom et al., 2004; Smither et al., 2007; Flashner et al., 2004).

Outer membrane proteins (OMPs) are in many cases virulence factors which participate in adherence and play an essential role in bacterial niche adaptation (Lin et al., 2002). One of the most potent OM antigens is the 80 kDa D15 protein, highly conserved in diverse *Haemophilus influenzae* strains. This antigen (as well as its related proteins) which was shown to confer immunoprotection, harbors a conserved domain denoted as POTRA (*Po*lypeptide-*Tr*ansport-*A*ssociated), a domain identified in the N-terminal region of protein families predominantly related to polypeptide transport (Sanchez-Pulido et al., 2003), and immunoprotection (Cameron et al., 2000; Ruffolo and Adler, 1996; Thomas et al., 2001). Another family of POTRA-containing proteins are transporters (TpsB) of the *T*wo *P*artner *S*ecretion (TPS) pathway, a pathway widely spread among Gram-negative bacteria (Jacob-Dubuisson et al., 2001). These transporters are located on the outer membrane, and translocate large, virulence-related exoproteins (up to ~500 kDa), denoted as TpsA. Here we report on a bioinformatic identification of 11 putative OMPs belonging to the TPS family that may be related to the pathogenesis of *Y. pestis*. The POTRA-containing, TpsB proteins were evaluated for their expression and immunogenicity both in vitro and in vivo following airway exposure of mice to a virulent *Y. pestis* strain.

17.2 Results and Discussion

17.2.1 Bioinformatic Mapping of Y. pestis *Outer Membrane Proteins and Selection of TpsBs*

A rational reductive strategy was employed in search for outer membrane proteins which could, by definition, be putatively involved in pathogenesis and be potential vaccine candidates. Of the total of 77 proteins predicted as OMPs (out of all 4,067 *Y. pestis* CO92 genome ORFs), 28 ORFs lack an ortholog in the closely related, nonpathogenic *E. coli* K12. Of these, 16 OMPs have no clear function, while most of the remaining proteins appear to be TPS components (six TpsB proteins, harboring a POTRA domain, and five TpsA; Table 17.1). We chose to evaluate first the antigenicity and immunoreactivity of the POTRA-containing TpsB proteins, which are known in several pathogens to be immunogenic and involved in virulence (Kurz et al., 2003; Nelson et al., 2001; Strauss et al., 1997).

17.2.2 Expression of rTpsBs in E. coli *and Evaluation of Their Imunoreactivity in Mice*

All six TpsBs were cloned from the fully virulent *Y. pestis* Kimberley53 strain and expressed in *E. coli* as full length proteins. The affinity-purified rTpsBs were

Table 17.1 Putative TPS components in the *Y. pestis* genome

TpsA (effector)			TpsB (transporter)		
Protein ID	Length (aa)	Annotation (Sanger[a])	Protein ID	Length (aa)	Annotation (Sanger[a])
YPO0599	3,296	Putative adhesion	**YPO0598**	563	Putative hemolysin activator
YPO2045	1,577	Putative hemolysin	**YPO2044**	558	Putative hemolysin activator
YPO2490	2,536	Putative hemolysin	**YPO2491**	563	Putative hemolysin activator
YPO3247[b]	1,911	Putative adhesion	**YPO3248**	597	Surface-exposed protein (RscB/HxuB)
YPO3721	1,636	Hemolysin	**YPO3720**	562	Hemolysin activator
–			**YPO4005**[c]	588	Putative hemolysin activator

[a]The annotation is according to the genomic sequence data (Parkhill et al., 2001), as deposited by the Sanger Institute: http://www.sanger.ac.uk/Projects/Y_pestis (except for YPO3248 which was determined in this study; HxuB from *H. influenzae* is translocating HxuA, the heme-hemopexin binding protein, via the TPS pathway)
[b]The effector YPO3247 was not included in the list of 28 OM selected ORFs since it was assigned as an extracellular protein localization by our subcellular prediction analysis
[c]YPO4005 does not have an identifiable adjacent "classical" effector in the genome, in term of size and function

observed as major product with apparent molecular weight of 58–59.5 kDa (Fig. 17.1a). Vaccination of mice with the six recombinant TpsBs generated relatively high titers of specific antibodies and a specific response (except for YPO2044; Fig. 17.1b). In addition, a cross reaction could be observed between the orthologous proteins YPO0598, YPO2491 and YPO3248 (Fig. 17.1b), designated as sero-group I, indicating the existence of common epitopes; the polypeptides belonging to sero-groups II (YPO3720) and III (YPO4005) do not cross react with any of the other TpsB members. The immunogenicity of the TpsB antigens during infection was further demonstrated by probing the different rTpsBs with antiserum collected from convalescent mice which were treated with α-F1 and α-V antibodies following an intranasal infection with the fully virulent *Y. pestis* Kimberley53 strain. Western Blot analysis reveals that this serum recognizes the recombinant proteins YPO0598, YPO2491 and YPO4005 and maybe also YPO3720 (Fig. 17.1c). Essentially, these findings provide the first evidence for actual expression of the TpsBs in *Y. pestis* during infection, in an environment encountered by the host.

Given these immunogenic properties of the POTRA-containing TpsB proteins in *Y. pestis*, we proceeded to assess their ability to confer immune protection against subcutaneous infection of fully virulent *Y. pestis*. Mice vaccinated with each of the various rTpsBs formulated in CFA and IFA, developed significant humoral response. Yet, upon challenge with 20 LD_{50} of the *Y. pestis* strain Kimberley53 strain (4 weeks after last boost), all vaccinated mice succumbed to the infection and their mean-time-to-death (MTTD) was not significantly different from that of the

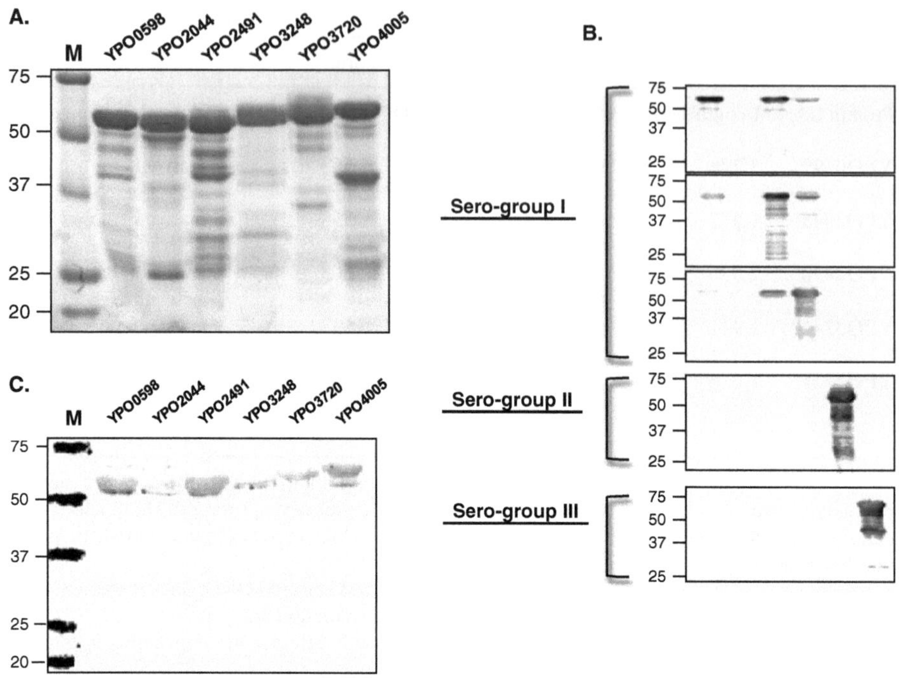

Fig. 17.1 Antigenicity and cross reactivity of the *Y. pestis* TpsB proteins. (**a**) Coomassie blue stained SDS-PAGE (10%) of affinity purified rTpsBs (5 μg/lane) and (**b**) Western blot analysis of resolved recombinant proteins using the corresponding serum generated in mice. (**c**) Western blot analysis of the six recombinant TpsB polypeptides probed with 1:100 diluted serum obtained from convalescent mice infected with the fully virulent *Y. pestis* Kimberley53

control mock vaccinated or naïve animals. To note, a recent study demonstrated that YPO3720 provides partial protection against challenge with a low dose of *Y. pestis* (Li et al., 2009).

17.2.3 Direct Evidence for Expression of TpsBs In Vivo in Y. pestis

In an attempt to directly assess expression in vivo of the TpsB proteins during infection, *Y. pestis* bacteria were collected from bronchoalveolar lavage fluids 24–48 h post airway infection (i.n) from animals inoculated with the virulent *Y. pestis* strain Kimberley53. This bacterial lysates exhibited a positive response with the anti-sera to the various rTpsBs tested (YPO0598, YPO03720 and YPO04005), indicating that the TpsB polypeptides belonging to the sero-reactive groups I, II and III are indeed "induced" in vivo following an airway infection. However, these in vivo expressed polypeptides appear as truncated approximately by 8.5–9.5 kDa (Fig. 17.2), implying that the TpsBs generated in vivo are susceptible to some

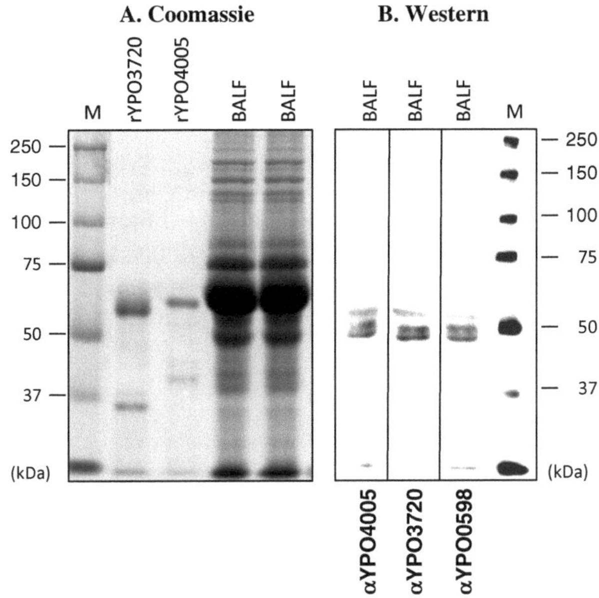

Fig. 17.2 Identification of in vivo expressed TpsBs from the BALF product. (**a**) Coomassie blue stained SDS-PAGE (7.5%) of recombinant TpsB proteins (*left lanes*), and the bronchoalveolar lavage fluid (BALF) isolated 48 h post i.n infection of mice with the 100 LD_{50} of the fully virulent *Y. pestis* Kimberely53 strain (*right lanes*, duplicated samples). (**b**) Western blot analysis of the bronchoalveolar lavage fluid (BALF), cross reacting with the specific anti-rTpsB sera, as indicated at the bottom of the panels

specific postranslation proteolytic processes. Comparison of the peptide map generated by MALDI-TOF-MS analysis of the in vivo product and the recombinant YPO0598 polypeptide revealed similar peptide maps except for two C-terminal peptides which were missing from the in vivo preparation (Fig. 17.3a). This difference would map the truncation of the latter to a region between amino acids 461–507 (Fig. 17.3b). Considering the apparent molecular weights of the in vivo expressed truncated YPO0598, we could further confine the region of the truncation to amino acids 477–486 at the C-terminal boundary of this TpsB. This region overlaps the L6 loop in the FhaC transporter of *B. pertussis*, a region proposed by the researchers to undergo a major conformational change that allows effector translocation (Clantin et al., 2007; Fig. 17.3b). It would therefore be tempting to speculate that the truncation of the TpsB isolated following infection is a reflection of the "open" conformation of TpsB, in which L6 is more susceptible to proteolysis, or alternatively, such a proteolytic process in this unique region facilitates the conformational change which is functionally important for the subsequent translocation of the TpsA effector.

To conclude, this study is the first report to show experimentally the expression of the TspB system in *Y. pestis* and its dependence on interaction with the host. Judging by the types of the TpsBs expressed in vivo and the specific pairwise interactions of TpsBs and TpsAs (Table 17.1), it may be inferred that in *Y. pestis*, following airway infection, both hemolysin and adhesion effectors could be translocated into host cells.

A. MALDI-TOF-based identification

A1. Recombinant YPO0598 (+HHHHHH)

1 ... 49

MSRFYYSILVAGVLLKIVAIPDASYGAELAPVQQSIHQQERQRALEER**LAPPTPDVRLSAPSAF**
FSRIIFPLETPCFVINRIK**ISGAEPLP**RWLPLQRIADQAQGQCLGAK**GINLLMSQMQNRLVDHG**
YVTTRVLAPQQDLNSGTLALNVVPGKIRGVELTPDSNR**YVTLFSAFPAR**AGTLLDLR**DIEQGLE**
NLQRVPTVQANMVLIPGSAPGETDIILNWQQRKMWRLAASLDDSGTRSTGR**YQGGATLFLDNPL**
SLSDLFYVSAGGALQRRGDKGTNNLTGHYSLPFGYWTAGMTASRYDYYQAVAGLNGDINYR**GES**
ENVAFQLSRLLHRNASQK**TTFTYDVLTR**SSK**NYINDTEVEVQR**RRTSAWRIGLQHR**HFISQAIL**
****
DAGISYQRGTRWFGAIPAQEEYFGEATALSKILRLNAQLDIPFVVMAQNLHYNLQYQR**QSTNTP**
****
LTPQDQFSIGGRWSVRGFNGERTLIADRGWWVR**NDIGWYLPLPGHELYVGVDYGEVGGR**SGAYL
LGRHLAGSAVGVRGNVLNTRYDLFAGKPLSKPNGFK**TDSLAVGFNLNWLY(HHHHHH)**

562

A2. *In vivo* expression - BAL product

1

MSRFYYSILVAGVLLKIVAIPDASYGAELAPVQQSIHQQERQRALEER**LAPPTPDVRLSAPSAF**
FSRIIFPLETPCFVINRIKISGAEPLPRWLPLQRIADQAQGQCLGAKGINLLMSQMQNR**LVDHG**
YVTTRVLAPQQDLNSGTLALNVVPGKIRGVELTPDSNR**YVTLFSAFPAR**AGTLLDLR**DIEQGLE**
NLQRVPTVQANMVLIPGSAPGETDIILNWQQRKMWRLAASLDDSGTRSTGR**YQGGATLFLDNPL**
SLSDLFYVSAGGALQRRGDKGTNNLTGHYSLPFGYWTAGMTASRYDYYQAVAGLNGDINYR**GES**
ENVAFQLSRLLHRNASQK**TTFTYDVLTR**SSK**NYINDTEVEVQR**RRTSAWRIGLQHR**HFISQAIL**
****
DAGISYQRGTRWFGAIPAQEEYFGEATALSKILRLNAQLDIPFVVMAQNLHYNLQYQR**QSTNTP**
****
LTPQDQFSIGGRWSVRGFNGERTLIADRGWWVRNDIGWYLPLPGHELYVGVDYGEVGGRSGAYL
LGRHLAGSAVGVRGNVLNTRYDLFAGKPLSKPNGFKTDSLAVGFNLNWLY

562

B. L6 suggested topology inferred from FhaC

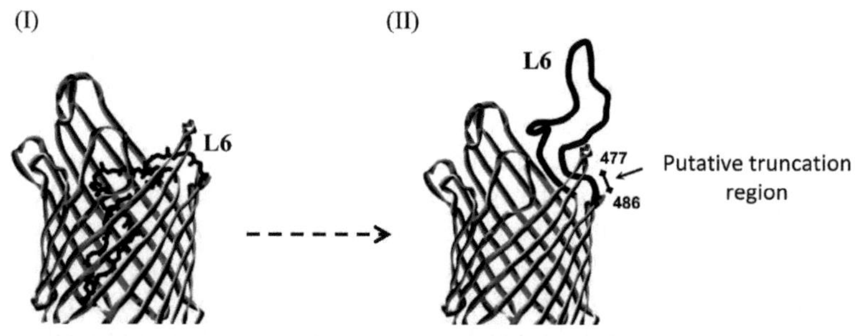

Fig. 17.3 Mass spectrometry identification of the recombinant and in vivo produced YPO0598. (**a**) Localization of identified peptides by MALDI-TOF-MS analysis of recombinant YPO0598 (A1) and in vivo BALF product (A2), on the protein sequence of YPO0598. Peptides identified are marked in *bold*, *underlined* and highlighted in *grey*. The sequence overlapping the FhaC loop designated as L6 (according to the FhaC *crystal structure*, Clantin et al., 2007) is marked by *asterisks*. The putative truncation region (amino acids 477–486) is marked by a *diamond-bound underline*. (**b**) *Ribbon representation* of the β-barrel region, based on the FhaC *crystal* structure ((I), Clantin et al., 2007), together with a schematic illustration of the suggested opening of the loop designated in FhaC as L6 (II). The location of the YPO0598 putative truncation region is indicated by an *arrow*

References

Anderson, G. W., JR., Heath, D. G., Bolt, C. R., Welkos, S. L. & Friedlander, A. M. (1998) Short- and long-term efficacy of single-dose subunit vaccines against *Yersinia pestis* in mice. *Am J Trop Med Hyg,* 58, 793–9.

Andrews, G. P., Strachan, S. T., Benner, G. E., Sample, A. K., Anderson, G. W., JR., Adamovicz, J. J., Welkos, S. L., Pullen, J. K. & Friedlander, A. M. (1999) Protective efficacy of recombinant *Yersinia* outer proteins against bubonic plague caused by encapsulated and nonencapsulated *Yersinia pestis. Infect Immun,* 67, 1533–7.

Benner, G. E., Andrews, G. P., Byrne, W. R., Strachan, S. D., Sample, A. K., Heath, D. G. & Friedlander, A. M. (1999) Immune response to *Yersinia* outer proteins and other *Yersinia pestis* antigens after experimental plague infection in mice. *Infect Immun,* 67, 1922–8.

Cameron, C. E., Lukehart, S. A., Castro, C., Molini, B., Godornes, C. & Van Voorhis, W. C. (2000) Opsonic potential, protective capacity, and sequence conservation of the *Treponema pallidum* subspecies *pallidum* Tp92. *J Infect Dis,* 181, 1401–13.

Clantin, B., Delattre, A. S., Rucktooa, P., Saint, N., Meli, A. C., Locht, C., Jacob-Dubuisson, F. & Villeret, V. (2007) Structure of the membrane protein FhaC: a member of the Omp85-TpsB transporter superfamily. *Science,* 317, 957–61.

Flashner, Y., Mamroud, E., Tidhar, A., Ber, R., Aftalion, M., Gur, D., Lazar, S., Zvi, A., Bino, T., Ariel, N., Velan, B., Shafferman, A. & Cohen, S. (2004) Generation of *Yersinia pestis* attenuated strains by signature-tagged mutagenesis in search of novel vaccine candidates. *Infect Immun,* 72, 908–15.

Garbom, S., Forsberg, A., Wolf-Watz, H. & Kihlberg, B. M. (2004) Identification of novel virulence-associated genes via genome analysis of hypothetical genes. *Infect Immun,* 72, 1333–40.

Heath, D. G., Anderson, G. W., JR., Mauro, J. M., Welkos, S. L., Andrews, G. P., Adamovicz, J. & Friedlander, A. M. (1998) Protection against experimental bubonic and pneumonic plague by a recombinant capsular F1-V antigen fusion protein vaccine . *Vaccine,* 16, 1131–7.

Inglesby, T. V., Dennis, D. T., Henderson, D. A., Bartlett, J. G., Ascher, M. S., Eitzen, E., Fine, A. D., Friedlander, A. M., Hauer, J., Koerner, J. F., Layton, M., Mcdade, J., Osterholm, M. T., O'Toole, T., Parker, G., Perl, T. M., Russell, P. K., Schoch-Spana, M. & Tonat, K. (2000) Plague as a biological weapon: medical and public health management. Working Group on Civilian Biodefense. *Jama,* 283, 2281–90.

Jacob-Dubuisson, F., Locht, C. & Antoine, R. (2001) Two-partner secretion in Gram-negative bacteria: a thrifty, specific pathway for large virulence proteins. *Mol Microbiol,* 40, 306–13.

Kurz, C. L., Chauvet, S., Andres, E., Aurouze, M., Vallet, I., Michel, G. P., Uh, M., Celli, J., Filloux, A., De Bentzmann, S., Steinmetz, I., Hoffmann, J. A., Finlay, B. B., Gorvel, J. P., Ferrandon, D. & Ewbank, J. J. (2003) Virulence factors of the human opportunistic pathogen *Serratia marcescens* identified by in vivo screening. *EMBO J,* 22, 1451–60.

Leary, S. E., Griffin, K. F., Galyov, E. E., Hewer, J., Williamson, E. D., Holmstrom, A., Forsberg, A. & Titball, R. W. (1999) *Yersinia* outer proteins (YOPS) E, K and N are antigenic but non-protective compared to V antigen, in a murine model of bubonic plague. *Microb Pathog,* 26, 159–69.

Li, B., Zhou, L., Guo, J. Y., Wang, X., Ni, B., Ke, Y., Zhu, Z., Guo, Z. & Yang, R. (2009) High-throughput identification of new protective antigens from a *Yersinia pestis* live vaccine by enzyme-linked immunospot assay. *Infect Immun,* 77(10), 4356–61.

Lin, J., Huang, S. & Zhang, Q. (2002) Outer membrane proteins: key players for bacterial adaptation in host niches. *Microbes Infect*, 4, 325–31.

Nelson, K. M., Young, G. M. & Miller, V. L. (2001) Identification of a locus involved in systemic dissemination of *Yersinia enterocolitica. Infect Immun,* 69, 6201–8.

Riedel, S. (2005) Plague: from natural disease to bioterrorism. *Proc (Bayl Univ Med Cent),* 18, 116–24.

Ruffolo, C. G. & Adler, B. (1996) Cloning, sequencing, expression, and protective capacity of the oma87 gene encoding the *Pasteurella multocida* 87-kilodalton outer membrane antigen. *Infect Immun,* 64, 3161–7.

Sanchez-Pulido, L., Devos, D., Genevrois, S., Vicente, M. & Valencia, A. (2003) POTRA: a conserved domain in the FtsQ family and a class of β-barrel outer membrane proteins. *Trends Biochem Sci,* 28, 523–6.

Smither, S. J., Hill, J., Van Baar, B. L., Hulst, A. G., De Jong, A. L. & Titball, R. W. (2007) Identification of outer membrane proteins of *Yersinia pestis* through biotinylation. *J Microbiol Methods,* 68, 26–31.

Strauss, E. J., Ghori, N. & Falkow, S. (1997) An *Edwardsiella tarda* strain containing a mutation in a gene with homology to shlB and hpmB is defective for entry into epithelial cells in culture. *Infect Immun,* 65, 3924–32.

Thomas, K. L., Leduc, I., Olsen, B., Thomas, C. E., Cameron, D. W. & Elkins, C. (2001) Cloning, overexpression, purification, and immunobiology of an 85-kilodalton outer membrane protein from *Haemophilus ducreyi*. *Infect Immun,* 69, 4438–46.

Williamson, E. D., Flick-Smith, H. C., Lebutt, C., Rowland, C. A., Jones, S. M., Waters, E. L., Gwyther, R. J., Miller, J., Packer, P. J. & Irving, M. (2005) Human immune response to a plague vaccine comprising recombinant F1 and V antigens. *Infect Immun,* 73, 3598–608.

Chapter 18
Genetic Detection of Vesicle Forming Pathogens by Arrayed Primer Extension (APEX)

Adi Beth-Din, Ofir Israeli, Ohad Shifman, Dana Stein, Einat Ben-Arie, Nir Paran, Shlomo Lustig, Anat Zvi, Naomi Ariel, and Arie Ordentlich

Abstract The Variola virus is the causative agent of smallpox disease. Accurate identification of Variola and differential diagnosis between Variola and other vesicle forming (smallpox-like) pathogens are both technically challenging and of great importance. We compiled a list of vesicle forming pathogens that are prone to be misdiagnosed as smallpox. Some of the pathogens (members of the Orthopoxvirus genus) are also genetically highly similar to Variola.

We established an approach for genetic identification and differential diagnosis between the closely related Orthopoxvirus genus members. This approach couples multiplex PCR with DNA microarray as molecular means for specific and sensitive pathogen identification. It consists of PCR amplification of conserved gene fragments, followed by sequence-based specific identification on a DNA microarray platform utilizing the Arrayed Primer Extension (APEX) technique. Our prototype array – the ChiPox, is based on genes that harbor variable regions flanked by highly conserved Orthopoxvirus sequences; a pattern supporting APEX-based diagnostics. This novel approach was successfully applied for the differential identification of six members of the Orthopoxvirus genus. The ChiPox is a powerful and sensitive tool for discrimination between closely related Orthopoxvirus species, and is the first step in the establishment of a comprehensive assay for genetic discrimination of vesicle forming pathogens.

Keywords APEX · Orthopoxvirus · Vesicle-forming pathogens · DNA microarray · Smallpox

---

A. Beth-Din (✉)
Departments of Biochemistry & Molecular Genetics, Israel Institute for Biological Research, P.O. Box 19, Ness-Ziona 74100, Israel
e-mail: adib@iibr.gov.il

A. Shafferman et al. (eds.), *The Challenge of Highly Pathogenic Microorganisms*,
DOI 10.1007/978-90-481-9054-6_18, © Springer Science+Business Media B.V. 2010

18.1 Introduction

18.1.1 Background

The group Orthopoxvirus (OPV) belongs to the family Poxviridae which comprises of over 70 members. The Poxviridae represent a family of large DNA genome viruses of approximately 200,000 bp (Massung et al., 1994) that encode about 200 proteins and replicate in the cytoplasm. This family is subdivided into two subfamilies, the Entomopoxvirinae and the Chordopoxvirinae, which infect insects and vertebrates, respectively. The Chordopoxvirinae have been further subdivided into eight genera (OPV, Avipoxvirus, Capripoxvirus, Leporipoxvirus, Molluscipoxvirus, Parapoxvirus, Suipoxvirus, and Yatapoxvirus) (Moss, 1996). Within the OPV genera, only four can infect humans, mainly via the airways epithelium through an aerosol or via the skin.

Variola virus, a member of the OPV genus, is the causative agent of the strictly human disease, smallpox, which is a major health and biosecurity concern (Henderson et al. 1999). Smallpox is a highly contagious disease with excessive mortality of up to 40% in unvaccinated human populations (Henderson, 2002). It is estimated that smallpox disease has killed more humans than all other infectious diseases combined (Moss, 1996). Smallpox is the first and only viral disease in human history to have been eradicated completely. In 1979, the World Health Organization (WHO) certified the eradication of smallpox virus following successful worldwide vaccination campaign. Since then, the only known smallpox virus samples have been stocked in repositories at the Centers for Disease Control and Prevention (Atlanta, Georgia) and the State Research Center of Virology and Biotechnology (Koltsovo, Russia).

Beside Variola virus, there are other three OPVs that can infect human: Vaccinia virus, Monkeypox virus and Cowpox virus. Recently, a few cases of human infections by Cowpox from household rodents were recorded in Germany (Becker et al., 2009).

18.1.2 The Significance of Accurate Identification of Vesicle Forming Pathogens

Following the eradication of smallpox and the development of vaccines for diseases with similar symptoms, like chickenpox and measles, there is a significant decrease in smallpox related morbidity and in the physician's skills for differential diagnosis between smallpox and diseases with similar symptoms, and therefore misdiagnosis is a major concern. In view of the fact that vaccination is efficient if given up to four days after infection, diagnosis of Variola virus at the earliest stage possible is critical for minimizing the casualties. Moreover, the high frequency of genomic recombination of OPVs, in conjunction with an exceptional ability to cross animal species barriers, substantially increases their pathogenic potential (Turner and Moyer, 1990; Esposito et al., 2006).

Development of a reliable and rapid method for sensitive detection of OPVs and accurate discrimination between the later and other vesicle forming pathogens is of immense importance. The lack of Variola virus and the extensive genetic similarity within the OPV group (over 95% similarity at the DNA level) (Arita and Tagaya, 1977) impairs the ability to develop a reliable diagnostic test. Several approaches are utilized for the identification of and differentiation between OPVs, including typical growth in chorioallantoic membranes (Obijeski et al., 1973), serology and mapping with DNA restriction enzymes (Esposito and Knight, 1985). These methods are fairly reliable but are time-consuming and require virus isolation and propagation.

Several genetic approaches were developed for rapid discrimination of OPVs, including conventional PCR assays with or without amplicon RFLP analyses, real-time PCR tests, and PCR cross-hybridization tests that use dot-blot or microarray for differential analysis (Meyer et al., 1993; Ropp et al., 1995; Kulesh et al., 2004; Ibrahim et al., 2003; Laassri et al., 2003; Nitsche et al., 2004; Wenli et al., 2004). These techniques can discriminate between known human infectious OPVs (Vaccinia, Variola, Cowpox or Monkeypox) to a reliable extent, but usually fail to pinpoint the vesicle forming pathogen, in a suspected sample, if it is not a human OPV pathogen. The aim of the current project was to provide a diagnostic tool for accurate and specific genetic detection of closely related species, based on DNA microarray.

18.2 Results and Discussion

18.2.1 Determination of Vesicle Forming Pathogens List

As the first step, a list of vesicle forming pathogens was assembled. The list includes 23 pathogen species, virus and bacteria, causing infectious diseases associated with vesicles; prone to be misdiagnosed as smallpox. Five members of the OPV genus that are genetically and morphologically highly similar to Variola were added to the list (Table 18.1).

The pathogens in the list were scored and prioritized according to the following criteria:

1. Clinical features and the resemblance to smallpox symptoms
2. Incidence of the disease
3. Severeness of the disease in human and mortality rate
4. Human infectivity

For the development of the detection array, 14 viruses from this list, with complete genomes available in databases were used. These 14 pathogens belong to two main families: Chordopoxvirus and Alphaherpesvirus, each one of them include subfamilies that group the pathogens. The pathogens in each group are closely related and share high degree of sequence similarity.

Table 18.1 The list of vesicle forming pathogens. The list integrates all the smallpox-like pathogens belonging to pox and herpes families and other infectious disease causing pathogens. In addition certain other animal infecting OPVs are included in the list

	Pathogen		Pathogen
1	Variola virus	15	Pseudocowpox virus
2	Monkeypox virus	16	Rickettsia akari
3	Vaccinia virus (Disseminated)	17	Volepox virus
4	Camelpox virus	18	Racoonpox virus
5	Cowpox virus	19	Skunkpox virus
6	Ectromelia virus	20	reindeerpox virus
7	Orf virus	21	Rubella virus
8	Bovine papular stomatitis virus	22	sealpox virus
9	Molluscum contagiosum virus	23	Measles virus (Rubeola)
10	Tanapox virus	24	Staph. Aureus (Impetigo)
11	Herpes simplex virus type 1 (Disseminated)	25	Strep. Pyogenes (Impetigo)
12	Herpes simplex virus type 2 (Disseminated)	26	Enterovirus 71 (Hand-foot-and-mouth disease)
13	Herpes B virus	27	Coxsackievirus A16 (Hand-foot-and-mouth disease)
14	Varicella - zoster virus	28	Treponema pallidum (pustular syphilis)

18.2.2 Establishment of an Approach for Diagnosis of Closely Related Species

The multi-pathogen identification test relies on diagnostic regions derived from common genes or sequences. A diagnostic region consists of high sequence homology in its margins that enables the use of common primers for PCR amplification of the region from all related pathogens in a particular subgroup. The center of the fragment harbors definite diagnostic bases that are utilized for specific identification of each pathogen. In-house computational tools (e.g. MUGDIR software) were developed for the array design based on these diagnostic markers. The software allows for color-coded visualization of common regions in multiple alignments of the group members, facilitating a quick and robust screen of potential diagnostic regions, and allowing for further down selection of the most suitable regions, for diagnostic purposes.

Following a group wise amplification, the pathogen related PCR product is specifically identified on a DNA microarray platform by an Arrayed Primer EXtension (APEX) Reaction (Fig. 18.1)

The probes on the array are covalently linked via their 5′ end to the matrix, leaving the 3′ end free for elongation. Each probe is complementary to a specific pathogen sequence ending one base upstream to a diagnostic base. Following hybridization, each complement probe is elongated by a single terminator base in a sequence dependent manner. Since the terminators are differentially labeled by four individual fluorophores, after the washing steps each relevant probe remains labeled and produces a signal indicative of the diagnostic base (=base call) (Tonisson et al., 2000, 2002; Kurg et al., 2000).

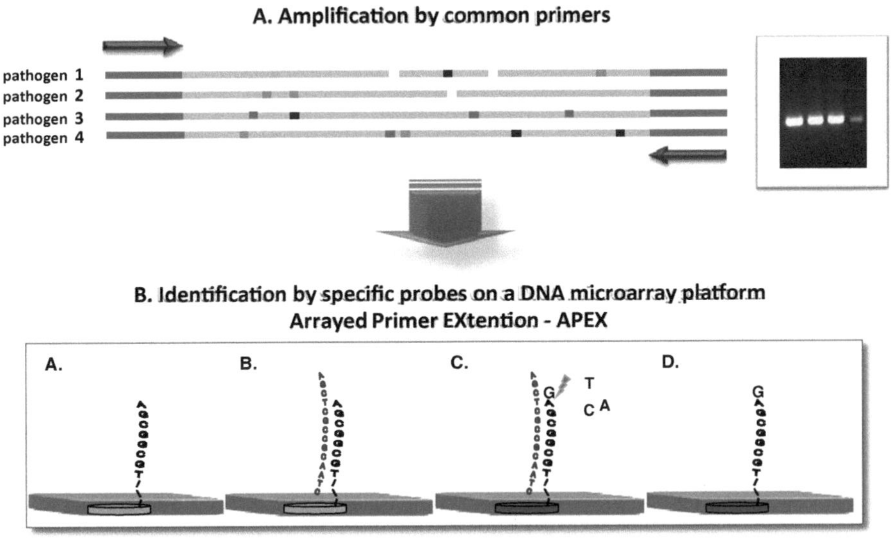

Fig. 18.1 The procedure for accurate genetic discrimination between closely related species. (**a**) PCR amplification of the diagnostic regions with universal primers, based on the common sequence patterns at the borders, shared by several pathogens (4 in this figure). After purification and fragmentation, the PCR product it is applied on the chip. (**b**) Parallel hybridization and primer extension. Each probe is labeled by a specific terminator nucleotide, in correspondence with the diagnostic base

The results from the probes of each pathogen compose a list of diagnostic bases which can be used as a fingerprint sequence of each pathogen (Syvanen, 2005; Podder et al., 2008; Pastinen et al., 2000). The readout sequence is subsequently compared to reference sequences of each pathogen in the array. For each comparison a score is calculated and normalized. The score corresponds to the percentage of true signals out of expected true probes for each pathogen. Threshold score values are: Score of 80 and up – positive, suspected sample 50–80 points and negative – less than 50.

18.2.3 Array Development Process

A process for the development of the array was established (Fig. 18.2). In each stage some principles for selection of the suitable regions and for the array probe design were defined.

The process was initiated with the outcome of a bioinformatic analysis, a pool of potential diagnostic regions. In the stage of primer design, selection was based on the following standards: Tm of 55–60°C, GC content of 25–75%, degeneracy level should be up to 4 (i.e. a maximum of two different bases for each primer), and the length of the region to be amplified shouldn't exceed 1 kb. In addition, primers accepted should be with maximal universality within the group members.

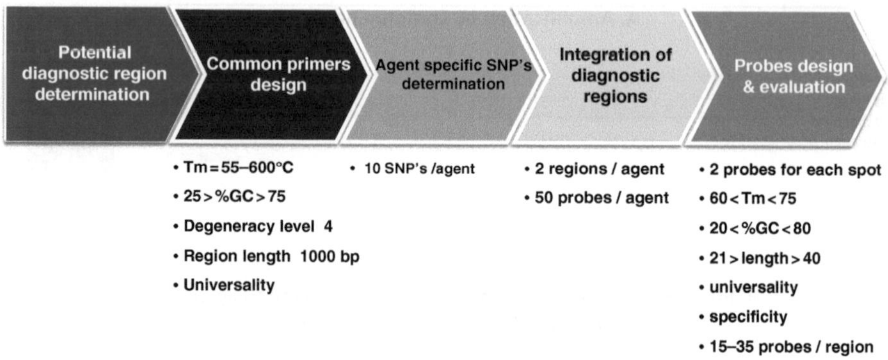

Fig. 18.2 Flowchart of the development of the DNA microarray test

The next step in selection of regions included assessment of pathogen specific SNP (single nucleotide polymorphism) content. Only regions containing at least 10 SNPs/pathogen were approved.

In the array integration stage, a minimal combination of regions based on two markers per agent with coverage of about 50 probes was assembled. A software (PROBES) was developed in-house for probes design and evaluation. This software uses a multiple sequence alignment of a diagnostic region as input, and automatically locates all the unique diagnostic bases. Next, the software designs probes for identification of these bases in the APEX reaction. The outcome of this analysis is a list of APEX probes and a table with predicted base extension of the probes in the list of various species.

18.2.4 The Development of the Array and Assessment of a Prototype Array- the ChiPox

The design protocol was applied to the list of 14 pathogens yielding an array of 9 diagnostic regions for the potential detection of all 14 pathogens; each one to be detected by at least 2 markers, with a coverage of about 50 probes (Table 18.2). When the computation identified more than 50 probes for a selected pathogen, the down selection for 50 probes was carried out based on prediction of probes efficiency and quality. Alongside the species-specific probes, a set of Orthopoxvirus universal probes (pan-ortho) was included in the array, in order to overcome the risk of the appearance of unknown /new OPV species, and to assure a positive result for each OPV group member.

A prototype array for the detection of OPV species was developed, and designated ChiPox. The ChiPox was assessed with DNA samples from OPV species, resulting in accurate identification of all five members examined (Fig. 18.3a). In order to determine the sensitivity of the ChiPox, a set of DNA samples from 10-fold

Table 18.2 The nine diagnostic regions selected for the ChiPox assay. The cells are colored to indicate the potential of the region to harbor suitable probes for identification of each pathogen (*darker*=higher number of potential probes). The selection of the regions was done based on their features and quality with preference to those which posses primers with low degeneracy level and relatively short amplicons

		Chordopox											Alphaherpes				
		Orthopox							Parapox		Mollus.	Yata.	Simplexvirus			Veri cello	
region name	region length	Pan Ortho	VAR	MPX	VAC	CPX	CMLP	ECT	ORF	BPSV	Mollus.	Tana	HSV1	HSV2	Herpes B	VZV	SUM
Chem	597	7	10	7	0	6	7	10									47
R60	638	20	21	13	12	33	16	28									143
R56	652	50	19	21	41	37	6	4									178
Y71	785	9	3	6	4	19	0	8				41					90
m8121L	1044	30	9	6	0	2	14	10				45					116
Mine	1017	0	10	2	0	50	8	23	50	50	50						243
B100	355								50	43	50						143
ORF45	530												33	27	25	50	135
ORF27	460												46	33	36	50	165

VAR variola, *MPX* monkeypox, *VAC* vaccinia, *CPX* cowpox, *CMLP* camelpox, *ECT* ectromelia, *BPSV* bovine papular stomatitis virus, *Mollus* molluscum contagiosum, *Tana* tanapox, *HSV1/2* herpes simplex virus ½, *VZV* varicella zoster virus, *Yata* yatapoxvirus

dilutions of Vaccinia virus was applied. Positive results were obtainable from as little as 100 pfu/ml (Fig. 18.3b).

The universality of the ChiPox was validated by the detection of five Vaccinia strains; all of them were indeed identified as Vaccinia with high scores (95–100). The specificity of the assay was examined with samples from closely related virus species as well as samples of human, monkey and mouse DNA, none of which gave false positive results (data not shown).

18.2.5 Evaluation of the ChiPox in Clinical Samples

The ChiPox was further evaluated, by detection of clinical samples. The disease model were rabbitpox infected rabbits. Animals were inoculated intra-nasal with 10^3 pfu/animal. Blood samples were collected from day 3–5 post inoculation, and sent for detection on ChiPox. All Vaccinia strains detected, attained high scores (87–100) and were clearly identified as Vaccinia (Fig. 18.4). All other OPV species didn't achieve the threshhold of 50 and were considered negative. Of note, the test allowed to distinguish between the Rabbitpox strains which displayed the highest score (100), and other Vaccinia strains showing lower but yet strictly positive scores (87–98). The level of detection (LOD) of the ChiPox in blood samples was determined to be 50–100 pfu/ml. Furthermore, this LOD is in agreement with results obtained from real-time PCR (data not shown).

In the same study, throat samples were collected on days 3–5 post infection and sent to detection on ChiPox. The level of detection of the ChiPox in those samples

A.

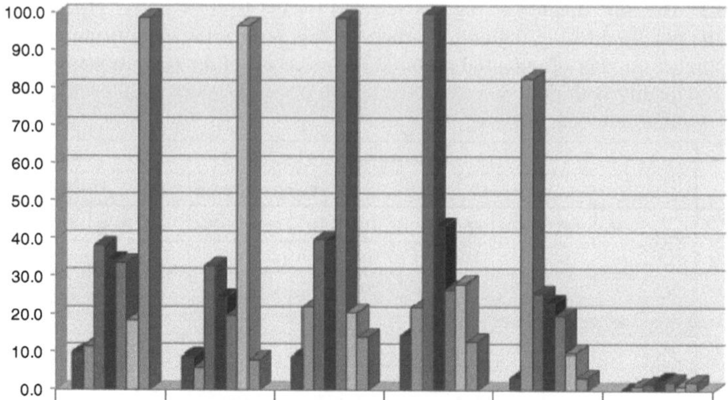

	Sample1	Sample2	Sample3	Sample4	Sample5	NTC
Variola	9.9	8.9	8.8	14.3	3.3	0.0
Monkeypox	11.5	6.0	22.2	22.1	82.8	1.2
Vaccinia	38.4	32.9	40.0	100.0	25.8	1.8
Cowpox A	34.2	24.7	39.7	43.7	23.3	2.7
Cowpox B	33.7	19.8	98.8	26.7	19.8	2.3
Camelpox	18.5	96.7	20.7	28.3	9.8	1.1
Ectromelia	98.9	8.0	14.0	12.8	3.4	2.3
	Ectromelia	***Camelpox***	***Cowpox* B**	***Vaccinia***	***Monkeypox***	***negative***

B.

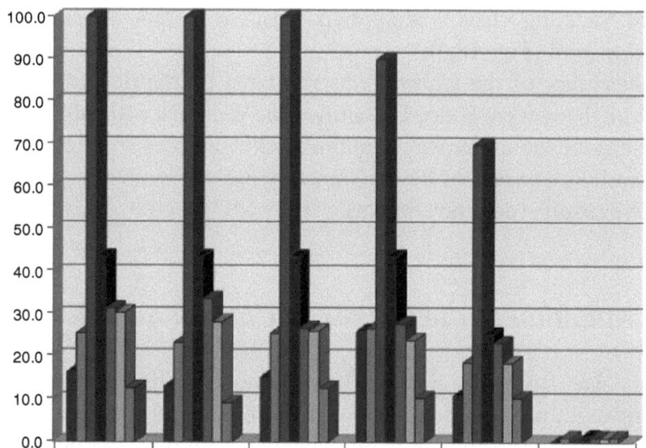

	100,000	10,000	1,000	100	10	NTC
Variola	16.5	13.2	15.4	26.1	11.0	0.0
Monkeypox	25.6	23.3	25.6	26.7	18.6	1.1
Vaccinia	100.0	100.0	100.0	90.0	70.0	0.0
Cowpox A	43.7	43.7	43.7	43.7	25.4	1.4
Cowpox B	31.4	33.7	26.7	27.9	23.3	1.2
Camelpox	30.4	28.3	26.1	23.9	18.5	1.1
Ectromelia	12.8	9.3	12.8	10.5	10.5	1.1

Fig. 18.3 Evaluation of the ChiPox with OPV species samples and assay sensitivity. (**a**) Evaluation of the ChiPox with five OPV species. Each column refers to one sample detected (sample identity is marked down). Results are presented as normalized score of each sample relative to the reference pathogens presented in the left column. The bars are colored according to the color index in the pathogens list. (**b**) Example of asay sensitivity for Vaccinia virus

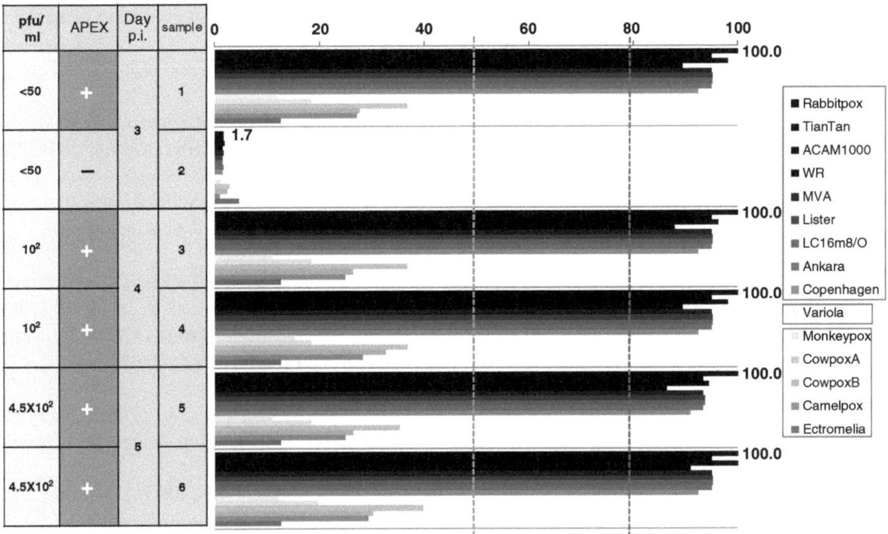

Fig. 18.4 Detection of Rabbitpox from blood samples. (**a**) Comparison of plaque assay determination and APEX results during the progression of the disease. (**b**) Scores calculated for the blood samples attained by APEX reaction. Rabbitpox score is represented by the top bar in each row of the chart, in each positive sample the highest score (100) was exhibited by Rabbitpox

was about 150 pfu/ml, again, very similar to real-time PCR sensitivity (data not shown).

18.3 Summary

The ChiPox array developed is a proof of concept to our approach for genetic detection based on group markers. It allowed to correctly identify all tested OPV species. The ChiPox appears also as a powerful and sensitive tool for discrimination not only between closely related OPV species but even between different strains of the same species. Moreover, the ChiPox was effective in disease detection in clinical samples. An expanded ChiPox array, which is now in development, will have the ability to discriminate between all 19 members listed as vesicle forming pathogens.

References

Arita, M. & Tagaya, I. (1977) Structural polypeptides of several strains of orthopoxvirus. *Microbiol Immunol,* 21(6), 343–6.

Becker, C., Kurth, A., Hessler, F., Kramp, H., Gokel, M., Hoffman, R., Kuczka, A. & Nitsche, A. (2009) Cowpox virus infection in pet rat owners. *Deutsches Arzteblatt Int,* 106(19), 329–34.

Esposito, J. J. & Knight, J. C. (1985) Orthopoxvirus DNA: a comparison of restriction profiles and maps. *Virology,* 143(1), 230–51.

Esposito, J. J., Sammons, S. A., Frace, A. M., Osborne, J. D., Olsen-Rasmussen, M., Zhang, M., Govil, D., Damon, I. K., Kline, R., Laker, M., Li, Y., Smith, G. L., Meyer, H., Leduc, J. W.

& Wohlhueter, R. M. (2006) Genome sequence diversity and clues to the evolution of variola (smallpox) virus. *Science,* 313, 807–12.

Henderson, D. A. (2002) Countering the post eradication threat of smallpox and polio. *Clin Infect Dis,* 34(1), 79–83.

Henderson, D. A., Inglesby, T. V. & Barlett, J. G. (1999) Smallpox as a biological weapon: medical and public health management. Working group on Civilian Biodefense. *JAMA,* 281(22), 2127–37.

Ibrahim, S. M., Kulesh, D. A., Saleh, S. S., Damon, I. K., Esposito, J. J., Schmaljohn, A. L. & Jahrling, P. B. (2003) Real-time PCR assay to detect smallpox virus. *J Clin Microbiol,* 41, 3835–9.

Kulesh, D. A., Baker, R. O., Loveless, B. M., Norwood, D., Zwiers, S. H., Mucker, E., Hartmann, C., Herrera, R., Miller, D., Christensen, D., Wasieloski, L. P., Huggins, J. & Jahrling, P. B. (2004) Smallpox and pan-orthopox virus detection by real-time 3-minor groove binder TaqMan assays on the Roche LightCycler and the Cepheid smart Cycler platforms. *J Clin Microbiol,* 42, 601–9.

Kurg, A., Tonisson, I., Georgiou, J., Shumaker, J., Tollett, A. & Metspalu, A. (2000) Arrayed primer extension: solid-phase four-color DNA resequencing and mutation detection technology. *Genetic Tests,* 4, 1–7.

Laassri, M., Chizhikov, V., Mikheev, M., Shchelkunov, S. & Chumakov, K. (2003) Detection and discrimination of orthopoxviruses using microarrays of immobilized oligonucleotides. *J Virol Methods,* 112, 67–78.

Massung, R. F., Liu, L. I., Qi, J., Knight, J. C. T., Yuran, E. A., Kerlavage, R., Parsons, J. M., Venter, J. C. & Esposito, J. J. (1994) Analysis of the complete genome of smallpox variola major virus strain Bangladesh-1975. *Virology,* 201, 215–40.

Meyer, H., Osterrieder, N. & Pfeffer, M. (1993) Differentiation of species of the genus Orthopoxvirus in a dot blot assay using digoxigenin-labeled DNA probes. *Vet Microbiol,* 34, 333–44.

Moss, B. (1996) Genetically engineered poxviruses for recombinant gene expression, vaccination, and safety. *Proc Natl Acad Sci USA,* 93(21), 11341–8.

Nitsche, A., Ellerbrok, H. & Pauli, G. (2004) Detection of orthopoxvirus DNA by real-time PCR and identification of Variola virus DNA by melting analysis. *J Clin Microbiol,* 42(3), 1207–13.

Obijeski, J. F., Palmer, E. L., Gafford, L. G. & Randall, C. C. (1973) Plyacrylamide gel electrophoresis of fowlpox and Vaccinia virus proteins. *Virology,* 51(2), 512–6.

Pastinen, T., Raitio, M., Lindroos, K., Tainola, P., Peltonen, L. & Syvanen, A. C. (2000) A system for specific, high-throughput genotyping by allele-specific primer extension on microarrays.*Genome Res,* 10(7), 1031–42.

Podder, M., Ruan, J., Tripp, B. W., Chu, Z. E. & Tebbutt, S. J. (2008) Robust SNP genotyping by multiplex PCR and arrayed primer extension. *BMC Med Genomics,* 5(1) 165–70.

Ropp, S. L., Jin, Q., Knight, J. C., Massung, R. F. & Esposito, J. J. (1995) PCR strategy for identification and differentiation of smallpox and other orthopoxviruses. *J Clin Microbiol,* 33, 2069–76.

Syvanen, A. C. (2005) Toward genome-wide SNP genotyping. *Nat Genet* 37(Suppl.), S5–10

Tonisson, N., Kurg, A., Kaasik, K., Lohmussaar, E. & Metspalu, A. (2000) Unravelling genetic data by arrayed primer extension. *Clin Chem Lab Med,* 38(2), 165–70.

Tonisson, N., Zernant, J., Kurg, A., Pavel, H., Slavin, G., Roomere, H., Meiel, A., Hainaut, P. & Metspalu, A. (2002) Evaluating the arrayed primer extension resequencing assay of TP53 tumor suppressor gene. *Proc Natl Acad Sci USA,* 99, 5503–8.

Turner, P. C. & Moyer, R. W. (1990) The molecular pathogenesis of poxviruses. *Curr Topics Microbiol Immunol,* 163, 125–51.

Wenli, M., Yan, W., Hongmin, W. & Wenling, Z. (2004) An oligonucleotide microarray for the detection of vaccinia virus. *Br J Biomed,* 61, 142–5.

Chapter 19
Subtyping Biothreat Bacterial Pathogens

Paul Keim and Talima Pearson

Abstract The precise identification of bacterial pathogens requires both robust methodologies and comprehensive population sampling. Methods have been evolving rapidly as genomic analyses have become faster and less expensive. The holy grail of this science is a complete and accurate whole genome sequence for every isolate of a pathogen. This is still unobtainable although it is possible to forecast its arrival in a few years. In the meanwhile, a combination of whole genome sequencing coupled to molecular assay development will prevail, with economics dictating the balance between the two. The best methods, however, are meaningless without the context of a population genetic database. All the genetic information about a single infective agent is much more valuable when compared to additional genomes. Ideally, genetic databases should be representative of the natural geographic, temporal and genetic distribution of the pathogen. Because of the plethora of different genotyping methodologies, it is also critical that methods employed in an investigation target information that is compatible with database content. It is also critical that the analysis of samples for both investigative and prior genetic databases be performed at a very high quality level. Erroneous results in an investigation will result in a false conclusion, but equally problematic are database errors that cloud the interpretation of results. Subtyping each biothreat pathogen is a unique exercise due to differences in biology, evolution and genome characteristics, so the population genetic structure and confidence statistics must be determined for each pathogen. The use of these methods and concepts for the study of natural outbreaks is an ideal way to train investigative teams and refine methods while building population genetic databases that are essential for epidemiological investigations.

Keywords Subtyping · CanSNPs · MLVA · Whole genome sequencing · *Bacillus anthracis*

P. Keim (✉)
Microbial Genetics and Genomics Center, Northern Arizona University, Flagstaff, AZ 86011-4073, USA; Pathogen Genomics Division, The Translational Genomics Research Institute, Phoenix, AZ, 85004, USA
e-mail: Paul.Keim@nau.edu

A. Shafferman et al. (eds.), *The Challenge of Highly Pathogenic Microorganisms*, 183
DOI 10.1007/978-90-481-9054-6_19, © Springer Science+Business Media B.V. 2010

19.1 History of Bacterial Strain Identification

19.1.1 Traditional Approaches

Microbiologists and epidemiologists have been refining their ability to differentiate among bacteria for more than a century. At first, phenotypic differences such as colony morphology, microscopic characteristics and metabolic capabilities were sufficient for identifying new genera and species. Among otherwise identical species, immunological differentiation was, at times, possible as different "strains" of bacteria elicited antibodies reacting with distinguishing antigens. Because serological analysis of infections was important for understanding immunity, it was often coupled with vaccine development. The precise identification of bacterial pathogens was also essential for understanding disease outbreaks and the identification of case clusters with a common source.

19.1.2 Molecular Approaches

Bacterial systematics and strain identification underwent a revolution with the development of DNA based methods. These included many different approaches that seemed to change annually over the last 20 years (Achtman, 1996). All were innovations that used changing molecular technologies to detect sequence based differences. The preponderances of approaches also lead to great confusion as to their relative strengths and weakness. The lack of data harmonization across methods also led to many "orphan databases" that had only limited utility for the general clinical and research communities. Notable exceptions include the 16S ribosomal RNA gene sequences, multiple locus sequence typing (MLST) (Maiden et al., 1998) and pulse-field gel electrophoretic (PFGE) separation of restriction fragment length polymorphism (RFLPs). The 16S sequence approach, in particular, was feasible for all bacteria as it is an essential gene for ribosomal production. However, its resolution was mostly limited to higher order relationships (e.g., families, genera and sometimes species) because of great sequence conservation. MLST and PFGE provided much greater resolution by examining less conserved genes and greater portions of the bacterial genome. Harmonization of DNA sequence data (16S and MLST) in databases is straightforward, allowing researchers and clinicians to easily add to and thus enhance established public databases. This is considerably more difficult for PFGE and it has only been through extensive cross-laboratory training, standardized methods and costly infrastructure programs that harmonization (sometimes imperfectly) of these data has occurred.

19.1.3 Hyper-Variable Locus Subtyping

In spite of these technologies, sufficient strain resolution has still been lacking in many recently emerged pathogens making it impossible to characterize outbreaks.

However, it was observed that bacterial genomes have hypervariable loci with many distinct alleles that could be used for distinguishing even among very closely related strains (Andersen et al., 1996). These proved to be variable number tandemly repeated (VNTR) regions that mutated at a relatively high frequency by insertion/deletion mechanisms. Conserved flanking regions can be used for primer design in order to support single locus PCR amplification. Extensive studies of natural populations and laboratory generated populations has led to an understanding of VNTR mutational rules, rates and products (Vogler et al., 2006, 2007; Girard et al., 2004). The use of multiple-locus VNTR analysis (MLVA) increased the probability of detecting an allelic difference and also decreased the potential for identical subtypes due to fortuitous independent but convergent mutations. While each species has a unique set of VNTR loci, MLVA has been developed for scores of pathogens (Vergnaud and Pourcel, 2009), illustrating both the need for and utility of this technology.

19.2 Whole Genome Sequencing and Subtyping

19.2.1 Escherichia coli

The "holy grail" of molecular subtyping is the use of complete genome sequences for differentiating strains. Even 5 years ago, this seemed to be an impossibility due to the high costs and long times needed to sequence a single genome. At that time, only a few dozen genomes had been completed and most researchers felt that one genome per species was adequate and would be the standard. These early genomes were extremely useful for identifying potential VNTRs and for the identification of other subtyping loci. Without this early work, few of the many species with MLVA systems would have been successful as finding VNTR loci using wet bench studies was extremely tedious (Schupp et al., 2000). The 2001 report of a second *E. coli* genome was seminal in shaping our concepts of bacterial genomic diversity as this strain and the previous one surprisingly differed by over 1.87 megabases (Perna et al. 2001). This is an incredible difference for an organism that contains only a total of 4–5 megabases in its genome. It was clear that we cannot even begin to describe the genomes of bacteria with only a single genome sequence per species. The use of whole genome sequencing for population genetics would require a great reduction in price, but this was occurring under the research and development pressure from the human genome project.

19.2.2 Bacillus anthracis

One of the first whole genome analyses of bacterial populations was reported in 2004 for *Bacillus anthracis* (Pearson et al., 2004). For this organism, rather than extensive gene differences as seen by Perna et al. (2001), there were very few, as *B. anthracis* is an example of a recently emerged pathogen (Keim et al., 1997). This study identified only a few thousand single nucleotide polymorphisms (SNPs) from five whole genome sequences. We selected strains for sequencing that were representative of diverse MLVA clades (Keim et al., 2000) in order to maximize the likelihood of finding differences due to their evolutionary separation. This strategic approach to SNP discovery was based on a large genetic database and used an "unbiased" genotyping method: MLVA. Because of the high cost of sequencing, rare SNPs discovered by whole genome sequencing were subsequently converted into single-locus genotyping assays and used to screen a larger set of strains.

19.3 Phylogenetic Discovery Bias and Subtyping

While our studies resulted in a very high-resolution estimate of the *B. anthracis* population structure, it also produced a strange phylogenetic topology due to a novel type of ascertainment bias (Worobey, 2005). Our phylogenetic tree had only branches leading to the sequenced genomes and all other strains resided on completely truncated branches. In fact, no strains except the "discovery strains" were on the branch tips, in contrast with other phylogenetic trees where all the strains are on the branch tips (Keim et al., 1997, 2000).

This highly unusual phylogenetic structure is due to "phylogenetic discovery bias," which particularly impacts phylogenies of clonal organisms (Pearson et al., 2004, 2009). As all of the SNPs were discovered by comparing the five original

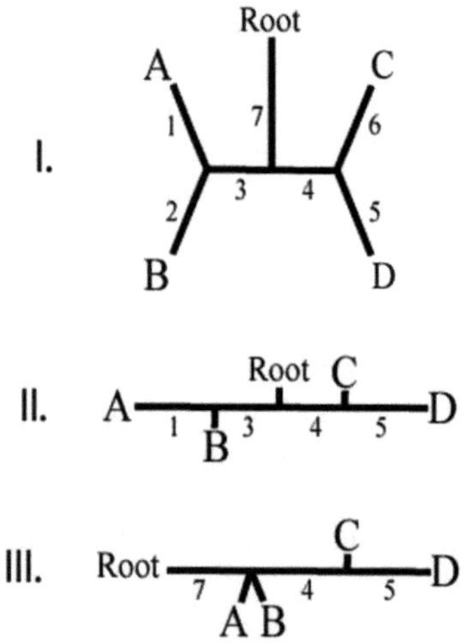

Fig. 19.1 The effect of discovery bias on phylogenetic reconstruction. A hypothetical five taxa tree (I) is presented to illustrate the effect of discovery (ascertainment) bias on phylogenetic reconstruction (Pearson et al., 2004, 2009). (II) SNP discovery by whole genome sequence comparisons of OTU-A and OTU-D, with subsequent genotyping of the other three OTU's. (III) SNP discovery by whole genome sequence comparisons of OTU-D, to the outgroup (root) genome sequence. This is a common first approach towards SNP development in a new species. The short branches in panels II and III would be zero in the absence of homoplasy due to recombination or errors in the data sets

genomes, we were naïve to much of the variation among other strains. In designing subsequent genotyping assays, we could only include SNPs discovered among the five genomes which provided only limited information regarding diversity and evolutionary patterns of other strains. This means that the SNPs found only in or among the un-sequenced genomes remained unknown and could not be included in the analyses. Hence, the branches leading to these strains are "collapsed" onto the phylogenetic backbone connecting the sequenced genomes (Fig. 19.1). Their positions on the backbone are extremely precise due to the large number of SNPs, but we can't estimate the branch lengths or topologies leading to them directly from these data. Figure 19.1 illustrates this phenomenon on a simple five OTU (operational taxonomic unit) tree. This biased discovery phenomenon is of little consequence if there are high levels of genetic exchanged (recombination) among isolates. In the case of *B. anthracis, Yersinia pestis* and *Francisella tularensis* however, there is no evidence of such genetic exchange since the evolutionary creation of these species (Achtman et al., 2004; Vogler et al., 2009).

19.4 Canonical SNPs for Efficient Subtyping

Currently, it is still prohibitive to perform whole genome sequencing on all isolates, thus other effective genotyping methods must be employed. Traditional methods such as MLST, PFGE and MLVA will continue to be used in the foreseeable future, but the approach of assaying SNPs discovered through whole genome comparisons against a broad panel of isolates will be especially valuable for phylogenetic accuracy (Pearson et al., 2004). A more economical and streamlined approach that capitalizes upon whole genome sequences and reduces the analytical complexity of thousands of SNPs has been proposed (Keim et al., 2004). This approach is based on selecting only a representative SNP from each phylogenetic branch. The *B. anthracis* phylogenetic tree, for example contains thousands of SNPs, yet the number of described branches is small (Van Ert et al., 2007a). Using only a few dozen strategically selected SNPs can thus be used to describe most of this species' diversity and provide rough phylogenetic classification of unknown isolates. We termed these SNPs canonical (canSNPs) as they define key phylogenetic positions. These canSNPs are analogous to "signature events" as described by Woese (Woese and Fox, 1977; Fox et al., 1980). As shown in Fig. 19.2, these can mark deep branches (Keim et al., 2004), terminal branches (Simonson et al., 2009) or even inter-species boundaries. CanSNPs on the terminal branches may represent strain specific markers, as were developed for the Ames strain (Van Ert et al., 2007b; Kenefic et al., 2008), while canSNPs on the deepest branch represent species specific markers such as the *plcR* SNP that defines *B. anthracis* (Easterday et al., 2005). For clinically or forensically important strains, canSNPs assays can greatly reduce the costs and time needed for identification.

The ability to differentiate among strains is a significant step towards genotyping recently emerged pathogens with little genetic variability, but interpreting results requires a thorough knowledge of the biology, genome characteristics and a reference database. Genotypic associations among samples may be due to a common

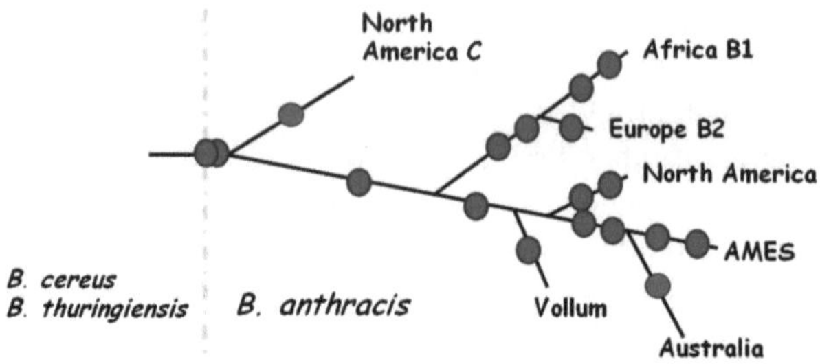

Fig. 19.2 *B. anthracis* phylogenetic tree with canonical SNPs. The phylogenetic structure of *B. anthracis* (Pearson et al., 2004) is represented along with the placement of canonical SNPs (canSNPs) represented as circles on the tree. Two canSNPs are at the species boundary between *B. cereus/thuringiensis* and *B. anthracis*

source, but they also may be due to relative lack of diversity in the markers chosen, or reflect limited diversity within the species such that a particular genotype is commonly found in nature. Therefore, population genetic studies are needed to support conclusions that are made concerning isolate attribution. Although a large and comprehensive genetic database is the theoretical goal, this may not be possible, especially *prior* to a biocrime or epidemiological event.

19.5 Summary

In summary, much progress in subtyping bacterial pathogens has occurred over the last decade, leading to whole genome sequencing. The availability of only a single genome has facilitated the identification of subgenomic genotyping loci such as VNTRs and MLST. Complementing the technology are growing databases with extensive global collections. The decreasing cost of whole genome sequencing has allowed comparisons of more than one genome, facilitating the discovery of SNPs which can be used to create highly accurate population and phylogenetic conclusions, although phylogenetic discovery bias must be thoroughly understood and considered. Further decreases in sequencing cost and an increased ability to handle and rapidly analyze large amounts of sequence data will allow us to reach the ultimate goal of sequencing every isolate in the near future. Large harmonized genetic databases are still critical, even with complete genome sequences

Acknowledgements The authors were supported by grants from the US National Institutes of Health (GM060795, AI070183), the Pacific-Southwest Region Center of Excellence (AI065359) and the Department of Homeland Security Science and Technology Directorate (NBCH2070001 and HSHQDC-08-C00158).

References

Achtman, M. (1996) A surfeit of YATMs? *J Clin Microbiol,* 34, 1870.
Achtman, M., Morelli, G., Zhu, P., Wirth, T., Diehl, I., Kusecek, B., Vogler, A. J., Wagner, D. M., Allender, C. J., Easterday, W. R., Chenal-Francisque, V., Worsham, P., Thomson, N. R.,

Parkhill, J., Lindler, L. E., Carniel, E. & Keim, P. (2004) Microevolution and history of the plague bacillus, *Yersinia pestis*. *Proc Natl Acad Sci USA,* 101, 17837–42.

Andersen, G. L., Simchock, J. M. & Wilson, K. H. (1996) Identification of a region of genetic variability among *Bacillus anthracis* strains and related species. *J Bacteriol,* 178, 377–84.

Easterday, W. R., Van Ert, M. N., Simonson, T. S., Wagner, D. M., Kenefic, L. J., Allender, C. J. & Keim, P. (2005) Use of single nucleotide polymorphisms in the plcR gene for specific identification of *Bacillus anthracis*. *J Clin Microbiol,* 43, 1995–7.

Fox, G. E., Stackebrandt, E., Hespell, R. B., Gibson, J., Maniloff, J., Dyer, T. A., Wolfe, R. S., Balch, W. E., Tanner, R. S., Magrum, L. J., Zablen, L. B., Blakemore, R., Gupta, R., Bonen, L., Lewis, B. J., Stahl, D. A., Luehrsen, K. R., Chen, K. N. & Woese, C. R. (1980) The phylogeny of prokaryotes. *Science,* 209, 457–63.

Girard, J. M., Wagner, D. M., Vogler, A. J., Keys, C., Allender, C. J., Drickamer, L. C. & Keim, P. (2004) Differential plague-transmission dynamics determine *Yersinia pestis* population genetic structure on local, regional, and global scales. *Proc Natl Acad Sci USA,* 101, 8408–13.

Keim, P., Kalif, A., Schupp, J., Hill, K., Travis, S. E., Richmond, K., Adair, D. M., Hugh-Jones, M., Kuske, C. R. & Jackson, P. (1997) Molecular evolution and diversity in *Bacillus anthracis* as detected by amplified fragment length polymorphism markers. *J Bacteriol,* 179, 818–24.

Keim, P., Price, L. B., Klevytska, A. M., Smith, K. L., Schupp, J. M., Okinaka, R., Jackson, P. J. & Hugh-Jones, M. E. (2000) Multiple-locus variable-number tandem repeat analysis reveals genetic relationships within *Bacillus anthracis*. *J Bacteriol,* 182, 2928–36.

Keim, P., Van Ert, M. N., Pearson, T., Vogler, A. J., Huynh, L. Y. & Wagner, D. M. (2004) Anthrax molecular epidemiology and forensics: using the appropriate marker for different evolutionary scales. *Infect Genet Evol,* 4, 205–13.

Kenefic, L. J., Pearson, T., Okinaka, R. T., Chung, W. K., Max, T., Trim, C. P., Beaudry, J. A., Schupp, J. M., Van Ert, M. N., Marston, C. K., Gutierrez, K., Swinford, A. K., Hoffmaster, A. R. & Keim, P. (2008) Texas isolates closely related to *Bacillus anthracis* Ames. *Emerg Infect Dis,* 14, 1494–6.

Maiden, M. C., Bygraves, J. A., Feil, E., Morelli, G., Russell, J. E., Urwin, R., Zhang, Q., Zhou, J., Zurth, K., Caugant, D. A., Feavers, I. M., Achtman, M. & Spratt, B. G. (1998) Multilocus sequence typing: a portable approach to the identification of clones within populations of pathogenic microorganisms. *Proc Natl Acad Sci USA,* 95, 3140–5.

Pearson, T., Busch, J. D., Ravel, J., Read, T. D., Rhoton, S. D., U'ren, J. M., Simonson, T. S., Kachur, S. M., Leadem, R. R., Cardon, M. L., Van Ert, M. N., Huynh, L. Y., Fraser, C. M. & Keim, P. (2004) Phylogenetic discovery bias in *Bacillus anthracis* using single-nucleotide polymorphisms from whole-genome sequencing. *Proc Natl Acad Sci USA,* 101, 13536–41.

Pearson, T., Okinaka, R. T., Foster, J. T., & Keim, P. (2009) Phylogenetic understanding of clonal populations in an era of whole genome sequencing. *Infect Genet Evol,* 9, 1010–9.

Perna, N. T., Plunkett, G., 3rd, Burland, V., Mau, B., Glasner, J. D., Rose, D. J., Mayhew, G. F., Evans, P. S., Gregor, J., Kirkpatrick, H. A., Posfai, G., Hackett, J., Klink, S., Boutin, A., Shao, Y., Miller, L., Grotbeck, E. J., Davis, N. W., Lim, A., DIMALANTA, E. T., Potamousis, K. D., Apodaca, J., Anantharaman, T. S., Lin, J., Yen, G., Schwartz, D. C., Welch, R. A. & Blattner, F. R. (2001) Genome sequence of *enterohaemorrhagic Escherichia coli* O157:H7. *Nature,* 409, 529–33.

Schupp, J. M., Klevytska, A. M., Zinser, G., Price, L. B. & Keim, P. (2000) vrrB, a hypervariable open reading frame in *Bacillus anthracis*. *J Bacteriol,* 182, 3989–97.

Simonson, T. S., Okinaka, R. T., Wang, B., Easterday, W. R., Huynh, L., U'ren, J. M., Dukerich, M., Zanecki, S. R., Kenefic, L. J., Beaudry, J., Schupp, J. M., Pearson, T., Wagner, D. M., Hoffmaster, A., Ravel, J. & Keim, P. (2009) *Bacillus anthracis* in China and its relationship to worldwide lineages. *BMC Microbiol,* 9, 71.

Van Ert, M. N., Easterday, W. R., Huynh, L. Y., Okinaka, R. T., Hugh-Jones, M. E., Ravel, J., Zanecki, S. R., Pearson, T., Simonson, T. S., U'ren, J. M., Kachur, S. M., Leadem-Dougherty, R. R., Rhoton, S. D., Zinser, G., Farlow, J., Coker, P. R., Smith, K. L., Wang, B., Kenefic, L. J., Fraser-Liggett, C. M., Wagner, D. M. & Keim, P. (2007a) Global genetic population structure of *Bacillus anthracis*. *PLoS One,* 2, e461.

Van Ert, M. N., Easterday, W. R., Simonson, T. S., U'ren, J. M., Pearson, T., Kenefic, L. J., Busch, J. D., Huynh, L. Y., Dukerich, M., Trim, C. B., Beaudry, J., Welty-Bernard, A., Read, T., Fraser, C. M., Ravel, J. & Keim, P. (2007b) Strain-specific single-nucleotide polymorphism assays for the *Bacillus anthracis* Ames strain. *J Clin Microbiol,* 45, 47–53.

Vergnaud, G. & Pourcel, C. (2009) Multiple locus variable number of tandem repeats analysis. *Methods Mol Biol,* 551, 141–58.

Vogler, A. J., Birdsell, D., Price, L. B., Bowers, J. R., Beckstrom-Sternberg, S. M., Auerbach, R. K., Beckstrom-Sternberg, J. S., Johansson, A., Clare, A., Buchhagen, J. L., Petersen, J. M., Pearson, T., Vaissaire, J., Dempsey, M. P., Foxall, P., Engelthaler, D. M., Wagner, D. M. & Keim, P. (2009) Phylogeography of *Francisella tularensis*: global expansion of a highly fit clone. *J Bacteriol,* 191, 2474–84.

Vogler, A. J., Keys, C., Nemoto, Y., Colman, R. E., Jay, Z. & Keim, P. (2006) Effect of repeat copy number on variable-number tandem repeat mutations in *Escherichia coli* O157:H7. *J Bacteriol,* 188, 4253–63.

Vogler, A. J., Keys, C. E., Allender, C., Bailey, I., Girard, J., Pearson, T., Smith, K. L., Wagner, D. M. & Keim, P. (2007) Mutations, mutation rates, and evolution at the hypervariable VNTR loci of *Yersinia pestis*. *Mutat Res,* 616, 145–58.

Woese, C. R. & Fox, G. E. (1977) Phylogenetic structure of the prokaryotic domain: the primary kingdoms. *Proc Natl Acad Sci USA,* 74, 5088–90.

Worobey, M. (2005) Anthrax and the art of war (against ascertainment bias). *Heredity*, 94, 459–60.

Chapter 20
A Rapid Method for *Bacillus anthracis* Genotyping

Haim Levy, Morly Fisher, Naomi Ariel, Shay Weiss, Anat Zvi, David Kobiler, and Zeev Altboum

Abstract Classification and differentiation of *Bacillus anthracis* isolates by genetic markers play an important role in the study of anthrax epidemiology. We have applied a PCR based method – Random Amplification of Polymorphic DNA (RAPD) to identify twenty-five *B. anthracis* genetic markers. These markers allowed for classification of the studied strains into five different groups. Three selected RAPD markers were cloned and sequenced. Typical integration of the three markers allowed for specific definition of the five RAPD derived genotypes. To test the universal power of these markers to discriminate between diverse *B. anthracis* strains, the nucleotide sequence of each marker was searched against all available *B. anthracis* genome sequences (both finished and unfinished). The three markers system could differentiate between strains belonging to the genetic groups Aβ, A1a, A1b, A4 and B1 (as defined by Keim et al., 2000; Maho et al., 2006) and gave rise to a unique combination for group A3, but couldn't distinguish between the sub groups A3a and A3b. In addition, this system could not distinguish between groups B2 and C (identical three marker combination). In an attempt to improve the resolution of this system we introduced a fourth marker. In silico analysis revealed that the resulting four markers system could now differentiate between group B2 and C, adding a second genotype to groups A1a, A3a and A4. This four marker system could potentially provide an accurate, simple, and inexpensive agarose-based system for classification *B. anthracis* strains in laboratories involved in research of this bacterium.

Keywords Genotyping · Phylogenetic tree · *Bacillus anthracis* · RAPD · GeneBank

H. Levy (✉)
Department of Infectious Diseases, Israel Institute for Biological Research,
P.O. Box 19, Ness-Ziona 74100, Israel
e-mail: haiml@iibr.gov.il

A. Shafferman et al. (eds.), *The Challenge of Highly Pathogenic Microorganisms*,
DOI 10.1007/978-90-481-9054-6_20, © Springer Science+Business Media B.V. 2010

20.1 Introduction

The spore forming *Bacillus anthracis* microorganism is the cause of Anthrax, and a major possible candidate to be used as a non-conventional weapon in acts of war or bio-terror. Methods that enable *B. anthracis* isolate classification are important for the identification of the bacterium both in cases of hostile dissemination and in routine research. Since *B. anthracis* is monomorphic, sensitive and accurate classification methods must be developed in order to identify and classify the possible origin of a specific isolate. For example, unequivocal identification of the source of the *B. anthracis* strain that was used in the 2001 envelope bio-terror event, required whole genome sequencing of isolated colonies (Enserink, 2008).

Up to recent years, a method for genetic typing of different isolates of this bacterium did not exist. The first genomic marker (vrrA) enabling the grouping of known *B. anthracis* strains was reported in 1996 by Andersen and co-workers (Andersen et al., 1996). Since then, three major approaches were undertaken to look for DNA based polymorphisims. The first and most extensive analysis was performed by P. Keim and his colleagues that was based on AFLP (amplified fragment length polymorphism) technique. This approach allowed for identification of 31 markers and to construct a phylogenetic tree (Keim et al., 1997). Based on these AFLP markers and the previously identified *vrrA,* Keim and coworkers (Keim et al., 2000) first proposed an eight- variable-number tandem repeats (VNTR) marker system. At a later stage additional seven were added (Matthew et al., 2007) which together served as Multiple-Locus VNTR Analysis (MLVA) 15 marker genotyping method. This method served to determine the genotype of about 1,000 *B. anthracis* strains, from all over the world.

Additional VNTR markers were identified by using bio-informatics tools to identify tandem repeats, documented in other species as rearrangements targets. Le Fleche et al. (2001) identified 14 such tandem repeats, and in combination with 6 of the MLVA markers, designed a phylogenetic tree which overall was similar to the one proposed by Keim et al. (2000), but had an additional genetic sub-group.

In an attempt to develop a simple classification system that will allow for typing of *B. anthracis* isolate without the need for a dedicated high cost equipment, we adapted the Random Amplified Polymorphic DNA (RAPD) method (Williams et al., 1990). RAPD is a PCR based method, based on short (10 nt) primers with an arbitrary sequence. Each PCR reaction is performed with a single primer which anneals to different binding sites in the genome. Using DNA from 7 different *B. anthracis* strains we noted polymorphisms in 25 out of 1,200 different RAPD primers that were identified (Levy et al., 2005). Using these markers we designed a phylogenetic tree which overall was similar to that proposed by Keim et al. (2000) and identical to the one described by Le Fleche et al. (2001). To simplify the typing method we selected and cloned three of the RAPD markers. Specific primers were designed, facilitating strain typing in a short rapid PCR reaction (Levy et al., 2005). Our three marker system could very simply and accurately differentiate between the strains in our collection and gave rise to a phylogenetic tree identical to the one generated by the 25 RAPD marker system (Levy et al., 2005). The aim of the present study was to test the universality of our system based on GenBank sequence data for selected *B. anthracis* strain.

20.2 Results

20.2.1 The RAPD Based Analysis System

All the markers represent direct repeats differing in length and number (Fig. 20.1). Primer pairs, specific for each of the three markers, were constructed according to the relevant sequence. For AA03 and AJ03 the primer sequence was based on the sequence of the RAPD primer and additional downstream sequences according to the bacterial genome. Since the AT07 RAPD marker was relatively large (~2 kbp), primer pairs flanking the multiple direct repeats consisting the site of polymorphysim were selected.

In an attempt to simplify the typing procedure, we tested the possibility to perform this analysis as a multiplex PCR. The predicted size of the PCR products of the primer pairs was around 400 nt for AJ03, 800 nt for AA03 and 600–1,000 nt for AT07. We combined the primer pairs of AJ03 and AA03 into a single multiplex reaction and performed the AT07 reaction separately. The results of the PCR reaction are presented in Fig. 20.2. In the upper gel the two markers divided the strains into two groups exhibiting a characteristic pattern of high or low MW bands. The marker in the lower gel divides the strains into three groups according to DNA band size (marked according to the number of direct repeats). Integrated results are presented as a table in Fig. 20.2. The results demonstrate that the three specific primers pairs allow for strain classification into 5 genotypes, in a manner identical to the outcome of the 25 RAPD marker system.

RAPD marker	No. of repeats	No. of alleles	Repeat size (nucleotides)	Location	Annotation
AA03	2–3	2	75	Upstream to gi/30256105	SNF symporter
AJ03	2–4	2	42	Between gi/30254841 gi/30254849	Intergenic
AT07	6–11	3	38	gi/30259155	*spoVID* homolog

Fig. 20.1 Structure of the RAPD based specific markers. The primers used to specifically amplify the AJ03 marker are (5′–3′): AJ035 – AGCACCTCGTTCATGCTCATAACGG and AJ036 – AGCACCTCGTCTACTTCATTTTGTGC; The AA03 marker are AA032 – TTAGCGCCCCCTTGCGTTCC and AA033 – TTAGCGCCCCTAGACCAATTGC: The AT07 marker are AT073 – CTCCTCAAATTACTAAAATGAAACC and AT074 – TTGGCATAGACGTATATTGCGGTCC

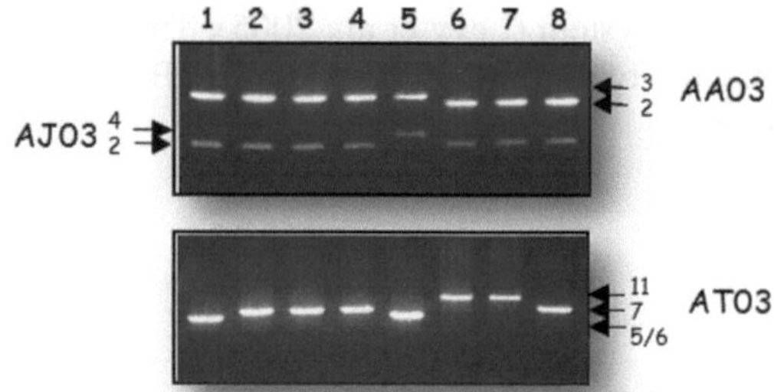

	Strain 1	2	3	4	5	6	7	8	Marker
Number of direct repeats	3	3	3	3	3	2	2	2	AA03
	2	2	2	2	4	2	2	2	AJ03
	6	7	7	7	5	11	11	7	AT03
	I	II	II	II	III	IV	IV	V	Genetic Group

Fig. 20.2 Development of a three marker system which enables the classification of the eight *B. anthracis* strains studied. The specific primers are marked as well as the number of direct repeats The AA and AJ primer are used in a multiplex PCR reaction. The strains analyzed are 1. Δ14185, 2. Sterne (10), 3. BA7001 (Sterne), 4. ATCC14186, 5. Vollum, 6. ATCC6605, 7. Pasteur, 8. BA8800. The analysis of the PCR products is summarized in the table above. The three marker system allowed for strain division into five subgroups

20.2.2 *Virtual Analysis of the* B. anthracis *Strains Present at the GeneBank Database*

To date, genome sequences of fourteen *B. anthracis* strains are available in the NCBI database. Grouping and exact genotyping of each strain, according to Keim's 8 or 15 VNTRs method, is noted in Fig. 20.3. These strains were analyzed using our three marker system and the result of this analysis is presented in Fig. 20.3. In order to enrich the strain collection in Fig. 20.4 two representative strains from Chad previously analyzed by the three marker system were added (Maho et al., 2006); as well as three strains form the IIBR collection (ATCC 14185, Pasteur and BA8800 (Levy et al., 2005)).

The results in Fig. 20.3 show that the three marker system represents a unique combination for Keim's groups B1 and A4 and actually grouped the previously untyped CDC 684 strain into the latter group. The Tsiankovskii (Russian vaccine strain) exhibits a unique marker combination where the A1a and Aβ groups are characterized by a unique marker combination for each group. On the other hand the system couldn't differentiate between the strains in groups B2 and C which had the same typing marker combination as was the case also for groups A3a and A3b. In

Strain	Group (VNTR 8)	Gen. (8/15)	AA	AJ	AT
Chad A1*	Aβ	/	2	2	5
Chad A8*	Aβ	/	2	4	5
WNA/A0174	A1a	3/205	2	2	11
A0193	A1a	10/216	2	2	11
6605/Pasteur♣	A1a	/	2	2	11
BA8800♣	A1a	/	2	2	7
Tsiankovskii	A1b?	/	2	4	7
ATCC 14185♣	A3a	45/107	3	2	6
Australia94/A0039	A3a	55/66	3	2	7
Sterne	A3b	61/50	3	2	7
A0389	A3b	63/45	3	2	7
A0248	A3b	68/62	3	2	7
A2012	A3b	/26	3	2	7
Ames/Ames Ancestor	A3b	62/24	3	2	7
Vollum/A0488	A4	77/116	3	4	5
CDC 684	A4?	/	3	4	5
Kruger B/A0442	B1	87/3	2	3	7
CNEVA	B2	79/17	3	3	7
A0465	B2	80/20	3	3	7
A1055	C	/1	3	3	7

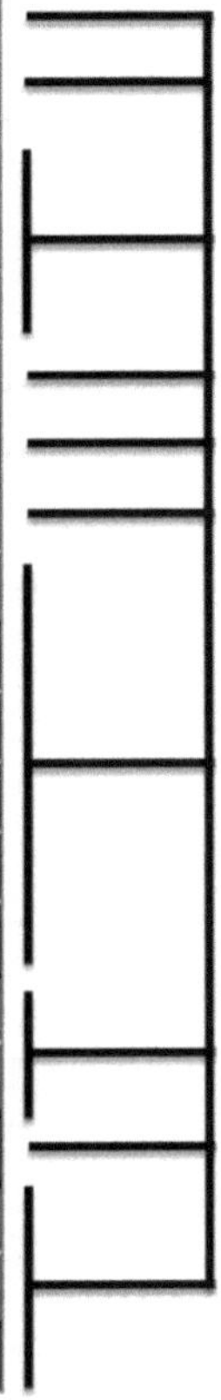

Fig. 20.3 Virtual analysis of the *B. anthracis* strains sequence available at the GeneBank database. The strains are marked according to their GeneBank name. Additional strains form the literature (Maho et al., 2006) are marked with * and strains from the IIBR collection (Levy et al., 2005) are marked with ♣

group A3a the ATCC 14185 strain exhibited a unique combination but the Australia 94 strain was indistinguishable from group A3b.

20.2.3 Addition of a Fourth Marker to the System – X06

To improve the resolution of the system we added a fourth marker. The RAPD marker X06 which like AT07 represented more than two genotypes was chosen. The relevant RAPD PCR product was cloned, sequenced and identified as part of the BA0871 gene, encoding for a cell wall anchored collagen adhesion homolog. The variable region consists of short (20 nt) direct repeats (up to 11). A virtual strain

Strain	Group (VNTR 8)	AA	AJ	AT	X
Chad A1	Aβ	2	2	5	/
Chad A8	Aβ	2	4	5	/
WNA/A0174	A1a	2	2	11	8
6605/Pasteur	A1a*	2	2	11	8
A0193	A1a	2	2	11	7
BA8800	A1a*	2	2	7	8
Tsiankovskii	A1b?	2	4	7	8
ATCC14185	A3a	3	2	6	8
Australia 94/A0039	A3a	3	2	7	11
Sterne	A3b	3	2	7	11
A0389	A3b	3	2	7	11
A0248	A3b	3	2	7	11
A2012	A3b	3	2	7	11
Ames/Ames Ancestor	A3b	3	2	7	11
Vollum/A0488	A4	3	4	5	8
CDC 684	A4?	3	4	5	11
Kruger B/A0442	B1	2	3	7	6
CNEVA	B2	3	3	7	9
A0465	B2	3	3	7	9
A1055	C	3	3	7	16

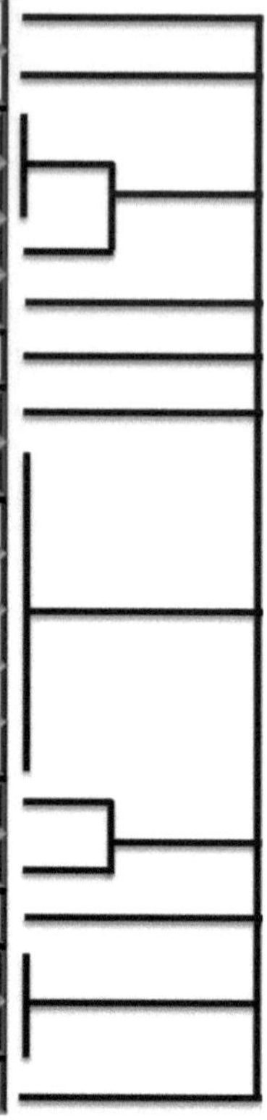

Fig. 20.4 Virtual analysis of the *B. anthracis* strain sequence available from the GeneBank database, using the four marker system. The X06 primers sequence is: X061- GTTGAGCA TGAGAGGTACCTTGTCCTTTTT, X062- AGTTCAAGCGCCAGAAGGTTATGAGTTATC

analysis employing the four markers systems is presented in Fig. 20.4. The fourth marker separates group B3 from group C, for which the new marker represents a novel allele of 16 repeats. Overall the fourth marker increases the number of genotypes typed from 10 (Fig. 20.3) to 13 (Fig. 20.4). Group A1a is now divided into 4 genotypes and each of the two strains in group A4 has its own genotype. However, the fourth marker did not allow for differentiation between the strain Australia 94 (A3b) and group A3a.

Recently Paul Keim and his coworkers (Matthew et al., 2007) published a new phylogenetic system based on selective (13) spontaneous point mutations that accumulated through evolution. Unlike VNTRs the single nucleotide polymorphisims (SNPs) are very stable and therefore can serve as a reliable genetic tool. A phylogenetic tree based on these SNPs is presented in Fig. 20.5. Using blast search we mapped the relevant SNPs onto the GenBank deposited strains sequences and placed

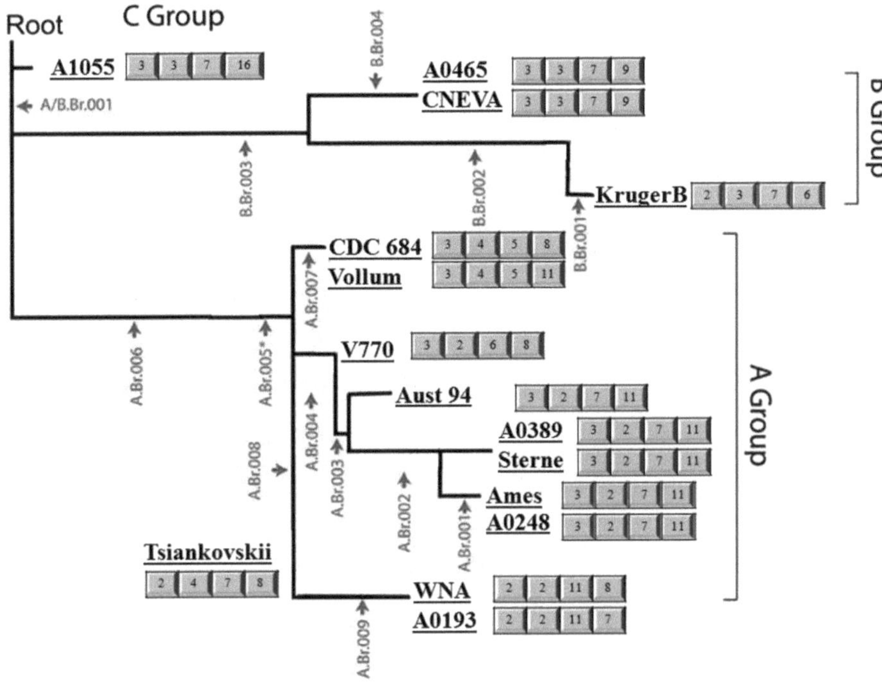

Fig. 20.5 Correlation between strain genotyping using the four marker system and the SNPs – based phylogenetic tree (van Erth et al., 2007)

them accordingly onto the phylogenetic tree (Fig. 20.5). The correlation between the four marker system and SNPs is relatively high. Groups C and B are well differentiated and the same applies to the different branches of the B group. In group A the correlation is very good for the branch of Tsiankovskii and the W/NA and A0193 where each of the three strains has its own marker combination, and the same applies to the CDC and Vollum strain branches. In the last branch, ATCC 14185 (V770) can be separated from the Australia 94 which resides on a separate branch, the same as in the results for the four marker analysis (Figs. 20.4 and 20.5). As was for the three marker system, the four marker system cannot differentiate between the strains Australia 94, A0389, Sterne, Ames and A0248.

20.3 Conclusion

Strain genotyping plays an important role in routine *B. anthracis* R&D' allowing for uniform nomenclature. A handy, simple and non expensive typing method can ensure/validate strain/isolate identity. In this work we described a simple PCR based method for general typing of *B. anthracis* isolates. This system shows good correlation with other established MLVA or SNP based typing methods. However, the limited sensitivity of the system still precludes the distinction between strains such as Sterne, Ames and Australia 94. In this case one should probably use additional markers such as the specific SNPs described by Matthew et al. (2007)

The four markers typing system offers a simple and rapid method for basic typing of known strains in the routine work of anthrax labs (QC) and enables a first general classification of unknown strains.

References

Andersen, G. L., Simchock, J. M. & Wilson, K. H. (1996) Identification of a region of genetic variability among *Bacillus anthracis* strains and related species. *J Bacteriol,* 178, 377–84.

Enserink, M. (2008) Anthrax investigation: Full-Genome sequencing paved the way from spores to a suspect. *Science,* 321, 5891–8.

Keim, P., Kalif, A., Schupp, J., Hill, K., Travis, S. E., Richmond, K., Adair, D. M., Hugh-Jones, M., Kuske, C. R. & Jackson, P. (1997) Molecular evolution and diversity in *Bacillus anthracis* as detected by amplified fragment length polymorphism markers. *J Bacteriol,* 179, 818–24

Keim, P., Price, L. B., Klevytska, A. M., Smith, K. L., Schupp, J. M., Okinaka, R., Jackson, P. J. & Hugh-Jones, M. E. (2000) Multiple-locus variable – number tandem repeat analysis reveals genetic relationships within *Bacillus anthracis. J Bacteriol,* 182, 2928–36.

Le Fleche, P., Hauck, Y., Onteniente, L., Prieur, A., Denoeud, F., Ramisse, V., Sylvestre, P., Benson, G., Ramisse, F. & Vergnaud, G. (2001) A tandem repeats database for bacterial genomes: application to the genotyping of *Yersinia pestis* and *Bacillus anthracis. BMC Microbiol,* 1, 2.

Levy, H., Fisher, M., Ariel, N., Altboum, Z., Kobiler, D. (2005) Identification of strain specific markers in Bacillus anthracis by Random amplification of polymorphic DNA FEMS. *Microbiology Letters*, 244, 199–205.

Maho, A., Rossano, A., Hachler, H., Holzer, A., Schelling, E., Zinsstag, J., Hassane, M. H., Toguebaye, B. S., Ayayi J. Akakpo, A. J., Van Ert, M, Keim, P., Kenefic, L., Frey, J. & Perreten, V. (2006) Antibiotic susceptibility and molecular diversity of *Bacillus anthracis* strains in chad: detection of a new phylogenetic subgroup. *J Clin Microbiol,* 44, 3422–5

Van Ert, M. N., Easterday, R. W, Huynh, L. Y., Richard T. Okinaka, R. T., Martin E. Hugh-Jones, M. E., Ravel, J., R. Zanecki, S. R., Pearson, T.,. Simonson T. S., U'ren, J. M., Kachur S. M., Leadem-Dougherty, R. R., Rhoton, S. D., Zinser, G., Farlow, J., Coker, P. R., Smith, K. L., Wang, B., Kenefic L. J., Fraser-Liggett, C. M., Wagner, D. M. & Keim, P. (2007) Global genetic population structure of *Bacillus anthracis. PLoS ONE,* | www.plosone.org 1 Issue 5 | , e461.

Chapter 21
Tularemia – A Disease with an Uncertain Impact on Public Health

Roland Grunow, Daniela Jacob, Andrea Kühn, and Herbert Nattermann

Abstract *Francisella tularensis* is the causative agent of tularemia and is considered as an agent with the potential to be used deliberately. The species includes subspecies and subtypes with different virulence for humans and occurs in certain areas of the Northern hemisphere. In North America, the virulence of *F. tularensis* subspecies *holarctica* is supposed to lie between the two subtypes A1 and A2 of *F. tularensis* subspecies *tularensis*. Clinical course and appearance depend on an early diagnosis and effective treatment. Detection and diagnosis require special laboratory tests including microbiological, molecular and immunological methods. During the last years, considerable substantial progress has been achieved to better understand the pathology and ecology of this zoonotic pathogen. However, the mechanisms of the obvious long-term persistence of this pathogen in the environment are not well known yet. In Germany, tularemia is a notifiable disease and occurs very rarely although it is probably underestimated. Interestingly, the cases are distributed almost over the entire territory of the country. Epidemiological studies will contribute to a better understanding of the reservoirs and ways of transmission of these bacteria. Outbreaks of tularemia are occasionally occurring in known and unknown endemic areas. The reasons and sources of such outbreaks are often not clarified. However, the intentional release of the agent must be excluded by all means. Therefore an algorithm has been developed to assess the probability of a biological attack and has come into use for tularemia outbreaks in Kosovo after the war in 1999. The results revealed that the unusual outbreak of tularemia in Kosovo most likely originated from a natural source supported by special ecological and hygienic conditions in a post-war situation.

Keywords Francisella · Tularemia · Epidemiology · Outbreak

R. Grunow (✉)
Director and Professor, Head of Division Highly Pathogenic Microorganisms (ZBS 2), Centre for Biological Security, Robert Koch-Institute, Berlin, Germany
e-mail: GrunowR@rki.de

A. Shafferman et al. (eds.), *The Challenge of Highly Pathogenic Microorganisms*,
DOI 10.1007/978-90-481-9054-6_21, © Springer Science+Business Media B.V. 2010

21.1 Microbiology and Clinical Characteristics of Tularemia

Francisella was isolated by McCoy and Chapin from a squirrel species in the county Tulare (route 99, between Fresno and Bakersfield) in California in 1911/1912 (isolation 1913). In 1925, Edward Francis identified the illness corresponding to the bacterium *Francisella tularensis* named tularemia. In recognition of his intensive work on this bacterium the name was later changed from *Pasteurella tularensis* to *Francisella* (*F.*).

At the moment, three species of *Francisella* are known: *F. tularensis, F. philomiragia, and F. noatunensis* or *piscicida* having been suggested for a new isolated species from salmon and other fish as well. Four subspecies of *F. tularensis* have been classified. The subspecies *tularensis* (or type A) includes the most virulent of the four known subspecies (biovar A1) and is associated with lethal pulmonary infections. Subspecies *holarctica* (or type B) is less virulent. The two other subspecies with low clinical relevance are ssp. *mediaasiatica* and ssp. *novicida*.

Francisella is a tiny, gram negative, non-motile, non-spore forming, capsuled bacterium. The cells are pleomorphic, typically appearing as short rods or coccoid forms. As a facultative intracellular pathogen it is also able to survive in ticks for more than 1 year. Cells are sensitive to heat and chemical disinfection and capable of surviving outside a mammalian host for weeks under humid and cool conditions.

Francisella is capable of infecting a large number of animal species such as hares, rabbits, mice and lemmings, and even fish, as more recently discovered. Birds are also known as carriers, but probably do not develop the disease themselves. On the other hand, *F. tularensis* is able to survive in the environment for weeks and has been found in water and grassland. The mechanisms for environmental resistance of these bacteria are not well understood, yet, protozoa like amoeba seem to play an important role as carriers or reservoir of *Francisella*. Biofilm formation could be another mechanism supporting the survival of *Francisella* outside host organisms.

The history shows that tularemia outbreaks were often associated with bad hygienic conditions especially in war and post-war situations. But also natural outbreaks occur in various endemic areas.

Humans acquire tularemia through contact with infected animals and/or vectors, by inhaling contaminated dust or aerosols, and by consumption of contaminated food or water. Various arthropods have been identified as being involved in the transmission of tularemia. In North America ticks are considered the most important vectors, whereas in Scandinavia mosquitoes probably play a major role. There is no evidence for human-to-human transmission. Mammals can be infected by a dose as low as 10–50 bacteria by intradermal or inhalational route of transmission. However, the oral uptake requires more than one million bacteria.

F. tularensis has been listed as a bioterrorism relevant category A agent (CDC classification). This may have different reasons: This pathogen can be transmitted by aerosols, is highly infectious, causes a lethality rate of 5–20% or even higher if untreated, shows same symptoms as by a natural infection, has an incapacitating effect with a long phase of reconvalescence and has been developed to become a biological weapon in the 1950s and 1960s.

The clinical manifestation depends on the route of infection and can be ulceroglandular or glandular, oropharyngeal, conjunctival, and pulmonal. The disease is called typhoidal if the way of infection cannot be identified. Common clinical symptoms are high fever and enlarged lymph nodes.

Most efficient treatment of tularemia, particularly in serious cases, is streptomycin. However, streptomycin is no longer used widely in human medicine because of its potential to cause vestibular toxicity and a frequent appearance of hypersensitive reactions. More often, other aminoglycosides, like gentamicin, or quinolone, like ciprofloxacin and doxycyclin, are used. For the treatment of tularemia meningitis, chloramphenicol is the most successfully used remedy in combination with streptomycin (World Health Organization, 2007).

21.2 Epidemiology

Tularemia has been reported in many countries of the Northern hemisphere, but not from the Southern hemisphere *to date*. At any rate, to our knowledge, no systematic studies on tularemia have been published, so far, concerning the Southern hemisphere. An isolate of *F. tularensis ssp. novicida* has been found in Australia, but without typical clinical presentation of tularemia. In the former Soviet Union, Scandinavia, the Balkan states, and North America endemic foci have existed for a long time. In Europe, a spread of the disease from North East to South West seems to take place.

During the last years, new insights in the ecology and epidemiology of *Francisella* could be achieved and several open questions have been identified, which should be considered for surveillance and human health. Some of these topics will be focussed in this article.

In a retrospective study of *Francisella* (samples from the USA and CDN, collected between 1964 and 2004) two genetically and geographically distinct biovars A1 and A2 of *F. tularensis* ssp. *tularensis* have been discovered by pulsed-field gel electrophoresis. These biovars are distributed to different host organisms and vectors, and, most important, they differ in their virulence for humans. A1 turned out to be most pathogenic, whereas A2 showed a lower virulence than *F. tularensis* ssp. *holarctica*. This observation is important regarding North America but also in terms of natural or intentional spread of the bacteria in other areas of the world (Petersen and Schriefer, 2005; Farlow, et al., 2005).

Some regions in Europe, like the UK and Iceland, are believed to be free of tularemia. In regions with rare or emerging and re-emerging tularaemia, the epidemiology and ecology of *Francisella* are of special interest. In such areas, the prevalence and transmission of pathogens and the conditions for their survival are only inadequately known.

21.3 Epidemiology of Tularemia in Regions with Probably Low Prevalence

Germany is an example for rare cases of human tularemia. Although tularemia is a notifiable infectious disease in Germany, only 3–5 cases have been reported on average per year for the decades. However, an elevated number of tularemia cases were registered between World War II and the mid 1950s. Most interestingly, the few cases reported in decades have been spread all over Germany. This observation was underlined by a more recent cross sectional study with more than 9,000 sera, revealing a seroprevalence of 0.2% throughout the German population, again distributed all over Germany. Assuming an adult population between 18 and 65 years and a titre persistence of over 20 years, an incidence of approximately 5,000 cases could be estimated for the whole German population, which is surprisingly high compared with the number of reported cases. However, this may reveal a probable underestimation of tularemia in Germany due to various possible reasons.

Calculating the prevalence of tularemia per 10^6 inhabitants for the different German States, based on data from 1974 to 2005, some regions showed a higher tularemia activity. In Mecklenburg-Western Pomerania, we could retrospectively find a high seroprevalence also in wild animals, like wild boars (1976–1989) 0.09% (own unpublished data, personal communication Dr. H. Nattermann) and, more recently, (1995/1996) 3.1% (Al Dahouk et al., 2005), as well as in hares (1976–1989) 10.7%. Other regions of interest may be deduced from these historical data in Saarland, Hesse and Saxony-Anhalt (Fig. 21.1).

In addition, more recent results revealed further tularemia regions. The above-mentioned regions are depicted in Fig. 21.1. Thus, a tularemia outbreak occurred in common marmosets (*Callitrix jacchus*) living in open cages with 5/62 deaths in Lower Saxony in 2004 (Splettstoesser et al., 2007). In 2005, 9/39 hunters got

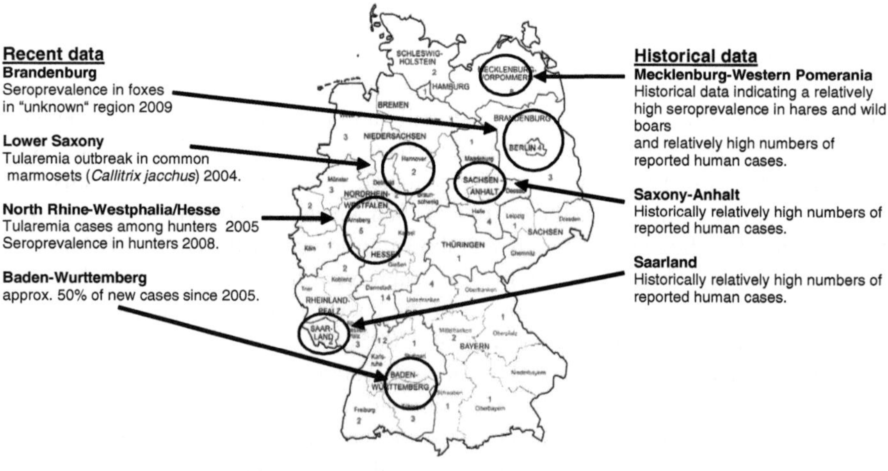

Fig. 21.1 Tularemia in Germany – geographical presentation

infected with *F. tularensis* after a hare hunting event (State Hesse). We investigated sera from hunters ($n = 313$) in North Rhine-Westphalia and Hesse, of which we found 6 positive (1.9%) (Jenzora et al., 2008). During the last 3 years, about half of all tularemia cases occurred in the South West of Germany in Baden-Wurttemberg. It seems that the number of reported tularemia cases increased over the last years since 2005 with the exception of one case only in 2006. Then, 15–21 cases have been registered which could mean that the higher numbers not only referred to improved surveillance and diagnostics. In this context, some peculiarities became obvious. Thus, for the first time in Germany, *F. tularensis* was detected in blood cultures and two cases of the pulmonary form were diagnosed, one of which was probably related to a mosquito bite (Splettstoesser et al., 2009). This way of transmission is rather suspected in Scandinavia. In 2009, we were able, for the first time, to describe a case probably caused by a tick (Lübbert et al., 2009).

Now, we were interested in studying the prevalence of tularemia in a region rather not suspected for the presence of tularemia. The aim was to verify that *Francisella* was spread over the entire German territory, including also fairly unexpected areas. We took advantage of material from wild foxes, which were killed in order to prevent overpopulation and which were investigated routinely for rabies.

Surprisingly, we detected a seroprevalence of 8.8% (31 of 351 foxes investigated so far) and obtained one *F. tularensis* isolate (Nattermann et al., 2009; Kühn et al., 2009). Interestingly, in November 2009, two human cases of tularemia were notified in this region. That would underline the assumption that *F. tularensis* is distributed in nature all over Germany. The main questions arising are: Where and how does the pathogen survive in nature? What is the real prevalence? And what circumstances can lead to an infection of humans?

21.4 Epidemiology in Regions with (re-)Emerging Tularemia

Kosovo could serve as an example of (re-)emerging tularemia. It has a territory of 10,908 km^2 with a population of approx. 2 million inhabitants, located in South East Europe.

By mid 1999, more than 10 years of political crisis and warfare resulted in environmental disruption, displacement of major parts of the population, and a breakdown of sanitation and hygiene.

It can be suspected that tularemia was present in this region during and/or after World War II but, officially, there are no registered tularemia cases since 1946. Although the disease was admitted to report in former Yugoslavia, local representatives stated that cases of tularemia were not detected before the war in Kosovo in 1998/1999. Thus, the first cases of tularemia were officially registered in Kosovo in 1999.

An international team recruited by WHO performed an intensive investigation of this first outbreak lasting from October 1999 to May 2000 (Reintjes et al., 2002). A second outbreak occurred in 2001/2002, which we investigated together with the Institute of Public Health in Pristina.

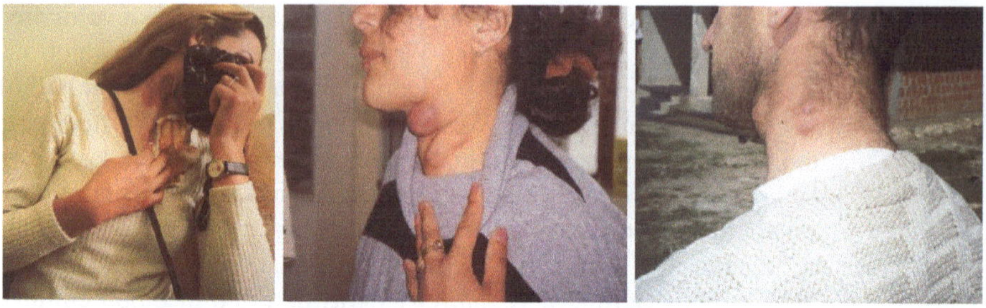

Fig. 21.2 Tularemia in Kosovo – clinical manifestation. Almost exclusively oropharyngeal Tularemia, cervical lymphadenitis, Kosovo, 2002

The clinical presentation of tularemia was almost exclusively oropharyngeal with cervical lymphadenitis (Fig. 21.2).

In both timely related outbreaks, 680 cases could be diagnosed by serology. The cases were probably spread along the main rivers coming from the mountains surrounding and dividing Kosovo. However, a correlation of numbers of cases with war afflicted areas could also be observed.

Francisella antigen could be detected in rodent faeces collected in food storages and in faeces of wild hares as well as in the tissue of dead rodents. Therefore, these animals were most likely the source of infection for humans. However, it was difficult to explain why the outbreak occurred with such intensity and in such a narrow time frame nearly all over Kosovo.

Possibly, the disease was enzootic before, but not known or it was imported by refugees or traders in food or exotic animals, birds, etc.

Another supposition is, that it resulted from deliberate release of *Francisella* during or after the war in Kosovo. Facing this question, we developed an approach to discriminate natural from intentional outbreaks of infectious diseases, which was applied by other authors for analysing of further outbreaks (Grunow and Finke, 2002; Dembek et al., 2007).

On the one hand, this approach considered the identification of "Conclusive Criteria" directly inferring that a biological attack has taken place. This could be the proof of the release of the agent by a biological weapon or the identification of the pathogen as a biological warfare agent. On the other hand, "Non-conclusive Criteria" included different aspects, such as political, military and social analyses of crisis afflicted regions, specific features of the pathogen, or the epidemiological and clinical characteristics of the epidemic. As these criteria may only indirectly indicate the usage of biological warfare agents, the likelihood of a biological warfare event must be estimated. Therefore, we used a scoring system, including weighting factors for the estimation of risks. The result of this assessment showed that the deliberate release of *Francisella* was doubtful.

The most probable explanation is that these new outbreaks in Kosovo occurred as a result of a typical post-war situation in autumn 1999. A spreading of tularemia among the overwhelming rodent population can be assumed. Unusually large numbers of rodents were reported by the local population. This was most likely the result of unharvested fields and a subsequent oversupply with food for these animals. Due to the close contact of animals with dwellings and food storages, a contamination of food and drinking water, followed by alimentary ingestion of the agent by humans, was likely. The spread of the infection over nearly the entire Kosovo can only be explained by assuming that the pathogen was already present in nature in most of the affected areas before. Ongoing regular cases and small outbreaks of tularemia show that the predicted development of an emerging or re-emerging endemic area for tularemia has taken place in Kosovo since the outbreaks in 1999 and 2001.

In summary, *Francisella* could epidemiologically remain static for many years although present in nature. The mechanisms of environmental resistance are not well known yet. Natural reservoirs like rodents, hares, amoeba, other water protozoa etc. are suspected but not yet confirmed. Reasons for epidemics and transmission to humans are poorly understood. Knowledge regarding the real prevalence of *Francisella* in the environment and factors causing epidemics, would contribute to estimate and control the impact of this pathogen on human and animal health.

References

Al Dahouk, S., Nöckler, K., Tomaso, H., Splettstoesser, W. D., Jungersen, G., Riber, U., Petry, T., Hoffmann, D., Scholz, H. C., Hensel, A. & Neubauer, H. (2005) Seroprevalence of brucellosis, tularemia, and yersiniosis in wild boars (*Sus scrofa*) from north-eastern Germany. *J Vet Med B, Infect Dis Vet Public Health,* 52(10), 444–55.

Dembek, Z. F., Kortepeter, M. G. & Pavlin, J. A. (2007) Discernment between deliberate and natural infectious disease outbreaks. *Epidemiol Infect,* 135(3), 353–71. Erratum in: *Epidemiol Infect,* 2007, 135(6), 1055.

Farlow, J., Wagner, D. M., Dukerich, M., Stanley, M., Chu, M., Kubota, K., Petersen, J. & Keim, P. (2005) *Francisella tularensis* in the United States. *Emerg Infect Dis,* 11(12), 1835–41.

Grunow, R. & Finke, E. J. (2002) A procedure for differentiating between the intentional release of biological warfare agents and natural outbreaks of disease: its use in analyzing the tularemia outbreak in Kosovo in 1999 and 2000. *Clin Microbiol Infect,* 8(8), 510–21.

Jenzora, A., Jansen, A., Ranisch, H., Lierz, M., Wichmann, O. & Grunow, R. (2008) Seroprevalence study of *Francisella tularensis* among hunters in Germany. *FEMS Immunol Med Microbiol,* 53(2), 183–9.

Kühn, A., Schulze, C., Kutzer, P., Ranisch, H., Bestek, S., Nattermann, H. & Grunow, R. (2009) Tularaemia seroprevalence studies among foxes and raccoon dogs in Germany. *6th International Conference on Tularemia 2009 Tularemia Network September 13–16, 2009, Berlin, Germany.* Abstract only. Available at: http://www.tularemia-network.com/ , P2-16.

Lübbert, C., Taege, C., Seufferlein, T. & Grunow, R. (2009) Prolonged course of tick-borne ulceroglandular tularemia in a 20-year-old patient in Germany-case report and review of the literature. *Deutsche Medizinische Wochenschrift,* 134(27), 1405–10.

Nattermann, H., Becker, S., Jacob, D., Kühn, A., Kutzer, P., Lochau, P., Schulze, C. & Grunow. R. (2009) First isolation of *Francisella tularensis* subspecies *holarctica* from Foxes *(Vulpes vulpes)* in Germany. *6th International Conference on Tularemia 2009 Tularemia Network September 13–16, 2009, Berlin, Germany.* Abstract only. Available at: http://www.tularemia-network.com/ , P1-43.

Petersen, J. M. & Schriefer, M. E. (2005) Tularemia: emergence/re-emergence. *Vet Res,* 36(3), 455–67.

Reintjes, R., Dedushaj, I., Gjini, A., Jorgensen, T. R., Cotter, B., Lieftucht, A., D'ancona, F., Dennis, D. T., Kosoy, M. A., Mulliqi-Osmani, G., Grunow, R., Kalaveshi, A., Gashi, L. &

Humolli, I. (2002) Tularemia outbreak investigation in Kosovo: case control and environmental studies. *Emerg Infect Dis,* 8(1), 69–73.

Splettstoesser, W. D., Mätz-Rensing, K., Seibold, E., Tomaso, H., Al Dahouk, S., Grunow, R., Essbauer, S., Buckendahl, A., Finke, E. J. & Neubauer, H. (2007) Re-emergence of *Francisella tularensis* in Germany: fatal tularaemia in a colony of semi-free-living marmosets (*Callithrix jacchus*). *Epidemiol Infect,* 135(8), 1256–65.

Splettstoesser, W. D., Piechotowski, I., Buckendahl, A., Frangoulidis, D., Kaysser, P., Kratzer, W., Kimmig, P., Seibold, E. & Brockmann, S. O. (2009) Tularemia in Germany: the tip of the iceberg? *Epidemiol Infect,* 137(5), 736–43.

World Health Organization. (2007) *WHO Guidelines on Tularemia.* Available at: WHO Library Cataloguing-in-Publication Data, Geneva, Switzerland. http://www.cdc.gov/tularemia/resources/whotularemiamanual.pdf

Chapter 22
Consequences of Antibiotic Treatment of *Francisella tularensis* Airways Infections

Erez Bar-Haim, Hila Cohen, Raphael Ber, Ofer Cohen, and Avigdor Shafferman

Abstract Antibiotics are considered as an effective treatment against *Francisella tularensis* (FT), the causative agent of Tularemia. In this study we compared the efficacy of two types of antibiotics: the bacteriostatic doxycycline and the bactericidic ciprofloxacin, for protection against intranasal FT-LVS in a mouse infection model. Antibiotic treatments were initiated 24–72 h post- bacterial infection, administered twice daily and for a period of 7 days (ciprofloxacin) or 10 days (doxycycline), respectively.

All treated mice survived the infection, even when the treatment was initiated after the appearance of disease symptoms (72 h). Bacterial clearance from the lungs, liver and spleen was more efficient in the ciprofloxacin-treated mice while in doxycycline-treated mice complete clearance was observed more than a week after cessation of antibiotic treatment. Nevertheless, no signs for relapse were observed in doxycycline-treated mice, probably due to the development of an immune response against the invading bacteria. Indeed, high antibody titers were observed in these mice, 10–50 fold higher than in the ciprofloxacin-treated mice.

Keywords *Francisella tularensis* · Antibiotic · Airways infection · LVS

22.1 Introduction

FT is the causative agent of tularemia, a highly infectious and potentially fatal disease in humans. Infection of humans can be established by a variety of exposure routs, with airway infection being the most severe form of the disease. For Type A strains, 30–60% mortality rates has been reported for respiratory disease in humans, if left untreated (Hepburn and Simpson, 2008; Sjostedt, 2006; Tarnvik and Berglund,

E. Bar-Haim (✉)
Department of Biochemistry and Molecular Genetics, Israel Institute for Biological Research, P.O. Box 19, Ness Ziona 74100, Israel
e-mail: erezb@iibr.gov.il

A. Shafferman et al. (eds.), *The Challenge of Highly Pathogenic Microorganisms*,
DOI 10.1007/978-90-481-9054-6_22, © Springer Science+Business Media B.V. 2010

2003). The low infection dose, severity of the disease, and the historic interest in FT as bioweapon, led to the classification of this pathogen as a category A bioterrorism agents. Due to the lack of efficient vaccines capable of conferring full protection against airway infections, antibiotics are still considered the treatment of choice for tularemia.

The most studied tularemia animal model is the LVS strain in mice, derived from *F. tularensis holarctica* subsp. LVS is highly virulent to mice and the LD_{50} of airways infection (i.n.) is approximately 10^3 cfu. An intranasal instillation dose of 10^5 cfu LVS ($100 \times LD_{50}$) results in a systemic disease, manifested by weight lost, low mobility and death by day 6–8 post infection (Fortier et al., 1991). Bacteria multiply in the lungs and disseminate to the liver and spleen from day 2 post infection (Bar-Haim et al., 2008; Fortier et al., 1991).

Antibiotic treatment options for tularemia, traditionally include streptomycin (and its oral alternative choice, gentamicin) and tetracycline (doxycycline). Fluoroquinolones are promising alternatives for treatment, having the advantage of bactericidal activity, and high in vitro potency (Hepburn and Simpson, 2008). Only few studies examined the efficacy of tularemia antibiotic treatment in a mouse infection model. In one study (Russell et al., 1998) ciprofloxacin and doxycyclin, administered intraperitoneally, were compared for efficacy against systemic (i.p.) infection of Schu S4 strain (*F. tularensis tularensis* subsp.). Both antibiotics were efficient when given prophylacticly (48 h before infection), or 1 day post infection (with a small advantage for ciprofloxacin over doxycycline in the post exposure treatment). However, delayed treatment or treatment against airway infection was not tested.

The current study is aimed to establish an experimental system for antibiotic treatment of *F. tularensis* LVS airways infections. This model is employed for analysis of treatment efficacy, delayed treatment, potential relapse, and acquisition of humoral immunity following recovery from antibiotic-treated infection.

22.2 Materials and Methods

Mice: All experiments were carried-out with 8–12 weeks-old, female BALB/c mice. The work was conducted in compliance with the guidelines of the animal use committee at the Israel Institute for Biological Research and is in accordance with the Animal Welfare Act.

Bacterial infection: LVS were grown to mid-log phase in TSB medium (Difco), added with 0.1% cysteine. Bacteria were washed and diluted in PBS up to a concentration of 4×10^6 cfu/ml. Mice anesthetized with ketamine/xylazine were infected intranasally with 25 μl of bacteria suspension, to give an infection dose of 10^5 cfu.

Antibiotic treatment: Mice (10 mice per group) were treated by i.p. injection twice daily with ciprofloxacin (50 mg/kg) for a period of 7 days or doxycycline (40 mg/kg) for 10 days. The treatment commenced 24, 48 or 72 h post infection. Mice were monitored for morbidity and mortality, for 28 days on a daily basis.

ELISA: Antibody titers were determined in serum samples withdrawn 28 days post infection from treated and untreated mice. Formalin-inactivated LVS served as an antigen for the ELISA.

22.3 Results and Discussion

22.3.1 Mortality, Morbidity and Bacterial Dissemination

Mice were intranasally exposed to a lethal dose of LVS (100 LD_{50}), followed by antibiotic treatment (ciprofloxacin, 50 mg/kg; doxycycline 40 mg/kg; administered twice daily i.p.) commencing 1, 2 or 3 days post infection. Mortality and morbidity were monitored for 28 days (Fig. 22.1). All antibiotic treated mice survived the LVS infection, in contrast to the control infected mice which succumbed to challenge. Morbidity was observed for all doxycycline-treated mice groups. For each group severity of weight loss was correlated with onset of treatment. Clear signs of morbidity were observed in the ciprofloxacin-treated animals when treatment was initiated 3 days post infection, and to a lesser extent in some of the mice treated from day 2 post infection (not shown). No relapse was observed during the 3 weeks follow-up after cessation of antibiotic treatment. The recovery of the doxycycline-treated mice was slower than that of the ciprofloxacin-treated mice, as reflected by the time required to reach the weight of the uninfected controls: 2–3 weeks for doxycycline treated mice as compared to 1–2 weeks for the ciprofloxacin treated group. These differences in disease progression and recovery are in line with the reported potency of ciprofloxacin treatment against LVS (Johansson et al., 2002).

Bacterial dissemination in the lung, liver and spleen was monitored during and after antibiotic treatment. The rapid effect of cypro is reflected by a lower bacterial burden in the lungs after one or two days of treatment (Fig. 22.2 *lower left panel*). Nearly complete clearance of bacteria was observed seven days post

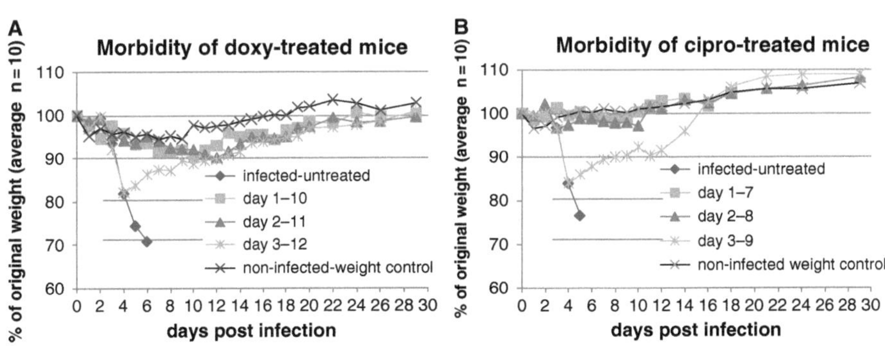

Fig. 22.1 Morbidity of antibiotic-treated mice. Mice were infected intranasally with 10^5 cfu of the LVS strain. Antibiotic treatment was initiated 1–3 days post infection, and lasted 10 days (doxycycline, **a**) or 7 days (ciprofloxacin, **b**)

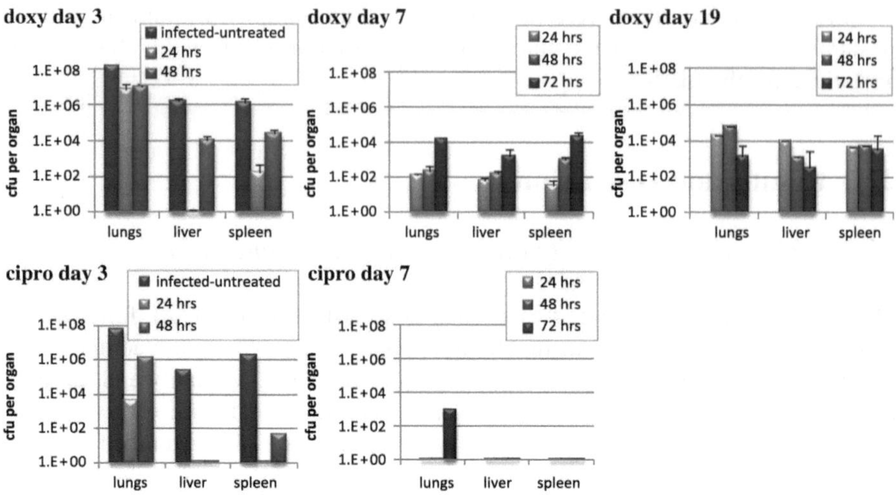

Fig. 22.2 Bacterial dissemination and multiplication. Groups of 3 mice were sacrificed at indicated time points post infection. Lung, liver and spleen were removed, homogenized, suspended in PBS, and serially diluted and plated onto CHA+1% hemoglobin. The presented data are an average of 3 mice

infection (Fig. 22.2 *lower right panel*). In the doxycycline treated group reduction rate in bacterial count is slow, (Fig. 22.2 *upper middle panel*), and live bacteria could be detected on day 19 post infection (7–9 days after cessation of doxycycline treatment). Complete clearance was observed only on day 28 post infection (not shown). These data are in line with the slower recovery of the doxycycline-treated mice (Fig. 22.1), and the bactericidal nature of doxycycline.

22.3.2 Development of Immune Response Following Antibiotic Treatment

Antibody titers against LVS were determined on day 28 post infection (Fig. 22.3). The titers of the doxycycline-treated mice were higher than those of the ciprofloxacin-treated animals (10–50 fold), and higher than the titers observed for control mice intranasally immunized with sub-lethal doses of LVS (10^2 cfu/mouse). The higher humoral response, following doxycycline-treatment, could be attributed to the slower bacterial killing kinetics by doxycycline as compared to cypro, and to incomplete bacterial clearance during doxycycline treatment (Fig. 22.2). Individual time point groups appear to exhibit dose dependence between antigen load and antibody titers, i.e. antibodies titers are lower in the 24-h group than the in the two other groups of doxycycline-treated mice where treatment was initiated later.

Upon re-challenged of the mice with 100 LD_{50} of LVS 4 or 5 weeks after the first challenge of ciprofloxacin-treated or doxycycline-treated mice respectively, we observed that most animals survived the challenge, indicating the

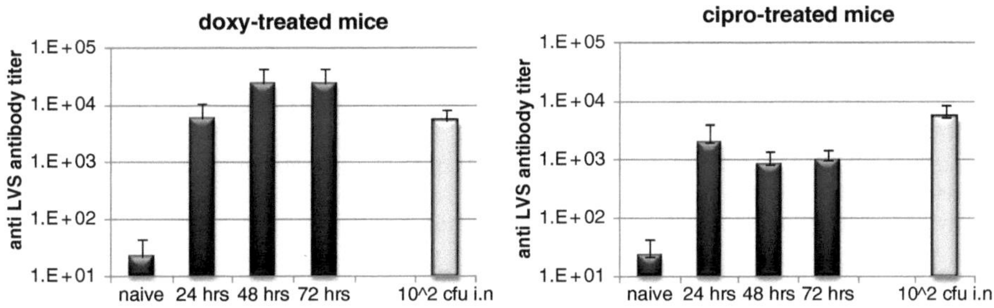

Fig. 22.3 Antibody response in the treated mice. Mice were bled 4 weeks post infection, while no signs of disease or live bacteria were observed in the mice. A control of intranasally-LVS vaccinated mice is included. The presented data are a geometric average of 3 mice

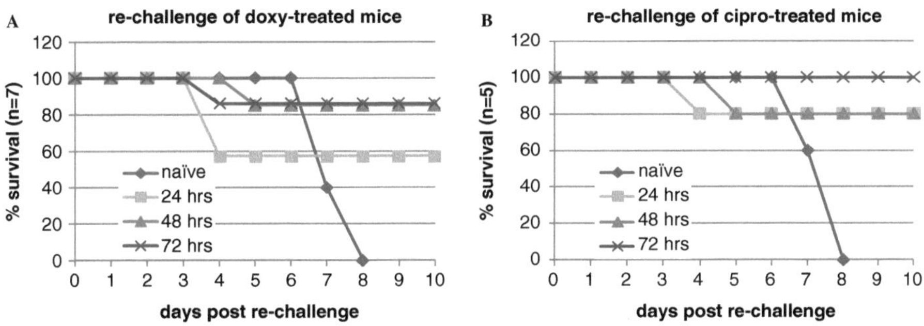

Fig. 22.4 Survival of re-challenged mice. Mice were challenged intranasally with 10^5 cfu of LVS ($100 \times LD_{50}$) 4 weeks (ciprofloxacin-treated; **a**) or 5 weeks (doxycycline treated; **b**) post initial infection

development of protective immune response following the antibiotic treatment. Surprisingly, the ciprofloxacin-treated mice which generated a less robust humoral immune response following the antibiotic treatment were better protected than the doxycycline-treated mice.

In summary, this study demonstrates that both antibiotics are effective for treatment of respiratory tularemia in the mouse model. Antibiotic treatments were efficient even when treatment commenced after manifestation of clinical symptoms. It appears that ciprofloxacin treatment is more advantageous than doxycycline, at least in this model system, based on the lower morbidity, quicker recovery and more efficient bacterial clearance observed for the ciprofloxacin vs. doxycycline treatment.

References

Bar-Haim, E., Gat, O., Markel, G., Cohen, H., Shafferman, A. & Velan, B. (2008) Interrelationship between dendritic cell trafficking and *Francisella tularensis* dissemination following airway infection. *PLoS Pathog,* 4, e1000211.

Fortier, A. H., Slayter, M. V., Ziemba, R., Meltzer, M. S. & Nacy, C. A. (1991) Live vaccine strain of *Francisella tularensis*: infection and immunity in mice. *Infect Immun,* 59, 2922–8.

Hepburn, M. J. & Simpson, A. J. (2008) Tularemia: current diagnosis and treatment options. *Exp Rev Anti Infect Ther,* 6, 231–40.
Russell, P., Eley, S. M., Fulop, M. J., Bell, D. L. & Titball, R. W. (1998) The efficacy of ciprofloxacin and doxycycline against experimental tularaemia. *J Antimicrob Chemother,* 41, 461–5.
Sjostedt, A. (2006) Intracellular survival mechanisms of *Francisella tularensis*, a stealth pathogen. *Microbes Infect,* 8, 561–7.
Tarnvik, A. & Berglund, L. (2003) Tularaemia. *Eur Respir J,* 21, 361–73.

Chapter 23
Novel Live Vaccine Candidates Against Airborne *Francisella tularensis*

Wayne Conlan and Anders Sjöstedt

Abstract *Francisella tularensis* is a facultative intracellular bacterial pathogen of humans and many other mammals. In particular, inhalation of small numbers of the *tularensis* subspecies has high mortality in the absence of effective clinical intervention. Current concerns about bioterrorism have renewed interest in the development of a licensable vaccine against the pathogen. An attenuated strain, LVS, of the less virulent *holarctica* subspecies was developed more than 60 years ago and demonstrated substantial, but not complete, protection in human volunteer studies. Issues concerning its safety, its method of attenuation and manufacture, and mechanism of action have been additional barriers to its licensure. In an attempt to address these issues, we have generated highly attenuated deletion mutants of subspecies *tularensis* strain, SCHU S4, that show better or worse protection than LVS against aerosol challenge with the wild-type strain in a murine model. By examining the immune responses elicited by vaccine strains with differing efficacies, a much more comprehensive understanding of protection mechanisms against the pathogen will emerge.

Keywords Francisella · Vaccine · Mutants · Aerosol

23.1 Introduction

Francisella tularensis is a zoonotic facultative intracellular coccobacillus, often arthropod-borne, and the etiological agent of a broad spectrum of infectious diseases collectively called, tularemia. Two subspecies, *F. tularensis* subps. *tularensis* and subsp. *holarctica* are pathogenic for humans. The latter is found throughout the

W. Conlan (✉)
National Research Council of Canada, Institute for Biological Sciences, Ottawa, ON K1C 2M7, Canada
e-mail: wayne.conlan@nrc-cnrc.gc.ca

A. Shafferman et al. (eds.), *The Challenge of Highly Pathogenic Microorganisms*,
DOI 10.1007/978-90-481-9054-6_23, © Springer Science+Business Media B.V. 2010

Northern hemisphere, and is a relatively common cause of infection in Scandinavia and Russia. In contrast, subsp. *tularensis* is confined to North America, and is a rare cause of human infection (reviewed in Sjostedt, 2007). Tularemia pathogenesis depends on the subspecies involved and its route of entry into the host; regardless of these criteria, the infectious dose is 25 CFU or less. Transdermal infection caused by various biting arthropods, or by handling infected animal carcasses, gives rise to ulceroglandular tularemia characterized by ulcer formation at the site of infection, and lymphadenopathy of the local draining lymph nodes. Combined these signs can be considered to be presumptively diagnostic (Sjostedt, 2007). Mosquito-borne ulceroglandular tularemia caused by subsp. *holarctica* accounts for the majority of natural cases of tularemia in Europe, and is never lethal. In contrast, transdermal infection with subsp. *tularensis* has a mortality rate of approximately 5% if left untreated. Respiratory tularemia is caused by inhalation of the pathogen, and it manifests with vague flu-like symptoms with or without overt respiratory involvement, and X-rays may remain clear. Inhalation of subsp. *holarctica* does not cause lethal human infection. However, inhalation of susbsp. *tularensis* can give rise to a disseminated infection, typhoidal tularemia, with a mortality rate of 30–60% without treatment (Sjostedt, 2007). This fact coupled with the low infectious dose led to the development by various countries including the USA and USSR of subspecies *tularensis* as an aerosolized biological warfare agent during the first part of the 20th century (Dennis et al., 2001). Although these programs have been verifiably dismantled, there is nowadays, renewed concern about this pathogen being used as a biothreat agent by terrorists. Concomitantly, there has been an upsurge in interest, primarily by the US government, in novel countermeasures against such a threat. In this regard, there are concerns that *F. tularensis* might be engineered to be resistant to the small number of effective antibiotics. This makes vaccination a preferred solution.

23.2 *Francisella tularensis* Live Vaccine Strain

Francisella tularensis live vaccine strain (LVS) is the sole tularemia preventative vaccine available in the West. It was developed in the 1950s from an attenuated strain of subspecies *holarctica*, strain S15,that was gifted to the USA from the USSR (reviewed in Ellis et al., 2002). LVS was extensively tested on human volunteers during the 1950s and 1960s. Given transdermally by scarification, it has an efficacy of 50–100% against intradermal or aerosol challenge with the prototypical susbsp. *tularensis* strain, SCHU S4 (McCrumb, 1961;Saslaw, et al., 1961; Hornick and Eigelsbach, 1966). Efficacy was dependent on the vaccine and challenge dose and the period between these two events. Because LVS was a pragmatically developed vaccine, regulatory concerns have been raised about (i), its method of attenuation and ability to revert; (ii), its method of manufacture; (iii) safety, especially for immunocompromised individuals; (iv), mechanism of action and correlates thereof (reviewed in Conlan and Oyston, 2007). Recent work has shown that the primary mutation responsible for attenuation is a deletion spanning two adjacent genes, *FTT0918* and *FTT0919* (Salomonsson et al., 2009). Interestingly, this generates a hybrid gene that encodes for a protein not found in natural strains of the pathogen (Twine et al., 2005). Additionally, a mutation of the gene *pilA*, encoding a putative type IV pilin, contributes a small degree of attenuation (Salomonsson et al., 2009). Progress has also been made by the development of a cGMP compliant manufacturing process for LVS (Pasetti et al., 2008). LVS is clearly safe for immunocompetent humans and experimental animals by transdermal routes. However, it is as virulent as clinical *F. tularensis* subsp. *holarctica* or subsp. *tularensis* strains for mice with defects in interferon gamma (IFNγ) or tumor necrosis factor alpha (TNFα) production or for neutropenic mice which raises concerns about its potential pathogenicity for immunocompromised humans (reviewed in Elkins et al., 2007). However, the aforementioned syndromes are clinically rare compared to defects in acquired immunity, and LVS is much more benign for mice lacking B- or T- cells (Elkins et al., 2007). Finally, at the time of the aforementioned human volunteer experiments, our knowledge of acquired immunity, especially cell-mediated immunity was, at best, rudimentary. Consequently, only relatively crude serological assays (hemagglutination and bacterial agglutination) were performed on volunteer samples (Saslaw and Carhart, 1961). This work demonstrated no correlation between antibody titres post-vaccination and protection post-challenge. LVS has been given to humans during the modern immunological era. Studies on these vaccinees revealed that LVS immunization generates antibodies to a diversity of *Francisella* antigens, as well as specific $CD4^+$ and $CD8^+$ T cells capable of secreting IFNγ (Conlan and Oyston, 2007). However, because the protection status of these individuals was never established, the utility of their immune responses to serve as correlates of protection remain unknown. According to murine studies, $CD4^+$ and $CD8^+$ T cells and IFNγ are critical for LVS-induced acquired immunity to intradermal or respiratory challenge with subsp. *tularensis*. However, some mouse strains are not protected by vaccination with LVS despite generating such immune responses. Thus, it would appear that merely generating specific T cells is not a suitable proxy for gauging protection. Instead, there must be more refined immunological parameters that govern post-vaccination protection status. However, their identity and ability to predict protection in the absence of pathogen challenge remain unknown.

23.3 Novel Live Vaccines

23.3.1 Mutants of **F. tularensis** *subsp.* **tularensis** *Strain SCHU S4 as Potential Vaccine Candidates*

We hypothesized that defined gene deletion mutants of *F. tularensis* subsp. *tularensis* might be better vaccines than LVS, not least because they express antigens unique to this subspecies that might provide additional useful protective epitopes (Twine et al., 2005). This required us to develop an appropriate gene deletion strategy, a task that had remained elusive to the field as methods devised for other pathogens failed to work with *F. tularensis*. Eventually, we developed a method for deleting genes from LVS that was based on homologous recombination in conjunction with a suicide vector which meant that the mutants contain unmarked, in-frame mutations (Golovliov et al., 2003). This was subsequently adapted to work with SCHU S4 (Twine et al., 2005). To date we have generated 56 unique single deletion mutants with ID LD_{50}s ranging from <10 CFU (same as wild-type) to >10^8 CFU as well as 18 double mutants all with LD_{50}s > 10^7 CFU (similar to LVS). The genotypes of many of these have been reported by us previously (Kadzhaev et al., 2009). Interestingly, highly attenuated mutants elicited varying degrees of protection in BALB/c mice against ID or respiratory challenge with SCHU S4. Some elicit no protection, some extend time to death, some protect against ID challenge only (like LVS), and some against both ID and aerosol challenge. Thus, these mutant strains are unique reagents with which to try and determine the immunological basis for protection against transdermal or respiratory challenge. Recently, we identified a mutant missing the heat shock protein gene, *clpB* that provided better protection than LVS against aerosol challenge, and a double deletion mutant missing the *FTT0918* gene that is disrupted in LVS and increases by 10,000-fold the ID LD_{50} SCHU S4, and the gene, *capB* that has homology to *Bacillus anthracis* capsule biosynthesis genes (Conlan et al., in press). The double mutant elicits protection intermediate between Δ*clpB* and LVS. All three strains have ID LD_{50}s > 10^7 CFU, and all three elicit complete protection against an ID challenge with 1,000 LD_{50} of SCHU S4.

For LVS immunized humans, protection against challenge with wild-type bacteria is the only correlate of protection that exists to date. Therefore, this seems to be the logical starting point for animal studies that aim to determine mechanisms and correlates of protection. To this end, we began by examining the course of infection in mice vaccinated ID with 10^5 CFU of one or other vaccine strain that were then challenged six weeks later with a low dose aerosol of SCHU S4. All three vaccine strains grew locally in the skin, and disseminated and grew in the liver, and spleen, but not the lungs. There were no overt differences in the sub-lethal infections caused by any of the three strains. Following aerosol challenge, mice immunized with Δ*clpB* survived longer than mice immunized with Δ*0918*Δ*capB* which, in turn, survived longer than mice immunized with LVS. Survival correlated with the degree to which immunized mice were able to control pulmonary and disseminated infection in the liver and spleen relative to control mice. LVS-immunized mice were able to slow down the course of infection such that lethal numbers were reached 2–3 days later than in control mice. Mice immunized with Δ*0918* Δ*capB* were able to further slow the rate of infection and survive an additional few days. Half of the mice immunized with Δ*clpB* were able to arrest bacterial growth after several days, then to eradicate the infection resulting in complete recovery. None of the 18 double mutants we have examined to date have been as effective as Δ*clpB;* all attempts to further attenuate Δ*clpB* have led to a severe loss of efficacy against aerosol challenge.

23.3.2 Correlates of Protection

It is extremely likely that qualitative or quantitative differences in immune response generated by vaccine strains explains their relative efficacies. Therefore, these immune responses have the potential to serve as correlates of protection. Following vaccination, changes in the levels of a panel of 21 cytokines and chemokines were monitored in the blood of BALB/c mice. Only IFNγ, TNFα, and interleukin- 6 levels were elevated for a few days post-vaccination, and the magnitude of the changes was similar for all three vaccine groups. By ELISA, sera taken 28 days post-vaccination had similar IgG and IgM titers to whole heat killed SCHU S4. A more detailed immunoproteomic analysis of these sera is underway, and detailed studies of the T cell phenotypes generated by immunization are planned. Finally, we examined the blood, lungs, and spleens of vaccinated and aerosol-challenged mice for changes in the same panel of 21 cytokines and chemokines throughout the course of infection. Most cytokines and chemokines rose in parallel with the rise in bacterial burden, and thus acted as correlates of infection. However, in the lungs only, preliminary evidence was obtained suggesting that IL-17 levels might correlate with protection, and confirmatory experiments are underway. Regardless, it is difficult to see how this might serve as a facile correlate of protection in humans since serum levels of IL-17 were similar in all three vaccinated and challenged groups.

23.4 Conclusions

The abuse of the potential biothreat agent, *F. tularensis* subsp. *tularensis*, has the capacity to cause a mass casualty event. This scenario could be averted by immunization with a safe and effective vaccine. The current vaccine, LVS, is hampered by several regulatory hurdles. Many of these are in the process of being addressed. However, the ability to define a mechanism and correlate of protection remains uncertain, but will almost certainly be required for full licensure. Moreover, from a clinical perspective, any correlate of protection will need to be measurable in the blood. Finally, LVS is not obviously 100% protective. We believe that many of the most contentious issues that continue to affect LVS can be addressed by using defined deletion mutants of SCHU S4 as live vaccines. This hypothesis is supported by our pre-clinical studies. Future studies will compare the immune responses of SCHU S4 mutants that differ in efficacy, thereby leading to a clearer understanding of the basis for protection against this highly virulent human pathogen.

References

Conlan, J. W. & Oyston, P. C. F. (2007) Vaccines against *Francisella tularensis*. *Ann NY Acad Sci*, 1105, 325–50.

Conlan, J. W., Shen, H., Golovliov, I. et al. (2010) Differential ability of novel attenuated targeted deletion mutants of *Francisella tularensis* subspecies *tularensis* strain SCHU S4 to protect mice against aerosol challenge with virulent bacteria: effects of host background and route of immunization. *Vaccine*, 28, 1824–31.

Dennis, D. T., Inglesby, T. V., Henderson, D. A. et al. (2001) Tularemia as a biological weapon. *JAMA*, 285, 2763–73.

Elkins, K. E, Cowley, S. C., & Bosio, C. M. (2007) Innate and adaptive immunity to *Francisella*. *Ann NY Acad Sci*, 1105, 284–324.

Ellis, J., Oyston, P. C. F., Green, M., & Titball, R. W. (2002) Tularemia. *Clin Microbiol Rev*, 15, 631–46.

Golovliov, I, Sjöstedt, A., Mokrievich, A. & Pavlov, V. (2003) A method for allelic replacement in *Francisella tularensis*. *FEMS Microbiol Lett*, 222, 273–80.

Hornick, R.B. & Eiglesbach, H. T. (1966) Aerogenic immunization of man with live tularemia vaccine. *Bact Rev* 30, 532–8.

Kadzhaev, K., Zingmark, C., Golovliov, I. et al. (2009) Identification of genes contributing to the virulence of *Francisella tularensis* SCHU S4 in a mouse intradermal infection model. *PLoS ONE* 4(5), e5463.doi:10.1371/journal.pone.0005463.

Mccrumb, F. R. (1961) Aerosol infection of man with *Pasteurella tularensis*. *Bact Rev*, 25, 262–7.

Pasetti, M. F., Cuberos, L., Horn, T. L. et al. (2008) An improved *Francisella tularensis* live vaccine strain (LVS) is well tolerated and highly immunogenic when administered to rabbits in escalating doses using various immunization routes. *Vaccine*, 26, 1173–85.

Salomonsson, E., Kuoppa, K., Forslund, A.-L. et al. (2009) Reintroduction of two deleted virulence loci restores full virulence to the live vaccine strain of *Francisella tularensis*. *Infect Immun*, 77, 3424–31.

Saslaw, S. & Carhart, S. (1961) Studies with tularemia vaccines in volunteers.III. Serological aspects following intracutaneous or respiratory challenge in both vaccinated and nonvaccinated volunteers. *Am J Med Sci*, 241, 689–99.

Saslaw, S., Eiglesbach, H. T., Prior, J. A. et al. (1961) Tularemia vaccine study II. Respiratory challenge. *Arch Int Med*, 107, 702–14.

Sjöstedt, A. (2007) Tularemia: history, epidemiology, pathogen physiology, and clinical manifestations. *Ann NY Acad Sci*, 1105, 1–29.

Twine, S.M., Bystrom, M., Chen, M. et al. (2005) A mutant of *Francisella tularensis* strain SCHU S4 lacking the ability to express a 58 kDa protein is attenuated for virulence and an effective live vaccine. *Infect Immun*, 73, 8345–52.

Chapter 24
Effect of Disruption of *mglA* on the Virulence and Immunogenicity of the *Francisella tularensis* Live Vaccine Strain (LVS)

Ofer Cohen, Galia Zaide, Erez Bar-Haim, Itai Mendelson, Orit Gat, Sharon Ehrlich, Gila Friedman, and Avigdor Shafferman

Abstract MglA (Macrophage Growth Locus A) is a pleiotropic transcription factor controlling the expression pattern of more than 100 genes in *Francisella novicida*, including all the ORFs located on the pathogenicity island. To further probe the role of MglA in the pathogenicity of the related strain *Francisella tularensis* LVS, we generated a mutant LVS strain in which the *mglA* locus was disrupted by insertion of a non-polar selectable marker cassette.

In vitro and in vivo analysis of the phenotype associated with the *mglA* null mutation in comparison to the wild-type parental strain and to a complementation strain (*ΔmglA:mglA*), established that the mutated strain *ΔmglA*: (i) cannot multiply in vitro in macrophages, (ii) is severely attenuated in a murine model of infection, exhibiting over 10,000 and 10,000,000 fold decrease in virulence by intranasal (IN) administration, and intraperitoneal (IP) route of infection respectively, (iii) unlike wild type LVS, the mutant strain cannot multiply in the lungs, liver and spleen of infected animals following IP administration, (iv) the Δ*mglA* mutant do not disseminate to target organs (e.g. liver and spleen) following IN administration, (v) Infection by Δ*mglA* bacteria elicits a significant humoral response, and systemic IP administration of high doses of Δ*mglA* cells results in full protection of animals against a subsequent IP challenge with high doses of the virulent wild-type strain.

These results indicate that MglA plays a major role in *F. tularensis* LVS virulence as previously shown for *F. novicida*. Our studies suggest that inactivation of MglA may serve as a platform for the development of an improved attenuated vaccine of virulent *F. tularensis* strains.

Keywords *Francisella* · Live vaccine strain · Pathogenicity island · Vaccine · MglA

O. Cohen (✉)
Department of Biochemistry and Molecular Genetics, Israel Institute for Biological Research, P.O. Box 19, Ness-Ziona 74100, Israel
e-mail: oferc@iibr.gov.il

A. Shafferman et al. (eds.), *The Challenge of Highly Pathogenic Microorganisms*,
DOI 10.1007/978-90-481-9054-6_24, © Springer Science+Business Media B.V. 2010

24.1 Introduction

Francisella tularensis, the aetiological agent of tularemia, is a Gram-negative, facultative intracellular bacterial pathogen. *F. tularensis* is considered highly infectious owing to the low infectious dose necessary for the onset of infection. The pneumonic form of tularemia has a high mortality rate and can be acquired when the organism is aerosolized (Ellis et al., 2002; Oyston et al., 2004). Although outbreaks of tularemia are thought to be rare, infections caused by *F. tularensis* have become a public health concern due to the potential use of this organism as a bioweapon. As of today, relatively little is known regarding *F. tularensis* pathogenesis and its repertoire of virulence factors, possibly because the *Francisellaceae* family represents a distinct phylogenetic group and virulence mechanisms active in other pathogenic bacteria may be irrelevant to the manifestation of its virulence (Forsman et al., 1994; Ritter, 1966). Following a variety of genomic studies, it was shown that *F. tularensis* exhibits a characteristic pathogenicity island (PI) spanning 16–19 genes over a 33.9 kb duplicated region in the chromosome, which is considered to be necessary for the infectious process (Larsson et al., 2005; Nano et al., 2004).

F. tularensis can invade and multiply in a wide range of cell types (Ben Nasr et al., 2006; Buddingh and Womack, 1941; Councilman, 1921; Francis, 1927; Lindemann et al., 2007; Shepard, 1959), but its ability to survive and replicate within macrophages appears to be a mandatory process for its pathogenesis (Fortier et al., 1994). Nevertheless, the genetic and molecular mechanisms that govern *Francisella* interactions with macrophages are still obscure. One of the few known genes that were shown to be involved in the survival of the bacteria in the macrophage was identified as a spontaneous mutation present in a variant of the subspecies *Francisella novicida,* characterized by a lack of acid phosphatase activity (Baron and Nano, 1998). This isolate was unable to replicate in macrophages, and the mutated genetic locus responsible for its phenotype was identified by a global-complementation strategy and coined *mglAB*, for macrophage growth locus (Baron and Nano, 1998).

MglA is a putative transcription regulator and an ortholog of the stringent starvation protein A (SspA) from *Escherichia coli*. It positively regulates the transcription of virulence-related genes found in the the *Francisella* pathogenicity island (FPI) as determined by RT-PCR, microarray and proteomic analyses (Brotcke et al., 2006; Guina et al., 2007; Lauriano et al., 2004). In the subspecies *novicida*, MglA controls the expression of ~100 genes with the vast majority being positively regulated. It was found that *F. novicida* MglA positively regulates all the genes encoded on the FPI as well as other genes located outside of the FPI, some of which may also be associated with virulence (Brotcke et al., 2006). Proteomic analysis indicated that MglA plays an important role in coordinating the stress response of *F. tularensis* and is essential for the survival in hostile environments (Guina et al., 2007). Recent studies in *F. tularensis* directed toward understanding the mechanism by which MglA regulates expression of its target genes, demonstrated that MglA forms a complex with another MglA-like protein annotated as SspA. This complex associates with RNA polymerase to positively control transcription of virulence genes critical for the intramacrophage growth and survival of the organism (Charity et al., 2007).

The *F. novicida mglA* mutant is highly attenuated in mice both by the intranasal and intraperitoneal routes of administration but it does not confer protection against a subsequent aerosol challenge with *F. novicida,* even at a low-dose of infection (West et al., 2008). In in vitro infection studies of macrophages, it was demonstrated that the Δ*mglA* strain is unable to escape the phagosome and reach the cytosol and does not promote apoptosis of the infected cells (Bonquist et al., 2008; Mariathasan et al., 2005; Santic et al., 2005). In this study we aimed to explore the possible role of MglA in the virulence of *F. tularensis* live vaccine strain (LVS) by characterizing the in vitro and in vivo phenotype exhibited by *mglA* mutated cells. We show that, as in the *F. novicida* strain, MglA is involved in manifestation of the bacterial virulence. In addition, we demonstrate that the LVS-*ΔmglA* mutant can confer effective protection against lethal *F. tularensis* challenge.

24.2 Results and Discussion

24.2.1 The LVS- ΔmglA Mutant Exhibits Altered Growth Characteristics in Culture and in Infected Macrophages

A Δ*mglA* mutant was constructed in LVS by insertion of a chloramphenicol resistance cassette replacing the wild-type allele by homologous recombination. The derived mutation was verified by PCR as illustrated in Fig. 24.1a, and the abrogation of MglA expression was confirmed by Western blot analysis using specific antibodies raised against MglA (Fig. 24.1b, *middle panel*). Consistent with the disruption of MglA we found that IglC, which is positively regulated by MglA (Lauriano et al., 2004), is not expressed in Δ*mglA* cells (Fig. 24.1b, *lower panel*).

In contrast to the WT, the Δ*mglA* mutant was unable to grow on GCHI (GC medium base, supplemented with 1% hemoglobin and 1% Iso-Vitalex) agar plates. Colonies obtained on CHA (cystine heart agar supplemented with 1% hemoglobin) agar plates were significantly smaller than those of the wild type even after an extended incubation time (more than 3 days). Culturing the Δ*mglA* mutant in TSBC (tryptic soy broth supplemented with 0.1% L-cystein) liquid media revealed a reduced growth rate in vitro (0.2 h^{-1} compared to 0.4 h^{-1} in the wild type strain). In addition, the mutated strain failed to reach the final turbidity of the WT even after 50 hr of growth (Fig. 24.1c). The impaired in vitro growth rate of the mutant is in line with previous results (Charity et al., 2007) and may be attributed to the pleiotropic nature of MglA which was shown to regulate the expression of at least 20 genes involved in metabolism of carbohydrates and amino acids (Brotcke et al., 2006). We note that these growth defects could be restored in the trans- complemented strain (Δ*mglA/pkk*-MglA).

The inability of the *F. novicida* and LVS-Δ*mglA* mutants to replicate in macrophages from various sources was previously reported (Baron and Nano, 1998; Bonquist et al., 2008; Lauriano et al., 2004; Santic et al., 2005). Here, we confirm this phenomenon for the LVS Δ*mglA* mutant in J774 cells in an experiment in which 2×10^7 mid-log bacterial culture were used to infect 1×10^5 cultured J774

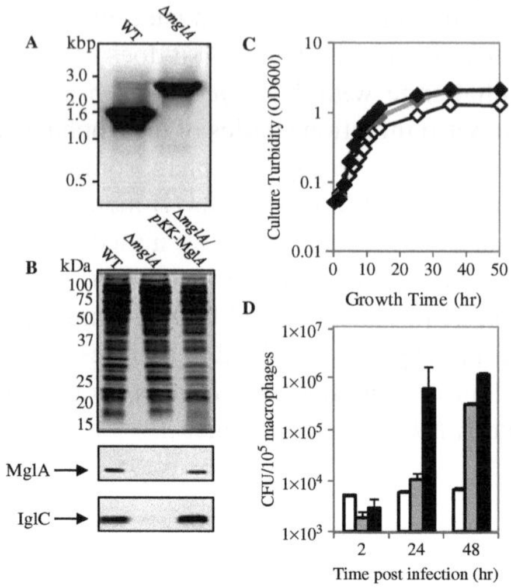

Fig. 24.1 Construction of the Δ*mglA* mutant. (**a**) PCR confirmation of the targeted integration of the chloramphenicol (Cm) cassette into the chromosomal *mglA* locus. The *mglA* null mutation was generated by allelic replacement with a chloramphenicol resistance (Cm^R) cassette disrupting the WT allele. Note that while a 1.6 kb fragment was amplified from the parental LVS chromosomal DNA, a 2.6 kb fragment was amplified from the Cm^R isolate, as expected from the targeted integration of the Cm cassette. (**b**) MglA gene-disruption abrogates expression of MglA. Coomassie blue stained SDS-PAGE gel and corresponding Western blots probed with specific mouse anti-MglA (*middle panel*) or anti-IglC (*bottom panel*) of WT, Δ*mglA* and the complemented (Δ*mglA* /*pKK*-MglA) strain. Bacterial pellets were collected from cultures grown overnight at 37°C in tryptic soy broth supplemented with 0.1% L-cystein (TSBC) liquid media and equal amounts of cell pellets were applied for each lane. (**c**) Growth curves of WT, Δ*mglA* and complemented strain (*black, open* and *gray diamonds respectively*) were carried out in TSBC liquid media at 37°C with vigorous shaking (200 rpm). Samples were drawn every 90 min for monitoring culture turbidity during the first 10 hr of growth and the final turbidity was determined after 50 h of growth. (**d**) Replication of Δ*mglA* in J774 cells. Cells were infected for 1 h ($t = -1$) with the WT(*black bars*), Δ*mglA* (*white bars*) or the complemented strain (*gray bars*) at an MOI of 200, washed ($t = 0$), incubated in the presence of gentamycin for 2 h, washed extensively ($t = 2$) and the total bacterial counts at 2, 24 and 48 h post infection were monitored

cells. Unlike wild type LVS which exhibited 24 hr post-infection at least 100 fold increase in intracellular bacterial count, the Δ*mglA* showed no increase in cell count (Fig. 24.1d). Most notably, the behavior of the complemented strain Δ*mglA/pkk*-MglA was indistinguishable from that of the wild type. Thus, MglA must be expressed for effective macrophages intracellular replication of *F. tularensis* LVS.

24.2.2 Δ*mglA* is Attenuated in a Murine Model of Infection

The lethal dose 50 (LD_{50}) values of the mutated strain were determined in mice both by intra-peritoneal (IP) and intra-nasal (IN) administration (Fig. 24.2). The mutated

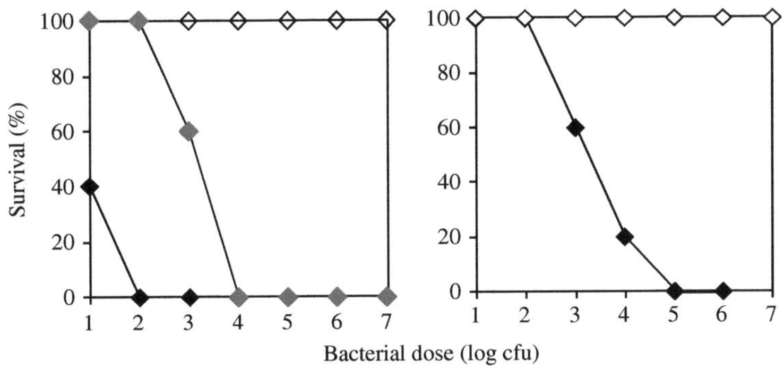

Fig. 24.2 Virulence of the WT (*black*), Δ*mglA* (*open diamonds*) and complemented strain (*gray*) following IP (*left*) or IN (*right*) administration. BABL/c mice were administrated with 10^1–10^7 cfu/ animal IP or 10^3–10^7 IN. Note that all mice infected with more than 10 cfu IP or more than 10^4 cfu IN of the WT strain succumbed, while all animals administrated with the mutant strain survived with no apparent signs of the disease

strain is severely attenuated in mice, exhibiting an increase in LD_{50} of at least 4 and 7 orders of magnitude in the IN and IP routes of infection respectively.

The virulence of the disrupted strain could be restored in the trans-complemented strain indicating that the phenotype associated with the Δ*mglA* mutation is indeed attributed to the absence of the MglA protein. However, only partial restoration of virulence was demonstrated, possibly because expression of MglA is not regulated from its native promoter, but rather from a constitutive promoter.

24.2.3 Δ*mglA* Cells Have Altered Dissemination and Target Organ-Propagation Profiles

F. tularensis does not produce any obvious exotoxins, and its lipopolysaccharide (LPS) is not endotoxic as is the case in other non-toxinogenic pathogens (Sandstrom et al., 1992). The virulence of *F. tularensis* appears to stem from its ability to induce a severe inflammatory response, possibly deleterious to the host, combined with a massive proliferation of bacteria within various host tissues and organs (Conlan and Oyston, 2007). In order to determine whether MglA is required for dissemination and replication of LVS to its target organs, BALB/c mice were infected IP or IN with 10^8 cfu/mouse of the mutant strain. At different time points post exposure, the bacteria recovered from the lungs, liver and spleen were quantified. We observed that following IP administration, the mutant cells disseminated to the liver, spleen and lungs, yet, unlike the WT, the mutant could not multiply in these organs (Fig. 24.3a). Furthermore, upon IN administration the mutant could neither replicate in the lungs nor disseminate to other organs (Fig. 24.3b).

Although unable to replicate in the target organs post IN administration, a significant number of the mutant bacteria were present in the lungs at least 3 days

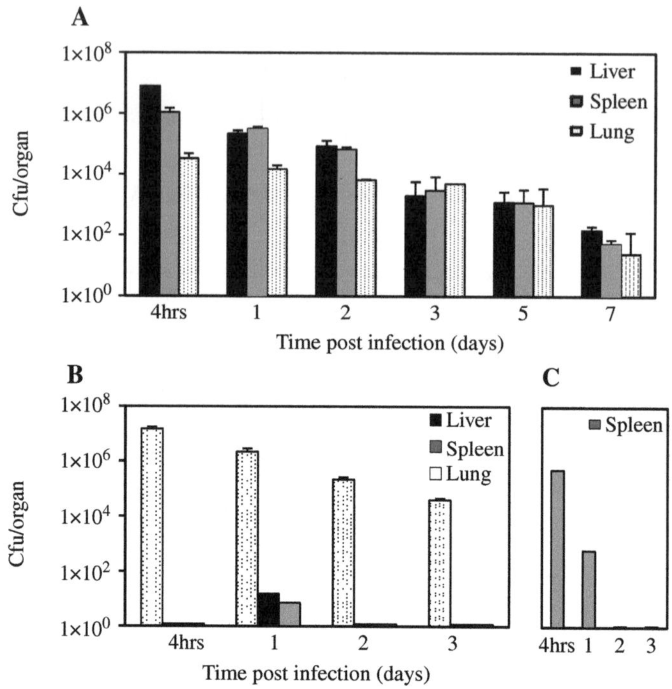

Fig. 24.3 Dissemination of *ΔmglA* cells to target organs. Mice were infected IP (**a**) or IN (**b**) with 10^8 cfu/mouse of the mutant strain and the bacterial load in the target organs was determined at the indicated times post infection by live count. The control WT strain was administered IN at 10^5 cfu/mouse. (**c**). Clearance rate of the inactivated WT strain following IP administration. Mice were administrated with 10^8 cfu/mouse of formalin-inactivated WT bacteria and the bacterial load in the spleen was determined by PCR

post-infection. This phenomenon of slow clearance was also observed following IP administration, were the mutated bacteria could be found in the liver, spleen and lungs 7 days post infection. We note that formaline-inactivated WT bacteria at doses as high as 10^8 cfu/mouse were efficiently cleared from the lungs to undetectable levels, as early as 48 hr post exposure (Fig. 24.3c).

24.2.4 Administration of the ΔmglA Elicits a Protective Immune Response

Since the Δ*mglA* mutant is highly attenuated, it can be administrated at high doses and therefore has the potential to serve as an improved live attenuated vaccine. In order to assess the ability of the mutant to confer protection against an LVS challenge, mice were immunized with a single dose (10^7 cfu/mouse) of the mutant strain by the IP or IN route. All animals elicited a significant humoral response (as determined by ELISA against total *F. tularensis* cell preparations). Four weeks

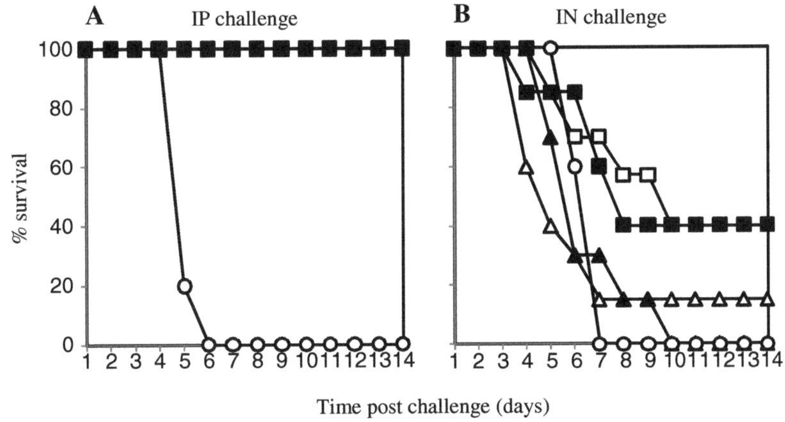

Fig. 24.4 Protective immunity conferred by the attenuated *ΔmglA* strain. Mice were immunized by IP (*squares*) or IN (*triangles*) administration of 10^7cfu/mouse. Four weeks post immunization, animals were challenged IP with 10 × LD_{50} (10 cfu/mouse) (**a**), or IN with 10 × LD_{50} (10^4 cfu/mouse; *empty symbols*) or 100 × LD_{50} (10^5 cfu/mouse; filled symbols) (**b**). Animals were monitored for illness and survival for 14 days. Control naïve animals (*empty circles*), were challenged with the lower doses

after the immunization, mice were challenged IP with 10 × LD_{50}/mouse (10 cfu) or IN with 10 × and 100 × LD_{50}/mouse (1 × 10^4 or 1 × 10^5 cfu, respectively) of the WT LVS strain. All animals survived the IP challenge, showing no signs of illness, regardless of the immunization route (Fig. 24.4a). On the other hand, protection against IN challenge was only partially achieved. Forty percent of the mice immunized IP survived both the 10 and 100 × LD_{50} challenge, and the rest exhibited an extended mean-time-to-death (MTTD) compared to the naïve control mice (Fig. 24.4b). Only 15% of the mice immunized *via* the IN route survived the low challenge and none of them survived the higher challenge (Fig. 24.4b). Thus, we conclude that IP immunization with the mutated strain is more efficient in promoting protection against IN challenge compared to protection promoted by IN immunization. These results appear to be consistent with the unique dissemination patterns exhibited by the mutated strain following the different routs of infection (see previous section).

To further explore the potential of *ΔmglA* to serve as a vaccine, we examined its ability to confer protection against higher challenge doses. Accordingly, mice were immunized IP with 10^4–10^6 cfu/mice and challenged IP with 50 × LD_{50} or 250 × LD_{50} of the WT strain four weeks after the immunization. All animals survived the low challenge dose (20% of the mice vaccinated with the low 10^4 cfu dose manifested disease symptoms). Full protection against the higher challenge dose was observed only for animals that were vaccinated with 10^5 cfu/mice or more. We note that protection against the low dose challenge was much less efficient using formaline-inactivated WT bacteria, compare to the protection afforded by the live mutant strain.

The LVS strain is considered to represent an efficacious live vaccine. Indeed, this strain has been historically used for human vaccination on a wide scale (Tigertt, 1962). Yet, as of today, due to its frequent reactogenicity, there is an increasing reluctance for the prophylactic use of this strain. In this study we demonstrate that bacteria of the *ΔmglA* strain, can be administrated at very high doses without causing any signs of disease. In addition, these high vaccine doses elicit an efficient protective response against a subsequent IP challenge in mice.

Antigens responsible for the protective immune response promoted by LVS against subsequent challenges with *F. tularensis* are unknown and identification of such protective antigens for vaccine development represents an objective of outmost importance. Here we show that the *mglA* mutant strain exhibits a slow clearance rate from the organs following IP or IN infection, and sufficient dissemination ability post IP infection. In addition, it confers partial protection against IN challenge. Thus, this strain can be employed as a platform strain in which the contribution of single/combination of selected antigens to protective immunity may be evaluated.

References

Baron, G. S. & Nano, F. E. (1998) MglA and MglB are required for the intramacrophage growth of *Francisella novicida. Mol Microbiol*, 29, 247–59.

Ben Nasr, A., Haithcoat, J., Masterson, J. E., Gunn, J. S., Eaves-Pyles, T. & Klimpel, G. R. (2006) Critical role for serum opsonins and complement receptors CR3 (CD11b/CD18) and CR4 (CD11c/CD18) in phagocytosis of *Francisella tularensis* by human dendritic cells (DC): uptake of *Francisella* leads to activation of immature DC and intracellular survival of the bacteria. *J Leukoc Biol*, 80, 774–86.

Bonquist, L., Lindgren, H., Golovliov, I., Guina, T. & Sjöstedt, A. (2008) MglA and Igl proteins contribute to the modulation of *Francisella tularensis* live vaccine strain-containing phagosomes in murine macrophages. *Infect Immun*, 76, 3502–10.

Brotcke, A., Weiss, D. S., Kim, C. C., Chain, P., Malfatti, S., Garcia, E. & Monack, D. M. (2006) Identification of MglA-regulated genes reveals novel virulence factors in *Francisella tularensis. Infect Immun*, 74, 6642–55.

Buddingh, G. J. & Womack, F. C. (1941) Observations on the infection of chick embryos with *Bacterium Tularense, Brucella,* and *Pasteurella Pestis. J Exp Med*, 74, 213–222.

Charity, J. C., Costante-Hamm, M. M., Balon, E. L., Boyd, D. H., Rubin, E. J. & Dove, S. L. (2007) Twin RNA polymerase-associated proteins control virulence gene expression in *Francisella tularensis. PLoS Pathog*, 3, e84.

Conlan, J. W. & Oyston, P. C. (2007) Vaccines against *Francisella tularensis. Ann NY Acad Sci*, 1105, 325–50.

Councilman, W. T. S. & Strong, R. P (1921) Plague-like infections in rodents. *Trans Assoc Am Phys.*, 36, 135–43.

Ellis, J., Oyston, P. C., Green, M. & Titball, R. W. (2002) Tularemia. *Clin Microbiol Rev*, 15, 631–46.

Forsman, M., Sandstrom, G. & Sjöstedt, A. (1994) Analysis of 16S ribosomal DNA sequences of *Francisella* strains and utilization for determination of the phylogeny of the genus and for identification of strains by PCR. *Int J Syst Bacteriol*, 44, 38–46.

Fortier, A. H., Green, S. J., Polsinelli, T., Jones, T. R., Crawford, R. M., Leiby, D. A., Elkins, K. L., Meltzer, M. S. & Nacy, C. A. (1994) Life and death of an intracellular pathogen: *Francisella tularensis* and the macrophage. *Immunol Ser*, 60, 349–61.

Francis, E. E. (1927) Microscopic changes of tularemia in the tick, *Dermacentor andersoni* and the bedbug, *Cimex lectularius. Pub Health Rep*, 42, 2763–72.

Guina, T., Radulovic, D., Bahrami, A. J., Bolton, D. L., Rohmer, L., Jones-Isaac, K. A., Chen, J., Gallagher, L. A., Gallis, B., Ryu, S., Taylor, G. K., Brittnacher, M. J., Manoil, C. & Goodlett, D. R. (2007) MglA regulates *Francisella tularensis* subsp. *novicida* (*Francisella novicida*) response to starvation and oxidative stress. *J Bacteriol*, 189, 6580–6.

Larsson, P., Oyston, P. C., Chain, P., Chu, M. C., Duffield, M., Fuxelius, H. H., Garcia, E., Halltorp, G., Johansson, D., Isherwood, K. E., Karp, P. D., Larsson, E., Liu, Y., Michell, S., Prior, J., Prior, R., Malfatti, S., Sjöstedt, A., Svensson, K., Thompson, N., Vergez, L., Wagg, J. K., Wren, B. W., Lindler, L. E., Andersson, S. G., Forsman, M. & Titball, R. W. (2005) The complete genome sequence of *Francisella tularensis*, the causative agent of tularemia. *Nat Genet*, 37, 153–9.

Lauriano, C. M., Barker, J. R., Yoon, S. S., Nano, F. E., Arulanandam, B. P., Hassett, D. J. & Klose, K. E. (2004) MglA regulates transcription of virulence factors necessary for *Francisella tularensis* intraamoebae and intramacrophage survival. *Proc Natl Acad Sci USA*, 101, 4246–9.

Lindemann, S. R., Mclendon, M. K., Apicella, M. A. & Jones, B. D. (2007) An in vitro model system used to study adherence and invasion of *Francisella tularensis* live vaccine strain in nonphagocytic cells. *Infect Immun*, 75, 3178–82.

Mariathasan, S., Weiss, D. S., Dixit, V. M. & Monack, D. M. (2005) Innate immunity against *Francisella tularensis* is dependent on the ASC/caspase-1 axis. *J Exp Med*, 202, 1043–9.

Nano, F. E., Zhang, N., Cowley, S. C., Klose, K. E., Cheung, K. K., Roberts, M. J., Ludu, J. S., Letendre, G. W., Meierovics, A. I., Stephens, G. & Elkins, K. L. (2004) A *Francisella tularensis* pathogenicity island required for intramacrophage growth. *J Bacteriol*, 186, 6430–6.

Oyston, P. C., Sjöstedt, A. & Titball, R. W. (2004) Tularaemia: bioterrorism defence renews interest in *Francisella tularensis*. *Nat Rev Microbiol*, 2, 967–78.

Ritter, D. B. G. & Gerloff, R. K. (1966) Deoxyribonucleic acid hybridization among some species of the genus *Pasteruella*. *J Bacteriol*, 92, 1838–9.

Sandstrom, G., Sjöstedt, A., Johansson, T., Kuoppa, K. & Williams, J. C. (1992) Immunogenicity and toxicity of lipopolysaccharide from *Francisella tularensis* LVS. *FEMS Microbiol Immunol* 5, 201–10.

Santic, M., Molmeret, M., Klose, K. E., Jones, S. & Kwaik, Y. A. (2005) The *Francisella tularensis* pathogenicity island protein IglC and its regulator MglA are essential for modulating phagosome biogenesis and subsequent bacterial escape into the cytoplasm. *Cell Microbiol*, 7, 969–79.

Shepard, C. C. (1959) Nonacid-fast bacteria and HeLa cells: their uptake and subsequent intracellular growth. *J Bacteriol*, 77, 701–14.

Tigertt, W. D. (1962) Soviet viable *Pasteurella tularensis* vaccines. *Bacterio Rev*, 26, 354–73.

West, T. E., Pelletier, M. R., Majure, M. C., Lembo, A., Hajjar, A. M. & Skerrett, S. J. (2008) Inhalation of *Francisella novicida* Delta *mglA* causes replicative infection that elicits innate and adaptive responses but is not protective against invasive pneumonic tularemia. *Microbes Infect*, 10, 773–80.

Chapter 25
Data Mining, Bioinformatic and Immunoinformatic Analyses of *Francisella tularensis* Schu S4 Genome in Search for Novel Vaccine Candidates

Anat Zvi, Naomi Ariel, and Avigdor Shafferman

Abstract The pathogenesis mechanism and the identity of major virulence factors of the highly infectious human pathogen *F. tularensis tularensis*, are still poorly characterized. The restricted efficacy of the sole available vaccine (the LVS attenuated strain) and residual toxicity has motivated extensive R&D efforts directed toward the identification of alternative tularemia vaccine formulations based on attenuated mutants, subunit vaccines and T-cell epitope-based vaccines. Identification of vaccine candidates by bioinformatic approaches is mostly driven by the availability of genome sequence data for numerous human-virulent as well as avirulent strains. In an attempt to select for *F. tularensis* Schu S4 potent antigens, we have developed a strategy based on genome-scale in silico analyses and data mining of published global experimental studies. The compiled information was divided into distinct biologically relevant categories. Protocols for qualitative and quantitative scoring for each of the categories, were developed. Together with implementation of a biological rationale, these served for ranking and prioritization of the putative antigens, providing a basis for subsequent selection of the top-ranking candidates to be evaluated experimentally.

Keywords *F. tularensis* · Bioinformatics · Immunoinformatics · Vaccine candidates

25.1 Introduction

Francisella tularensis is a Gram-negative, facultative intracellular bacterial pathogen capable of causing severe disease (tularemia) in a wide range of mammals, including humans, and is classified as a category A biodefense agent. The pathogen

A. Zvi (✉)
Department of Biochemistry and Molecular Genetics, Israel Institute for Biological Research, P.O. Box 19, Ness-Ziona 74100, Israel
e-mail: anatz@iibr.gov.il

A. Shafferman et al. (eds.), *The Challenge of Highly Pathogenic Microorganisms*,
DOI 10.1007/978-90-481-9054-6_25, © Springer Science+Business Media B.V. 2010

colonizes macrophages as well as other phagocytic (dentritic and neutrophils) and non-phagocytic (hepatocytes and endothelial) cells. As few as 10 cells have been found to cause respiratory tularemia, with ensuing mortality rate of 30–60% (in the most virulent strain) if infection is left untreated. In the absence of known, classical virulence factors such as exotoxins, attempts to identify traits which play a role in the virulence of *F. tularensis* by global transcriptomics or proteomic screens as well as targeted mutagenesis, are conducted (see references in Table 25.1).

Table 25.1 Literature-based sources for knowledge dataset

Category	Reference	Description/experimental model	Method
Expression	Gilmore et al. (2004)	Identification and in vitro localization of surface-associated proteins, Schu S4	TnphoA mutant library
	Hubalek et al. (2003)	Comparison of in vitro expressed strain-specific proteins (*holarctica* & *mediasiatica*)	Proteomics (2D&Mass)
	Pavkova et al. (2005)	Membrane proteins, LVS	Proteomics (2D&Mass)
	Twine et al. (2005)	Membrane proteins, LVS & FSC033	Proteomics (2D&Mass)
	Twine et al. (2006a)	Culture vs. in vivo/oxidative stress	in vivo proteomics
	Pavkova et al. (2006)	Membrane associated proteins (Schu S4, *holarctica*, LVS)	Proteomics (2D&Mass), PCR
	Lee et al. (2006)	major extracellular proteins (in vitro, LVS)	metabolic labelling, 2D
	Waldo et al. (2007)	Whole-cell extract, top-ranked ordered proteome, Schu S4	Proteomics (2D, IEF, mRNA microarray)
	Rohmer et al. (2008)	in vitro expressed proteins – exponential phase, *novicida* U112	Proteomics, codon-adaptation, cellular localization
Growth & Survival/Virulence	Tempel et al. (2006)	Tissue dissemination, attenuation & protection, MF & mouse studies, cytotoxicity, *F. novicida* U112	Tn5 mutagenesis
	Qin and Mann (2006)	in vitro survival (MF, hepatic cell-line), Schu S4	EZ:TN mutagenesis

Table 25.1 (continued)

Category	Reference	Description/experimental model	Method
	Gallagher et al., (2007)	Growth in rich medium, *novicida* U112	Near-saturation transposon library
	Maier et al. (2007)	in vitro/in vivo attenuation & protection, LPS phenotype, LVS	Himar1 transposon mutagenesis
	Weiss et al. (2007)	Genes essential for growth or survival in vivo, phenotype-pyroptosis, F. *novicida* U112	Tn-5 library, microarray-based negative selection
	Su et al. (2007)	Essential genes for establishment of lung infection, growth index, attenuated phenotype confirmation (CI), LVS	STM
	Tempel et al. (2006)	Attenuation & protection in macrophage and mouse, *novicida* U112	Tn5
	Horzempa et al. (2008)	Response and attenuation in chicken-embryo/mammalian cell, temperature control, LVS	in vitro growth, microarray, Q-PCR
	Kadzhaev et al. (2009)	Virulence genes, mouse intra-dermal infection model (survival & attenuation), Schu S4	Transposon & targeted mutagenesis
	Wehrly et al. (2009)	Time-dependent transcriptional profiling inside MF – virulence determinants, targeted mutagenesis, mouse infection model (attenuation), Schu S4	Microarray, allelic replacement
Immuno-reactivity	Havlasova et al. (2002)	Cell lysate, human sera, LVS	Proteomics + SERPA

Table 25.1 (continued)

Category	Reference	Description/experimental model	Method
	Havlasova et al. (2005)	Antigenic profile in murine model, compared to human proteome, disease time-scale, LVS	Proteomics (2D&Mass) blot analysis, predictions
	Twine et al. (2006b)	Antigens recognized by antisera from LVS immunized mice (protected vs. unprotected)	Proteomics (2D&Mass) blot analysis
	Twine et al. (2006a)	Proteome expressed in the host environment (mouse spleen) vs. culture, FSC033 (Type A)	Proteomics (2D&Mass) blot analysis
	Janovska et al. (2007a)	Membrane fractions, LVS	Proteomics (2D&Mass) blot analysis, bioinformatics
	Janovska et al. (2007b)	Membrane fractions, human serum in type A strain(s)	Proteomics + SERPA
	Eyles et al. (2007)	Identification of immunodominant antigens; mice immunized with LVS and challenged with Schu S4	Protein microarray
	Sundaresh et al. (2007)	Probed with sera of infected & healthy individuals	Protein microarray
Regulation	Brotcke et al. (2006)	MglA regulation, isogenic mutant; targeted mutagenesis, MF replication, cytotoxicity & attenuation, *F. novicida* U112	Microarray
	Guina et al. (2007)	MglA regulation, isogenic mutant, survival under starvation & oxidative stress, *F. novicida* U112	Proteomics

Table 25.1 (continued)

Category	Reference	Description/experimental model	Method
	Mohapatra et al. (2007)	FTT1557c regulation, mutant, virulence gene expression, attenuation and protection, *F. novicida* U112	Microarray, qRT-PCR
	Charity et al. (2007)	sspA and mglA regulation, mutants, gene expression, LVS	DNA microarray
	Meibom et al. (2009)	Hfq regulation – stress tolerance & virulence of a virulent clinical strain FSC200 and its Hfq mutant	Targeted mutagenesis, transcriptomics
T-cell	Lee et al. (2006)	Protein profiles of culture filtrate & cell lysates, T-cell proliferation against filtrate proteins, LVS & a virulent clinical isolate	Proteomics (2D), N-terminal sequencing, PCR, lymphocyte proliferation, cytokine staining)
	McMurry et al. (2007)	T-cell epitopes recognized by T lymphocytes, Schu S4	Immunoinformatics
	Harndahl et al. (2007)	Epitopes binding human MHC class I alleles, LVS	MHC binding assays
Iron-regulation	Deng et al. (2006)	Up/down expression in LVS in vitro iron restricted conditions	DNA microarray, qRT-PCR
	Lenco et al. (2007)	Up/down expression LVS in vitro iron restricted conditions	Proteomics
Comparative genomics	Broekhuijsen et al. (2003)	Definition of regions of difference (Schu-like, *holarctica*, *mediasiatica*)	DNA microarray
	Rohmer et al. (2006)	Potential source for LVS attenuation (LVS vs. *holarctica* FSC200)	Whole-genome comparison

Table 25.1 (continued)

Category	Reference	Description/experimental model	Method
	Rohmer et al. (2007)	Three-way genome comparison (Schu S4, LVS, *novicida* U112)	Bioinformatics
	Petrosino et al. (2006)	Chromosome rearrangement and diversification of *F. tularensis holarctica* OSU18 Identification of putative virulence determinants	Bioinformatics
	Beckstrom-Sternberg et al. (2007)	Complete genomic characterization of a pathogenic AII strain (WY96) vs. AI (Schu S4)	Bioinformatics
	Chaudhuri et al. (2007)	Comparison of a european type A isolate (FSC198) to Schu S4	Bioinformatics
	Murthy et al. (2007)	A full genomic sequence verified protein coding gene collection for *F. tularensis*	Bioinformatics

To date, eight *Francisella* genomes, including the three subspecies *F. tularensis tularensis*, *holarctica* and *mediasiatica*, as well as the most ancestral species *F. tularensis novicida*, have been fully sequenced (Barabote et al., 2009; Larsson et al., 2005; Petrosino et al., 2006; Rohmer et al., 2007). All are genetically highly similar (97–99% nucleotide identity between and within subspecies), but exhibit numerous type-dependent DNA rearrangements. *F. tularensis tularensis* (Type A, two genetically distinct clades) and *F. tularensis holarctica* (Type B) are the primary biovars associated with human disease, where the latter cause a milder disease. The genetic basis for the difference in virulence between the subspecies remains unknown. Moreover, *Francisella* has no close human pathogenic relatives and is distantly related to the human pathogens *Coxiella burnetii* and *Legionella pneumophila*. It is therefore not surprising that ~40% of total ORF products within an *F. tularensis* isolate are still considered hypothetical proteins with unknown function.

The potential use of *F. tularensis* as a biothreat agent has motivated studies directed to development of a vaccine as a countermeasure. While killed vaccines were highly reactogenic, poorly immunogenic and failed to protect against type A infection, the live attenuated vaccine (LVS) based on a derivative of a type B strain, has been used as a fairly reactive human vaccine for the past 50 years (Mann and Ark, 2009; Wayne Conlan and Oyston, 2007). However, the unknown nature of its attenuation and residual virulence following vaccination by the aerosol route has limited its licensing and overall utility. Early attempts to identify protective antigens, as components of subunit vaccines, were based on the experimental identification of immunoreactive surface-associated proteins and other surface antigens (e.g. carbohydrates and LPS; (Pechous et al., 2009)). However, these studies have generally met with limited success. It was soon realized that although specific antibodies are readily detectable in sera upon *F. tularensis* infection, their importance to immunity remains unclear and humoral response must probably be coupled with an effective cellular immune response to confer efficient protection (Sebastian et al., 2009).

The availability of full genome sequences of numerous human pathogens concomitant with the development of high-throughput technologies such as functional and structural genomics completely revolutionized vaccine design. The first evolving paradigm, coined "reverse vaccinology", relies on whole genome-based in silico analysis of the entire repertoire of the pathogen genome encoded genes. This approach is mainly directed towards the identification of putative virulence factors and surface anchored/secreted antigens, having the potential to elicit a humoral response. It was first described by Rappuoli and his collaborators (Pizza et al., 2000), for the analysis of *Neisseria meningitidis* genome and has been since successfully applied to a variety of human pathogens (Bambini and Rappuoli, 2009), including *B. anthracis* (Ariel et al., 2003; Gat et al., 2006). Such an approach is expected to have a limited applicability for the search of vaccine candidates in the case of intracellular pathogens, where the cellular response is known to dominate over the antibody-based response.

In a previous study we have described the design and application of an alternative genome-based approach for rational selection of vaccine candidates for an intracellular pathogen (Zvi et al., 2008). Putative vaccine candidates were selected by integration of cross-matched data from documented global experimental studies, in silico analyses, a rational prioritization scheme and biological rationale.

In this study we report on the construction of a *Francisellae* compiled knowledge dataset, development of dedicated scoring protocols and preliminary ranking of the 1603 *F. tularensis tularensis* Schu S4 ORF products, for selection of putative vaccine candidates to be evaluated experimentally . Combined with genome-scale CTL epitope mapping, this dataset constitutes a platform for selection of novel antigens to be evaluated as subunit vaccines, targets for directed mutagenesis or antigens harboring prominent T-cell epitopes.

25.2 Results and Discussion

The availability of the genome sequence of the human virulent Schu S4 strain, as well as that of other strains (of varying human pathogenicity), permitted to apply an in silico, genome-scale approach in search for novel vaccine candidates. Yet, the evolutionary distance between this organism and other human pathogens, together with the poor knowledge of the pathogen secretory mechanisms and the absence of most of the well established virulence-related secretion pathways known in other Gram negative bacteria, render the application of the classical "reverse vaccinology" strategy not as straightforward as for other pathogens. In addition, the intracellular life style of *F. tularensis* accentuates the major putative importance of the T-cell immunity arm. All these characteristics led us to apply a different whole genome-based reductive methodology, based on a strategy recently developed and established by us in the analysis of *Mycobacterium tuberculosis* for selection of vaccine candidates (Zvi et al., 2008). Description of the methodology and its preliminary application to the selection of vaccine candidates from the *F. tularensis* Schu S4 1603 ORF genome, are detailed in the next paragraphs.

25.2.1 Data Mining and Compilation of Literature-Based Evidence Derived from Global Analyses

Extensive inspection of experimental data derived from comprehensive global in vivo & in vitro analyses, conducted with human virulent strains as well as LVS or the model strain novicida, was conducted. Retrieved evidences were compiled in accordance with the pathogen's route of entry, intracellular niche, establishment in different organs, survival and spread. The data was grouped in aspects related to both pathogenesis and vaccine development (such as: expression, growth, virulence, immunoreactivity, pleotropic transcription regulation and iron regulation) and compiled into a single knowledge dataset. Based on these aspects, the data was divided into 7 categories (Table 25.1).

25.2.2 Bioinformatic Analyses of F. tularensis Schu S4 1603 ORF Products

In an attempt to extend our knowledge on some of the characteristics of the organism ORF products, we conducted a revision of the existing functional annotation which, according to the originally deposited data, was limited to ~ 60% of the total ORF products, while the remaining ~40% were classified as proteins of unknown function (Larsson et al., 2005). This revision was based on sequence similarity searches, together with domain and motifs assignment and resulted in assignment of a putative function for additional 6% of the proteins (Fig. 25.1). To note, a similar procedure carried out for *M. tuberculosis* sequences resulted in an increase of 20% in the number of proteins for which a function could be assigned (Zvi et al., 2008). As annotation is mostly inferred by homology, the relative minor outcome of the re-annotation effort may be attributed yet again to the phylogenetic divergence between *F. tularensis* and other organisms. With regard to cellular localization, our main effort was directed towards the identification of proteins secreted by non

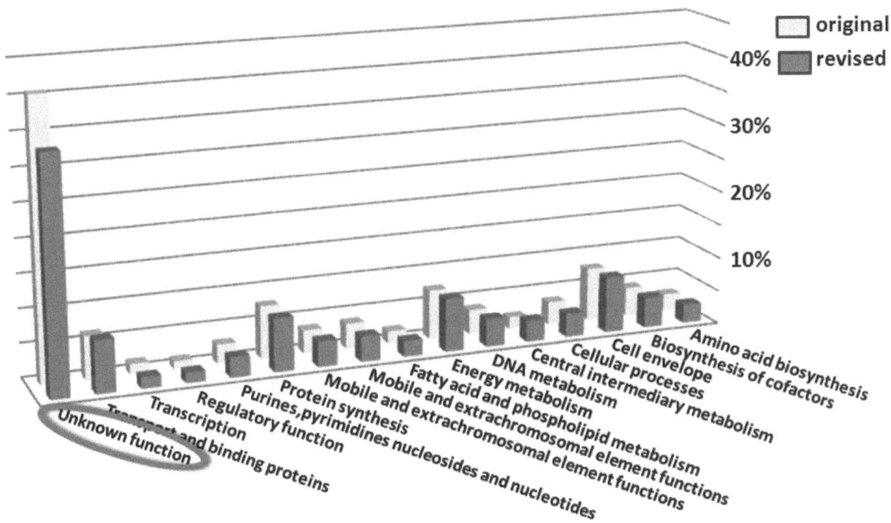

Fig. 25.1 Re-Distribution of *F. tularensis* genes in functional categories according to revised annotation. The functional categories are adapted from the TIGR Comprehensive Microbial Resource (http://cmr.jcvi.org/tigr-scripts/CMR/CmrHomePage.cgi). The distribution of genes according to the original annotation (Larsson et al., 2005) and after revisiting the annotation (this study) is presented in light grey and dark grey, respectively

classical, non-signal triggered pathways, applying the SecretomeP analysis which produces ab initio predictions of non-signal triggered secreted proteins, based on sequence-derived features with discriminatory value (Bendtsen et al., 2005). This analysis revealed 125 ORF products predicted as secreted, of which ~40 could be virulence-related (data not shown). This data, together with information resulting from classical analysis of protein subcellular localization conducted with the PsortB tool (Gardy et al., 2005), were added to the compiled dataset.

Based on evidence on the contribution of cellular immunity to protection, and in view of the knowledge gap on *F. tularensis* experimentally identified CTL epitopes eliciting an immune response (as few as ~200 documented epitopes, originating in individual proteins, (Peters et al., 2005) www.immuneepitope.org), a whole-genome immunoinformatic analysis was conducted in order to map all putative human MHC class I binders. Our analysis predicts CTL epitopes of 9, 10 and 11-mer long peptides and covers ~120 different HLA alleles represented by 12 supertypes, clustering alleles which have similar binding motifs (Lundegaard et al., 2008). This analysis resulted in mapping of 50,508 putative binders, out of > 1.6×10^7 peptides with 9–11 amino acid long, spanning the whole genome. This information was added to the compiled dataset, and may be used either as a yet another measure for evaluating and ranking the antigens by their predicted immunogenicity potential or for a standalone selection of most potent CTL epitopes by applying various filtering considerations (i.e. density of binders/ORF product, allelic coverage, binding affinities etc.).

25.2.3 Assignment of a 'Qualitative' Score for Each ORF Product According to the Accumulating Evidence

The resulting knowledge dataset, which combines the literature-based data together with results obtained from the bioinformatic analyses (see above), is compiled into an excel spreadsheet containing a row for each ORF product, whereas each column represents data extracted from the global experimental screen or computational analyses. This dataset forms the basis for antigen ranking. The advantage in prioritizing *all antigens* (as compared to the basic reverse vaccinology methodology, where each step in the analysis is used as a filtering criteria to down select for a remaining subset of antigens), is that we benefit from all the information accumulated for all 1603 protein sequences. For each ORF product, a "+" sign is assigned in each of the categories where evidence exists, and the arithmetic summation of all "+" signs generates an unweighted score. Inspection of the range and distribution of the calculated "qualitative" scores discloses that 72% (1150 of all ORF products) have any recorded evidence (qualitative score of 1–15), while scores of >= 6 or >= 10 are attributed to only 5% or 1% of the ORF products, respectively. This qualitative score, although simplistic and enabling a preliminary ranking of the antigens, necessitates refining measures in order to overcome the following inherent limitations: (1) large clusters of antigens harbor the same qualitative score, rendering it, to some extent, non discriminatory; (2) the different criteria that contribute to the total score are unweighted for their relevance to vaccine/pathogenesis; (3) numerical values of the actual data are totally ignored. All these features are addressed by employing a "quantitative" score, as detailed below.

25.2.4 Assignment of a 'Quantitative' Score for Each ORF Product

With the intention of using a more discriminative measure for antigen ranking, we introduced a scoring scheme based on: (1) a total score for each category, which may vary to reflect the significance/relevance of a specific category (2) internal scores within each category, which are designed to reflect relative result intensity for a particular category and/or different parameters relevant to the particular category (Table 25.2). These internal scores compose the total quantitative score mentioned above. The numerical scheme is dependent on the nature of the data and on the scientific question in hand, as will be exemplified below.

Compilation of the knowledge dataset and establishment of an antigen ranking scheme according to the total accumulated data, paves the way for either selecting a predetermined number of antigens by relying solely on their relative ranking and setting an arbitrary cutoff, or for tailoring antigen selection according to a specific scientific question. Deciding on the relevant categories together with up- or down-weighting of selected categories by applying different total quantitative scores are the two means by which one can cross-examine the dataset and direct the search towards the desired antigens according to the specific question in hand. Such

Table 25.2 Numerical scheme for calculation of qualitative scores

Category	Criteria	Reference[a]	Internal score[b]	Total score
Growth & Survival/Virulence	No evidence		0	4
	Persistence in lungs	Su et al. (2007) Weiss et al. (2007) Kraemer et al. (2009)	3	
	Survival in macrophages	Tempel et al. (2006)	0.5	
	Dissemination in spleen&lung&liver	Kraemer et al. (2009)	0.5	
	Inhibition growth index (<0.7)	Su et al. (2007)	−1	
Specific mutant	No evidence		0	4
	Partial attenuation	(Multiple)	2	
	Full attenuation	(Multiple)	4	
	Not attenuated	(Multiple)	−3	
Protection	No evidence		0	2
	Protective	(Multiple)	2	
	Not protective	(Multiple)	−2	
Immunoreactivity	No SERPA disease	Sundaresh et al. (2007) Eyles et al. (2007)	0	3
	SERPA disease low	Sundaresh et al. (2007) Eyles et al. (2007)	2	
	SERPA disease high	Sundaresh et al. (2007) Eyles et al. (2007)	3	
Iron-regulation	No evidence		0	2
	Low (<2)	Deng et al. (2006) Lenco et al. (2007)	1	
	High (>2)	Deng et al. (2006) Lenco et al. (2007)	2	

[a]Multiple references: Originate from publications on specific targeted mutants
[b]Negative scores are given as "penalty"

a primary selection process for the identification of vaccine candidates targeted to mutagenesis and/or attenuation was followed, as described below.

25.2.5 Selection of Putative Vaccine Candidates

We adapted our genome-based screen to exploit the available information in aspects pertinent to virulence, immunoreactivity and protection, aiming to focus on targets where mutation or deletion may potentially result in attenuation of virulence.

To this end, we chose to further examine a list of over 500 antigens having either a total qualitative score equal to or smaller than 6, provided that evidence ensuing from the categories of growth/virulence and iron regulation contributed to the score. This list was further subjected to additional filtering steps, and ORFs were removed according to the following practical considerations: (a) ORFs which are part of the pathogenicity island (PAI), which exists in two almost identical copies in the genome, were removed, to avoid possible complementation by the paralogous copy of the gene; (b) ORFs which were already studied as specific mutants in Schu S4 or in the closely related vaccine strain LVS; (c) Ribosomal proteins or transposases. This filtering, in addition to manual curation, resulted in a list of 40 putative antigens. These 40 antigens were ranked according to the calculated qualitative scores. To allow for a further resolved prioritization, based on numerical measures, we calculated a qualitative score for each of the 40 antigens, based on the numerical scheme presented in Table 25.2. All "+" signs in the different categories, given for existing evidence, were replaced by an appropriate numerical score, according to the numerical scheme, and the arithmetic sum of the individual numerical scores for each ORF product generated the total qualitative score. The process resulted in total scores ranging from 2 to 13.5 and demonstrates the benefit of using these refined scores as a discriminatory tool for further trimming down the number of antigens for experimental evaluation. Most of the recently studied promising virulence-related antigens, such as ClpB (Meibom et al., 2009), DsbB (Qin and Mann, 2006), FTT0918 (Bonquist et al., 2008), capB (Su et al., 2007) and ggt (Meibom et al., 2009), were originally among the top-ranking antigens (quantitative score of 5–13.5) however they were removed from the final list following the filtering step (b) mentioned above. Among the top-ranking antigens in the list (score 5 and above), several housekeeping genes and chaperone proteins are included, proteins which may be relevant to the mechanisms of adaptation of the pathogen to its specific niche. Remarkably, 42% of the highest ranking antigens are ORFs with unassigned function. This observation illustrates the advantage of the methodology, by which unknown antigens may surface as promising candidates as a consequence of the added value of cross matching all documented evidence and computed data, for a particular ORF.

References

Ariel, N., Zvi, A., Makarova, K. S., Chitlaru, T., Elhanany, E., Velan, B., Cohen, S., Friedlander, A. M. & Shafferman, A. (2003) Genome-based bioinformatic selection of chromosomal *Bacillus anthracis* putative vaccine candidates coupled with proteomic identification of surface-associated antigens. *Infect Immun,* 71, 4563–79.

Bambini, S. & Rappuoli, R. (2009) The use of genomics in microbial vaccine development. *Drug Discov Today,* 14, 252–60.

Barabote, R. D., Xie, G., Brettin, T. S., Hinrichs, S. H., Fey, P. D., Jay, J. J., Engle, J. L., Godbole, S. D., Noronha, J. M., Scheuermann, R. H., Zhou, L. W., Lion, C. & Dempsey, M. P. (2009) Complete genome sequence of *Francisella tularensis* subspecies *holarctica* FTNF002-00. *PLoS One,* 4, e7041.

Beckstrom-Sternberg, S. M., Auerbach, R. K., Godbole, S., Pearson, J. V., Beckstrom-Sternberg, J. S., Deng, Z., Munk, C., Kubota, K., Zhou, Y., Bruce, D., Noronha, J., Scheuermann, R. H., Wang, A., Wei, X., Wang, J., Hao, J., Wagner, D. M., Brettin, T. S., Brown, N., Gilna, P. & Keim, P. S. (2007) Complete genomic characterization of a pathogenic A.II strain of *Francisella tularensis* subspecies *tularensis*. *PLoS One*, 2, e947.

Bendtsen, J. D., Kiemer, L., Fausboll, A. & Brunak, S. (2005) Non-classical protein secretion in bacteria. *BMC Microbiol,* 5, 58.

Bonquist, L., Lindgren, H., Golovliov, I., Guina, T. & Sjostedt, A. (2008) MglA and Igl proteins contribute to the modulation of *Francisella tularensis* live vaccine strain-containing phagosomes in murine macrophages. *Infect Immun,* 76, 3502–10.

Broekhuijsen, M., Larsson, P., Johansson, A., Bystrom, M., Eriksson, U., Larsson, E., Prior, R. G., Sjostedt, A., Titball, R. W. & Forsman, M. (2003) Genome-wide DNA microarray analysis of *Francisella tularensis* strains demonstrates extensive genetic conservation within the species but identifies regions that are unique to the highly virulent *F. tularensis* subsp. *tularensis*. *J Clin Microbiol*, 41, 2924–31.

Deng, K., Blick, R. J., Liu, W. & Hansen, E. J. (2006) Identification of *Francisella tularensis* genes affected by iron limitation. *Infect Immun*, 74, 4224–36.

Chaudhuri, R. R., Ren, C. P., Desmond, L., Vincent, G. A., Silman, N. J., Brehm, J. K., Elmore, M. J., Hudson, M. J., Forsman, M., Isherwood, K. E., Gurycova, D., Minton, N. P., Titball, R. W., Pallen, M. J. & Vipond, R. (2007) Genome sequencing shows that European isolates of *Francisella tularensis* subspecies *tularensis* are almost identical to US laboratory strain Schu S4. *PLoS One*, 2, e352.

Gallagher, L. A., Ramage, E., Jacobs, M. A., Kaul, R., Brittnacher, M. & Manoil, C. (2007) A comprehensive transposon mutant library of *Francisella novicida*, a bioweapon surrogate. *Proc Natl Acad Sci USA,* 104, 1009–14.

Gardy, J. L., Laird, M. R., Chen, F., Rey, S., Walsh, C. J., Ester, M. & Brinkman, F. S. (2005) PSORTb v.2.0: Expanded prediction of bacterial protein subcellular localization and insights gained from comparative proteome analysis. *Bioinformatics,* 21, 617–23.

Gat, O., Grosfeld, H., Ariel, N., Inbar, I., Zaide, G., Broder, Y., Zvi, A., Chitlaru, T., Altboum, Z., Stein, D., Cohen, S. & Shafferman, A. (2006) Search for *Bacillus anthracis* potential vaccine candidates by a functional genomic-serologic screen. *Infect Immun,* 74, 3987–4001.

Gilmore, R. D., Jr., Bacon, R. M., Sviat, S. L., Petersen, J. M. & Bearden, S. W. (2004) Identification of *Francisella tularensis* genes encoding exported membrane-associated proteins using TnphoA mutagenesis of a genomic library. *Microb Pathog,* 37, 205–13.

Harndahl, M., Lamberth, K., Justesen, S., Roder, G., Madsen, M., Nielsen, M., Lundegaard, C., Larsen, M. V., Tang, S., Brunak, S., Lund, O. & Buus, S. (2007) Large scale analysis of peptide-HLA class I interactions. Submitted to IEDB: http://www.immuneepitope.org

Horzempa, J., Carlson, P. E., Jr., O'dee, D. M., Shanks, R. M. & Nau, G. J. (2008) Global transcriptional response to mammalian temperature provides new insight into *Francisella tularensis* pathogenesis. *BMC Microbiol,* 8, 172.

Hubalek, M., Hernychova, L., Havlasova, J., Kasalova, I., Neubauerova, V., Stulik, J., Macela, A., Lundqvist, M. & Larsson, P. (2003) Towards proteome database of *Francisella tularensis*. *J Chromatogr B Analyt Technol Biomed Life Sci,* 787, 149–77.

Kadzhaev, K., Zingmark, C., Golovliov, I., Bolanowski, M., Shen, H., Conlan, W. & Sjostedt, A. (2009) Identification of genes contributing to the virulence of *Francisella tularensis* SCHU S4 in a mouse intradermal infection model. *PLoS One,* 4, e5463.

Kraemer, P. S., Mitchell, A., Pelletier, M. R., Gallagher, L. A., Wasnick, M., Rohmer, L., Brittnacher, M. J., Manoil, C., Skerett, S. J. & Salama, N. R. (2009) Genome-wide screen in *Francisella novicida* for genes required for pulmonary and systemic infection in mice. *Infect Immun,* 77, 232–44.

Larsson, P., Oyston, P. C., Chain, P., Chu, M. C., Duffield, M., Fuxelius, H. H., Garcia, E., Halltorp, G., Johansson, D., Isherwood, K. E., Karp, P. D., Larsson, E., Liu, Y., Michell, S.,

Prior, J., Prior, R., Malfatti, S., Sjostedt, A., Svensson, K., Thompson, N., Vergez, L., Wagg, J. K., Wren, B. W., Lindler, L. E., Andersson, S. G., Forsman, M. & Titball, R. W. (2005) The complete genome sequence of *Francisella tularensis*, the causative agent of tularemia. *Nat Genet,* 37, 153–9.

Lee, B. Y., Horwitz, M. A. & Clemens, D. L. (2006) Identification, recombinant expression, immunolocalization in macrophages, and T-cell responsiveness of the major extracellular proteins of *Francisella tularensis. Infect Immun,* 74, 4002–13.

Lenco, J., Hubalek, M., Larsson, P., Fucikova, A., Brychta, M., Macela, A. & Stulik, J. (2007) Proteomics analysis of the *Francisella tularensis* LVS response to iron restriction: induction of the *F. tularensis* pathogenicity island proteins IglABC. *FEMS Microbiol Lett,* 269, 11–21.

Lundegaard, C., Lamberth, K., Harndahl, M., Buus, S., Lund, O. & Nielsen, M. (2008) NetMHC-3.0: Accurate web accessible predictions of human, mouse and monkey MHC class I affinities for peptides of length 8-11. *Nucleic Acids Res,* 36, W509–12.

Maier, T. M., Casey, M. S., Becker, R. H., Dorsey, C. W., Glass, E. M., Maltsev, N., Zahrt, T. C. & Frank, D. W. (2007) Identification of *Francisella tularensis* Himar1-based transposon mutants defective for replication in macrophages. *Infect Immun,* 75, 5376–89.

Mann, B. J. & Ark, N. M. (2009) Rationally designed tularemia vaccines. *Expert Rev Vaccines,* 8, 877–85.

Meibom, K. L., Forslund, A. L., Kuoppa, K., Alkhuder, K., Dubail, I., Dupuis, M., Forsberg, A. & Charbit, A. (2009) Hfq, a novel pleiotropic regulator of virulence-associated genes in *Francisella tularensis. Infect Immun,* 77, 1866–80.

Mcmurry, J. A., Gregory, S. H., Moise, L., Rivera, D., Buus, S. & De groot, A. S. (2007) Diversity of *Francisella tularensis* Schu4 antigens recognized by T lymphocytes after natural infections in humans: identification of candidate epitopes for inclusion in a rationally designed tularemia vaccine. *Vaccine,* 25, 3179–91.

Murthy, T., Rolfs, A., Hu, Y., Shi, Z., Raphael, J., Moreira, D., Kelley, F., Mccarron, S., Jepson, D., Taycher, E., Zuo, D., Mohr, S. E., Fernandez, M., Brizuela, L. & Labaer, J. (2007) A full-genomic sequence-verified protein-coding gene collection for *Francisella tularensis. PLoS One*, 2, e577.

Pavkova, I., Hubalek, M., Zechovska, J., Lenco, J. & Stulik, J. (2005) *Francisella tularensis* live vaccine strain: proteomic analysis of membrane proteins enriched fraction. *Proteomics,* 5, 2460–7.

Pavkova, I., Reichelova, M., Larsson, P., Hubalek, M., Vackova, J., Forsberg, A. & Stulik, J. (2006) Comparative proteome analysis of fractions enriched for membrane-associated proteins from *Francisella tularensis* subsp. tularensis and *F. tularensis* subsp. *holarctica* strains. *J Proteome Res,* 5, 3125–34.

Pechous, R. D., Mccarthy, T. R. & Zahrt, T. C. (2009) Working toward the Future: Insights into *Francisella tularensis* pathogenesis and vaccine development. *Microbiol Mol Biol Rev,* 73, 684–711.

Peters, B., Sidney, J., Bourne, P., Bui, H. H., Buus, S., Doh, G., Fleri, W., Kronenberg, M., Kubo, R., Lund, O., Nemazee, D., Ponomarenko, J. V., Sathiamurthy, M., Schoenberger, S., Stewart, S., Surko, P., Way, S., Wilson, S. & Sette, A. (2005) The immune epitope database and analysis resource: from vision to blueprint. *PLoS Biol,* 3, e91.

Petrosino, J. F., Xiang, Q., Karpathy, S. E., Jiang, H., Yerrapragada, S., Liu, Y., Gioia, J., Hemphill, L., Gonzalez, A., Raghavan, T. M., Uzman, A., Fox, G. E., Highlander, S., Reichard, M., Morton, R. J., Clinkenbeard, K. D. & Weinstock, G. M. (2006) Chromosome rearrangement and diversification of *Francisella tularensis* revealed by the type B (OSU18) genome sequence. *J Bacteriol,* 188, 6977–85.

Pizza, M., Scarlato, V., Masignani, V., Giuliani, M. M., Arico, B., Comanducci, M., Jennings, G. T., Baldi, L., Bartolini, E., Capecchi, B., Galeotti, C. L., Luzzi, E., Manetti, R., Marchetti, E., Mora, M., Nuti, S., Ratti, G., Santini, L., Savino, S., Scarselli, M., Storni, E., Zuo, P., Broeker, M., Hundt, E., Knapp, B., Blair, E., Mason, T., Tettelin, H., Hood, D. W., Jeffries, A. C., Saunders, N. J., Granoff, D. M., Venter, J. C., Moxon, E. R., Grandi, G. &

Rappuoli, R. (2000) Identification of vaccine candidates against serogroup B *meningococcus* by whole-genome sequencing. *Science,* 287, 1816–20.

Qin, A. & Mann, B. J. (2006) Identification of transposon insertion mutants of *Francisella tularensis* tularensis strain Schu S4 deficient in intracellular replication in the hepatic cell line HepG2. *BMC Microbiol,* 6, 69.

Rohmer, L., Brittnacher, M., Svensson, K., Buckley, D., Haugen, E., Zhou, Y., Chang, J., Levy, R., Hayden, H., Forsman, M., Olson, M., Johansson, A., Kaul, R. & Miller, S. I. (2006) Potential source of *Francisella tularensis* live vaccine strain attenuation determined by genome comparison. *Infect Immun,* 74, 6895–906.

Rohmer, L., Fong, C., Abmayr, S., Wasnick, M., Larson Freeman, T. J., Radey, M., Guina, T., Svensson, K., Hayden, H. S., Jacobs, M., Gallagher, L. A., Manoil, C., Ernst, R. K., Drees, B., Buckley, D., Haugen, E., Bovee, D., Zhou, Y., Chang, J., Levy, R., Lim, R., Gillett, W., Guenthener, D., Kang, A., Shaffer, S. A., Taylor, G., Chen, J., Gallis, B., D'argenio, D. A., Forsman, M., Olson, M. V., Goodlett, D. R., Kaul, R., Miller, S. I. & Brittnacher, M. J. (2007) Comparison of *Francisella tularensis* genomes reveals evolutionary events associated with the emergence of human pathogenic strains. *Genome Biol,* 8, R102.

Rohmer, L., Guina, T., Chen, J., Gallis, B., Taylor, G. K., Shaffer, S. A., Miller, S. I., Brittnacher, M. J. & Goodlett, D. R. (2008) Determination and comparison of the *Francisella tularensis* subsp.novicida U112 proteome to other bacterial proteomes. *J Proteome Res,* 7, 2016–24.

Sebastian, S., Pinkham, J. T., Lynch, J. G., Ross, R. A., Reinap, B., Blalock, L. T., Conlan, J. W. & Kasper, D. L. (2009) Cellular and humoral immunity are synergistic in protection against types A and B *Francisella tularensis. Vaccine,* 27, 597–605.

Su, J., Yang, J., Zhao, D., Kawula, T. H., Banas, J. A. & Zhang, J. R. (2007) Genome-wide identification of *Francisella tularensis* virulence determinants. *Infect Immun,* 75, 3089–101.

Tempel, R., Lai, X. H., Crosa, L., Kozlowicz, B. & Heffron, F. (2006) Attenuated *Francisella novicida* transposon mutants protect mice against wild-type challenge. *Infect Immun,* 74, 5095–105.

Twine, S. M., Mykytczuk, N. C., Petit, M., Tremblay, T. L., Lanthier, P., Conlan, J. W. & Kelly, J. F. (2005) *Francisella tularensis* proteome: low levels of ASB-14 facilitate the visualization of membrane proteins in total protein extracts. *J Proteome Res,* 4, 1848–54.

Twine, S. M., Mykytczuk, N. C., Petit, M. D., Shen, H., Sjostedt, A., Wayne Conlan, J. & Kelly, J. F. (2006) In vivo proteomic analysis of the intracellular bacterial pathogen, *Francisella tularensis,* isolated from mouse spleen. *Biochem Biophys Res Commun,* 345, 1621–33.

Waldo, R. H., Cummings, E. D., Sarva, S. T., Brown, J. M., Lauriano, C. M., Rose, L. A., Belland, R. J., Klose, K. E. & Hilliard, G. M. (2007) Proteome cataloging and relative quantification of *Francisella tularensis* tularensis strain Schu4 in 2D PAGE using preparative isoelectric focusing. *J Proteome Res,* 6, 3484–90.

Wayne Conlan, J. & Oyston, P. C. (2007) Vaccines against *Francisella tularensis. Ann N Y Acad Sci,* 1105, 325–50.

Weiss, D. S., Brotcke, A., Henry, T., Margolis, J. J., Chan, K. & Monack, D. M. (2007) In vivo negative selection screen identifies genes required for *Francisella* virulence. *Proc Natl Acad Sci U S A,* 104, 6037–42.

Zvi, A., Ariel, N., Fulkerson, J., Sadoff, J. C. & Shafferman, A. (2008) Whole genome identification of *Mycobacterium tuberculosis* vaccine candidates by comprehensive data mining and bioinformatic analyses. *BMC Med Genomics,* 1, 18.

Chapter 26
Modified Vaccinia Virus Ankara (MVA) based Vaccines – Immunostimulatory and Protective Capacity

Michael H. Lehmann, Melanie Kremer, Yasemin Suezer, and Gerd Sutter

Abstract Modified vaccinia virus Ankara (MVA) is a highly attenuated and replication-deficient vaccinia virus (VACV) which serves as promising candidate replacement vaccine against smallpox. Moreover, genetically modified MVA viruses are investigated as vector vaccines against various infectious diseases and cancer. A long-standing observation is that non-replicating MVA vaccines are paradoxically immunogenic in comparison to immunizations with fully replication-competent VACV. Here, we present evidence from recent experiments supporting the notion that MVA has particular immunostimulatory properties. Immunizations in mouse models revealed the in vivo synthesis of type I IFN shortly after MVA vaccine administration, and the activation of dendritic cells by both TLR-dependent and TLR-independent pathways. In addition, rapid immigration of leukocytes into the lung of mice is triggered upon intranasal immunization with MVA and not with other VACV strains. Importantly, MVA infection efficiently induced the expression of several chemokines (CCL2, CCL3, CCL4, CXCL10), and we identified CCL2 as the main CC chemokine attracting monocytes after MVA infection in vitro and in vivo. We conclude that the failure of MVA to prevent early activation of type I interferon, chemokine expression, and early immigration of leukocytes contributes to its potential as a protective vaccine, and supports the suitability of MVA vaccines to serve for efficient prophylaxis against suddenly emerging infectious diseases.

Keywords Poxvirus · Smallpox · Biothreat · Immunomodulation · Interferon · Chemokine · MCP-1

G. Sutter (✉)
Institut für Infektionsmedizin und Zoonosen, Ludwig-Maximilians-Universität, Veterinärstr. 13, 80539 München, Germany
e-mail: gerd.sutter@lmu.de

A. Shafferman et al. (eds.), *The Challenge of Highly Pathogenic Microorganisms*,
DOI 10.1007/978-90-481-9054-6_26, © Springer Science+Business Media B.V. 2010

26.1 Origin, Attenuation, and Early Use of VACV MVA

MVA was developed as an attenuated strain of VACV for use as safer vaccine during the last decades of smallpox eradication campaign. It originates from chorioallantois vaccinia virus Ankara (CVA), a VACV virus strain historically used at the Turkish vaccine institute in Ankara as basis for smallpox vaccine production. At the Institute for Microbiology and Infectious Diseases of Animals of the University of Munich CVA was chosen as candidate VACV for serial passage in tissue cultures to investigate the possibility to restrict the broad host range of vaccinia virus as an attempt to mimic the evolution of host range restricted Orthopoxviruses such as variola virus the causative agent of human smallpox. Indeed, Mayr & Munz (Mayr and Munz, 1964) reported that CVA had developed particular growth characteristics on the chorioallantois membrane, in tissue cultures and in laboratory animals after 371 passages on chicken embryo fibroblasts (CEF). Upon further passage a new highly attenuated phenotype could be confirmed and from the 516th CEF passage the VACV strain was renamed modified vaccinia virus Ankara (MVA) (Mayr et al., 1975). Concurrently, MVA was chosen and used by the Bavarian State Vaccine Institute in Munich as candidate vaccine for safer immunization against smallpox. Until 1988 a licensed MVA vaccine has been used in more than 100 000 humans without documentation of any of the severe adverse reactions associated with the application of conventional VACV vaccines (Mahnel and Mayr, 1994; Stickl et al., 1974). A few years later the first molecular analysis of the MVA genome revealed that during attenuation the virus genome had suffered large deletions totalling to a loss of about 15% of genetic information compared to the anchestor virus CVA (Meyer et al., 1991). The complete sequence of the MVA genome demonstrated that deletions and mutations affected many genes considered as important regulators of VACV-host interaction genome representing the likely genetic basis for avirulence and host cell restriction (Antoine et al., 1998). The molecular characterization of MVA infection showed that while unable to propagate in human and many other mammalian cells MVA can efficiently express viral and recombinant genes supporting its usefulness as viral vector (Sutter and Moss, 1992). The first recombinant MVA vaccine was engineered to deliver the hemagglutinin and the nucleoprotein of influenza virus as foreign antigens. Its application in mice stimulated antigen-specific humoral and cellular immunity and solidly protected against a highly lethal influenza virus challenge (Sutter et al., 1994) initiating the development of many new MVA candidate vaccines. For reviews see Acres and Bonnefoy (2008), Drexler et al. (2004), Gomez et al. (2008), Paran and Sutter (2009), Rimmelzwaan and Sutter (2009), Sutter and Staib (2003).

26.2 Immune Stimulatory Capacity of MVA and Induction of Interferon

Anton Mayr and co-workers had already observed immune stimulatory activities associated with experimental MVA inoculations (Mayr et al., 1975). Intraperitoneal application of virus to mice enhanced the in vivo clearance of carbon marker particles from the blood of the animals suggesting an increased phagocytic activity of immune cells 2 days after MVA inoculation. In addition, intranasal delivery of MVA to rabbits resulted within hours in the induction of "serum interferons" that efficiently inhibited sindbis virus replication in an in vitro infection assay. Today, type I and type II interferons (IFN) are well known as important mediators of versatile anti-viral response mechanisms and it has been shown that VACV is capable to antagonize IFN action by secretion of soluble viral IFN-binding proteins from infected cells (Alcami and Smith, 1995; Symons et al., 1995; Colamonici et al., 1995). However, recently it has been demonstrated that infection with MVA but not with other VACV induces type I IFN in mice as well as in bone marrow-derived plasmacytoic dendritic cells (Waibler et al., 2007).

26.3 MVA Induces Chemokine Expression In Vitro and In Vivo

After the activation of pattern recognition receptors by viral infection, the response of the host immune system includes also the production of inflammatory cytokines and chemokines. For review see Mogensen, 2009. Hereby, chemokine gene expression can again be enhanced by type I interferons and chemokines coordinate the migration of leucocytes by chemotaxis. This activity has a crucial role in bridging the innate and adaptive immune defence and is considered a prerequisite for acquiring protective immunity.

To test the capacity of MVA to induce chemokines, we used a protein array, and detected several chemokines being upregulated including the CC chemokine ligand 2 (CCL2). Upregulation of chemokine expression by MVA was further confirmed by ribonuclease protection assay, RT-PCR and ELISA in the human monocytic cell line THP-1 and primary human monocytes. Intranasal infection of mice with

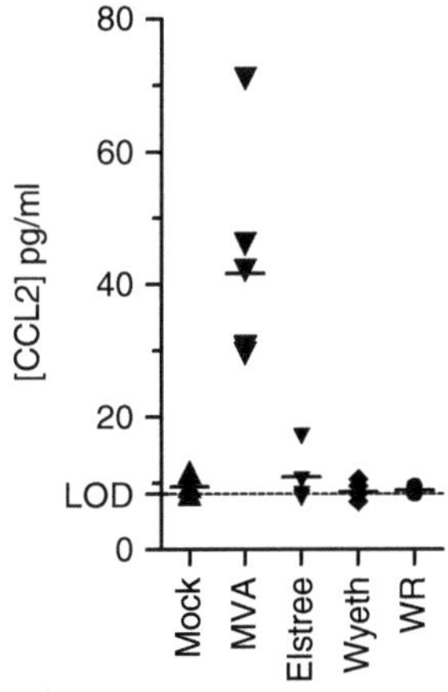

Fig. 26.1 MVA exclusively induces CCL2 production in vivo. Mice were intranasally infected with MVA or with VACV Elstree, Wyeth or WR. Bronchoalveolar lavage was performed 24 h post infection. CCL2 concentration in bronchoalveolar fluids of mice was determined by multiplex bead-based Luminex® technology (Invitrogen). Mice that inhaled equal volumes of PBS served as control (Mock). Limit of detection (LOD)

MVA, in contrast to infection with other VACV strains, confirmed the expression of CCL2 (Fig. 26.1). Further, we have shown that UV-inactivated MVA cannot trigger chemokine expression in vitro and in vivo, which demonstrates the requirement for an activated molecular viral life cycle and categorically excludes that MVA exploits a cell membrane receptor for signalling to exert this effect (Lehmann et al., 2009). These results are in contrast to the suggestion that MVA infection stimulates TOLL-like receptor (TLR) 2 mediated immune activation including chemokine expression (Barbalat et al., 2009; Delaloye et al., 2009; Zhu et al., 2007). We could not reproduce the relevance of TLR2 mediated signalling and rather prefer a model where MVA triggers chemokine production through activation of intracellular viral nucleic acid detectors.

26.4 MVA Induces Immune Cell Migration In Vitro and In Vivo

Active immigration of immune cells to the site of immunization is an urgently sought action of vaccine adjuvants. Infection of monocytes with MVA but not with other VACV strains induces the expression of chemokines including a chemotactic

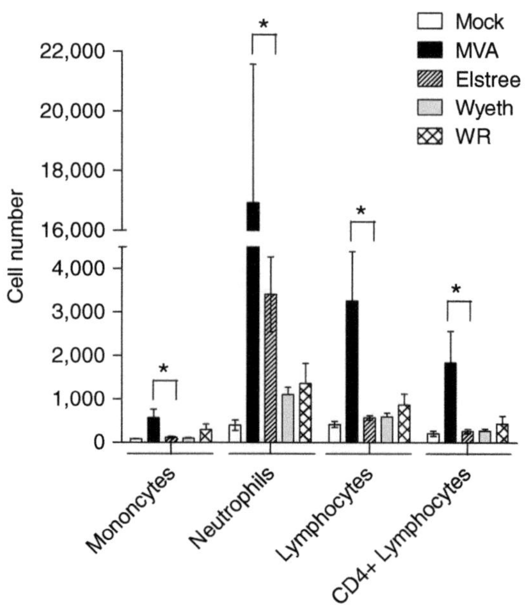

Fig. 26.2 MVA, but not VACV Elstree, Wyeth or WR triggers early respiratory infiltration of leukocytes. Mice were intranasally infected with MVA, or VACV Elstree, Wyeth or WR. An equal volume of PBS was used as control (Mock). Cells in the lung were recovered 24 h post infection by bronchoalveolar lavage and analyzed by flow cytometry. Monocytes, neutrophils and lymphocytes were identified by their surface expression profile of CD11b, Gr-1 (Ly-6C+Ly-6G), CD3 or CD4, respectively. At least five mice per group were used. Columns represent the mean cell number of each individual cell population ± SEM. $^{**}P<0.01$

factor for human monocytes. Using an in vitro chemotaxis assay, we confirmed that infection with MVA results in production of a soluble chemotactic factor for monocytes, T cells and NK cells. By applying neutralizing antibodies against CCL2, we identified CCL2 as the exclusive factor secreted by MVA-infected cells capable of attracting THP-1 cells and primary human monocytes.

Intranasal infection of mice with MVA, but not with VACV Elstree, Wyeth and Western Reserve (WR) triggers immigration of leukocytes into the lung. Immigrating cells were identified as monocytes, neutrophils and CD4+ lymphocytes by flow cytometry and could be detected 24 h post infection and reached peak numbers 48 h post infection (Fig. 26.2). Using CCL2-deficient mice we could demonstrate that CCL2 plays a key role in MVA-triggered respiratory immigration of leukocytes (Lehmann et al., 2009). Further, MVA-induced CCL2 may also be involved in forming bronchus-associated lymphoid tissue constituting a part of the adaptive mucosal immune system of the lung, which is induced in mice after intranasal infection with MVA (Halle et al., 2009).

Of note, heat- and UV-treated MVA did not trigger respiratory immigration of leukocytes which nicely corresponds to the inability of UV-treated MVA to induce chemokines.

26.5 Immunogenicity and Protective Capacity of MVA – Does Immune Stimulation Compensate for Replication Deficiency?

A surprising first observation has been that non-replicating MVA vectors seemed to be paradoxically immunogenic in comparison to fully replicating recombinant VACV, which should amplify and deliver much higher amounts of antigen upon administration in vivo (Hirsch et al., 1996; Sutter et al., 1994). Moreover, we and others have shown that immunizations with MVA effectively protect mice against concurrent lethal infections with orthopoxviruses (Paran et al., 2009; Samuelsson et al., 2008; Staib et al., 2006), and in a monkeypox virus challenge model it could be shown that also in non-human primates protective immunity developed more rapidly after vaccination with MVA compared to the conventional Dryvax vaccine (Earl et al., 2008).

How can we explain this stunning efficacy of the non-replicating MVA vaccines? Efficient activation and quality of IFN, cytokine and chemokine coordinated innate responses may be prerequisites for the induction of protective immunity upon vaccination. Pathogenic poxviruses including conventional VACV succeed to counteract the activity of IFNs, cytokines and chemokines through expression of soluble binding proteins and receptor antagonists (Seet et al., 2003). In addition, VACV exploits stunning intracellular synthesis of other viral immune evasion proteins, such as A52 and A46, to antagonize signalling pathways and to block the production of interferons, inflammatory cytokines and chemokines already at the level of transcriptional activation (Bowie et al., 2000; Harte et al., 2003; Stack et al., 2005). Notably, of all

VACV genes suspected to be involved in immune evasion most are inactivated or truncated in the MVA genome (Antoine et al., 1998).

However, MVA infection, even under non-replicating conditions, produces normal amounts of viral RNA, DNA and proteins, and thus MVA must be expected to deliver the key pathogen associated molecular patterns for an optimal immune stimulation. Consequently, we suggest that the failure of MVA infection to early interfere with key immune defence mechanisms including interferon and chemokine activities may significantly contribute to its attenuated phenotype and to its particular potential as vaccine.

References

Acres, B. & Bonnefoy, J. Y. (2008) Clinical development of MVA-based therapeutic cancer vaccines. *Expert Rev Vaccines,* 7, 889–93.

Alcami, A. & Smith, G. L. (1995) Vaccinia, cowpox, and camelpox viruses encode soluble gamma interferon receptors with novel broad species specificity. *J Virol,* 69, 4633–9.

Antoine, G., Scheiflinger, F., Dorner, F. & Falkner, F. G. (1998) The complete genomic sequence of the modified vaccinia Ankara strain: Comparison with other orthopoxviruses. *Virology,* 244, 365–96.

Barbalat, R., Lau, L., Locksley, R. M. & Barton, G. M. (2009) Toll-like receptor 2 on inflammatory monocytes induces type I interferon in response to viral but not bacterial ligands. *Nat Immunol,* 10, 1200–7.

Bowie, A., Kiss-Toth, E., Symons, J. A., Smith, G. L., Dower, S. K. & O'neill, L. A. (2000) A46R and A52R from vaccinia virus are antagonists of host IL-1 and toll-like receptor signaling. *Proc Natl Acad Sci U S A,* 97, 10162–7.

Colamonici, O. R., Domanski, P., Sweitzer, S. M., Larner, A. & Buller, R. M. (1995) Vaccinia virus B18R gene encodes a type I interferon-binding protein that blocks interferon alpha transmembrane signaling. *J Biol Chem,* 270, 15974–8.

Delaloye, J., Roger, T., Steiner-Tardivel, Q. G., Le Roy, D., Knaup Reymond, M., Akira, S., Petrilli, V., Gomez, C. E., Perdiguero, B., Tschopp, J., Pantaleo, G., Esteban, M. & Calandra, T. (2009) Innate immune sensing of modified vaccinia virus Ankara (MVA) is mediated by TLR2-TLR6, MDA-5 and the NALP3 inflammasome. *PLoS Pathog,* 5, e1000480.

Drexler, I., Staib, C. & Sutter, G. (2004) Modified vaccinia virus Ankara as antigen delivery system: how can we best use its potential? *Curr Opin Biotechnol,* 15, 506–12.

Earl, P. L., Americo, J. L., Wyatt, L. S., Espenshade, O., Bassler, J., Gong, K., Lin, S., Peters, E., Rhodes, L., Jr., Spano, Y. E., Silvera, P. M. & Moss, B. (2008) Rapid protection in a monkeypox model by a single injection of a replication-deficient vaccinia virus. *Proc Natl Acad Sci USA,* 105, 10889–94.

Gomez, C. E., Najera, J. L., Krupa, M. & Esteban, M. (2008) The poxvirus vectors MVA and NYVAC as gene delivery systems for vaccination against infectious diseases and cancer. *Curr Gene Ther,* 8, 97–120.

Halle, S., Dujardin, H. C., Bakocevic, N., Fleige, H., Danzer, H., Willenzon, S., Suezer, Y., Hammerling, G., Garbi, N., Sutter, G., Worbs, T. & Forster, R. (2009) Induced bronchus-associated lymphoid tissue serves as a general priming site for T cells and is maintained by dendritic cells. *J Exp Med,* 206, 2593–601.

Harte, M. T., Haga, I. R., Maloney, G., Gray, P., Reading, P. C., Bartlett, N. W., Smith, G. L., Bowie, A. & O'neill, L. A. (2003) The poxvirus protein A52R targets Toll-like receptor signaling complexes to suppress host defense. *J Exp Med,* 197, 343–51.

Hirsch, V. M., Fuerst, T. R., Sutter, G., Carroll, M. W., Yang, L. C., Goldstein, S., Piatak, M., Jr., Elkins, W. R., Alvord, W. G., Montefiori, D. C., Moss, B. & Lifson, J. D. (1996) Patterns of viral replication correlate with outcome in simian immunodeficiency virus (SIV)-infected macaques: effect of prior immunization with a trivalent SIV vaccine in modified vaccinia virus Ankara. *J Virol,* 70, 3741–52.

Lehmann, M. H., Kastenmuller, W., Kandemir, J. D., Brandt, F., Suezer, Y. & Sutter, G. (2009) Modified vaccinia virus ankara triggers chemotaxis of monocytes and early respiratory immigration of leukocytes by induction of CCL2 expression. *J Virol,* 83, 2540–52.

Mahnel, H. & Mayr, A. (1994) [Experiences with immunization against orthopox viruses of humans and animals using vaccine strain MVA]. *Berl Munch Tierarztl Wochenschr,* 107, 253–6.

Mayr, A., Hochstein-Mintzel, V. & Stickl, H. (1975) Abstammung, Eigenschaften und Verwendung des attenuierten Vaccinia-Stammes MVA. *Infection,* 3, 6–14.

Mayr, A. & Munz, E. (1964) Veränderungen von Vaccinevirus durch Dauerpassagen auf Hühnerembryofibroblasten-Kulturen. *Zentralbl Bakteriol Orig A,* 195, 24–35.

Meyer, H., Sutter, G. & Mayr, A. (1991) Mapping of deletions in the genome of the highly attenuated vaccinia virus MVA and their influence on virulence. *J Gen Virol,* 72(Pt 5), 1031–8.

Mogensen, T. H. (2009). Pathogen recognition and inflammatory signaling in innate immune defenses. *Clin Microbiol Rev,* 22, 240–73.

Paran, N., Suezer, Y., Lustig, S., Israely, T., Schwantes, A., Melamed, S., Katz, L., Preuss, T., Hanschmann, K. M., Kalinke, U., Erez, N., Levin, R., Velan, B., Lower, J., Shafferman, A. & Sutter, G. (2009) Postexposure immunization with modified vaccinia virus Ankara or conventional Lister vaccine provides solid protection in a murine model of human smallpox. *J Infect Dis,* 199, 39–48.

Paran, N. & Sutter, G. (2009) Smallpox vaccines: New formulations and revised strategies for vaccination. *Hum Vaccin,* 5, 824–31.

Rimmelzwaan, G. F. & Sutter, G. (2009) Candidate influenza vaccines based on recombinant modified vaccinia virus Ankara. *Expert Rev Vaccines,* 8, 447–54.

Samuelsson, C., Hausmann, J., Lauterbach, H., Schmidt, M., Akira, S., Wagner, H., Chaplin, P., Suter, M., O'keeffe, M. & Hochrein, H. (2008) Survival of lethal poxvirus infection in mice depends on TLR9, and therapeutic vaccination provides protection. *J Clin Invest,* 118, 1776–84.

Seet, B. T., Johnston, J. B., Brunetti, C. R., Barrett, J. W., Everett, H., Cameron, C., Sypula, J., Nazarian, S. H., Lucas, A. & Mcfadden, G. (2003) Poxviruses and immune evasion. *Annu Rev Immunol,* 21, 377–423.

Stack, J., Haga, I. R., Schroder, M., Bartlett, N. W., Maloney, G., Reading, P. C., Fitzgerald, K. A., Smith, G. L. & Bowie, A. G. (2005) Vaccinia virus protein A46R targets multiple Toll-like-interleukin-1 receptor adaptors and contributes to virulence. *J Exp Med,* 201, 1007–18.

Staib, C., Suezer, Y., Kisling, S., Kalinke, U. & Sutter, G. (2006) Short-term, but not post-exposure, protection against lethal orthopoxvirus challenge after immunization with modified vaccinia virus Ankara. *J Gen Virol,* 87, 2917–21.

Stickl, H., Hochstein-Mintzel, V., Mayr, A., Huber, H. C., Schafer, H. & Holzner, A. (1974) [MVA vaccination against smallpox: clinical tests with an attenuated live vaccinia virus strain (MVA) (author's transl)]. *Dtsch Med Wochenschr,* 99, 2386–92.

Sutter, G. & Moss, B. (1992) Nonreplicating vaccinia vector efficiently expresses recombinant genes. *Proc Natl Acad Sci U S A,* 89, 10847–51.

Sutter, G., Ramsey-Ewing, A., Rosales, R. & Moss, B. (1994) Stable expression of the vaccinia virus K1L gene in rabbit cells complements the host range defect of a vaccinia virus mutant. *J Virol,* 68, 4109–16.

Sutter, G. & Staib, C. (2003) Vaccinia vectors as candidate vaccines: the development of modified vaccinia virus Ankara for antigen delivery. *Curr Drug Targets Infect Disord,* 3, 263–71.

Symons, J. A., Alcami, A. & Smith, G. L. (1995) Vaccinia virus encodes a soluble type I interferon receptor of novel structure and broad species specificity. *Cell,* 81, 551–60.

Waibler, Z., Anzaghe, M., Ludwig, H., Akira, S., Weiss, S., Sutter, G. & Kalinke, U. (2007) Modified vaccinia virus Ankara induces Toll-like receptor-independent type I interferon responses. *J Virol,* 81, 12102–10.

Zhu, J., Martinez, J., Huang, X. & Yang, Y. (2007) Innate immunity against vaccinia virus is mediated by TLR2 and requires TLR-independent production of IFN-beta. *Blood,* 109, 619–25.

Chapter 27
Therapeutic Potential of Vaccinia Hyper Immune Sera in Mouse Models of Lethal Orthopoxviruses Infection

Sharon Melamed, Nir Paran, Tomer Israely, Noam Erez, Shaul Reuveny, Arie Ordentlich, and Shlomo Lustig

Abstract Vaccinia Immune Globulin (VIG) is currently used to treat severe complications of smallpox vaccines. In this study we compare the therapeutic potential of vaccinia virus rabbit hyper immune sera (RHIS) with that of human VIG. The clearance rate of RHIS from mouse circulation is only slightly slower than that of VIG ($t_{1/2}$=10 and 7.5 days respectively). Like VIG, passively administered RHIS can protect mice against lethal respiratory and dermal Ectromelia virus (ECTV) challenge. Administration of both homologous (anti ECTV) and heterologous (anti VACV-WR or VACV-Lister) anti-sera conferred efficient protection against a subsequent lethal respiratory ECTV challenge. These observations formed the basis for passive cross protection studies against ECTV, conducted in mice. RHIS conferred better protection as compared to VIG as a result of its better specific activity which is about 100 folds higher than that of VIG, allowing for significant protection even if administered 5 days post infection. This study emphasizes the advantage of a hyper immune product and validates the potential use of VIG and other antibody based therapeutics, not only as prophylactic measures against post-vaccination complications but also for post-exposure treatment of smallpox disease.

Keywords ECTV · Hairless mice · Hyper immune sera · Smallpox · Vaccinia Immune Globulin (VIG)

27.1 Introduction

Smallpox is an acute highly contagious disease, caused by Variola virus, a member of the orthopoxviruses. Success of the global eradication campaign through intensive worldwide vaccination [initiated in 1967 by WHO (World et al., 1980)] resulted in gradual suspension of smallpox vaccination programs. As a result, the

S. Melamed (✉)
Department of Infectious Diseases, Israel Institute for Biological Research, P.O. Box 19, Ness-Ziona 74100, Israel
e-mail: sharonm@iibr.gov.il

A. Shafferman et al. (eds.), *The Challenge of Highly Pathogenic Microorganisms*,
DOI 10.1007/978-90-481-9054-6_27, © Springer Science+Business Media B.V. 2010

worldwide number of unvaccinated individuals is constantly on the rise. The potential reintroduction of variola virus as a bioterror threat agent led to renewed interest in countermeasures to orthopoxviruses.

Smallpox vaccines contain live vaccinia virus, closely related to the disease causingagent (Variola virus). These vaccines though efficacious, exhibit a high incidence of adverse effects, which can be life threatening (Fulginiti et al., 2003, Wittek, 2006).

Several antiviral drugs have been recently evaluated as a post-exposure antipoxvirus treatment measures (Smee and Sidwell, 2003; Baker et al., 2003; De Clercq, 2002; Quenelle et al., 2003, 2007; Yang et al., 2005). However, none has been approved so far. Since smallpox has been eradicated, any attempt to develop new countermeasures depends on appropriate animal models employing closely related viruses (Jordan and Hruby, 2006; Smee, 2008; Smee and Sidwell, 2003). In the absence of Variola virus and a reliable animal disease model, data compiled from diverse experimental models is used to evaluate and approve new countermeasures.

Several studies support the concept of serum therapy for Orthopoxvirus infections (Goldsmith et al., 2004; Law et al., 2005; Shearer et al., 2005). Recently, we showed that VIG can protect mice against ECTV, a natural highly virulent and contagious mouse pathogen and a member of the Orthopox family. The similarities in Variola and ECTV pathogenesis and mechanisms utilized by both viruses to evade the host immune response, render ECTV infections of small animals a relevant surrogate model for smallpox infections (Panchanathan et al., 2008; Esteban and Buller, 2005; Paran et al., 2009). Using this model, we have demonstrated the ability of VIG to protect mice against an ECTV lethal challenge as a prophylactic measure as well as in a post-exposure scenario (Lustig et al., 2009).

Here we show the protective effect of rabbit hyper immune sera (RHIS) in treatment of lethal respiratory ECTV infections and dermal manifestations of ECTV, in mice. The improved specific activity of RHIS over VIG allows for efficient protection even when administered 4–5 days post exposure.

27.2 Results

27.2.1 In-Vitro Characteristics and Pharmacokinetics of Vaccinia Virus RHIS

Three RHIS preparations were prepared by consecutive sub-cutaneous injections of live ECTV, VACV-WR or VACV-Lister. For relative efficacy evaluation, all sera were "normalized" to equal antibody titers [ELISA against VACV-IHD-J, (6×10^5/injection dose)] allowing for comparison of antibodies homologous to the challenge strain (ECTV in our case) versus antibodies originating in other species (heterologous to the challenge strain) (Table 27.1). Specific activity of the rabbit hyper immune sera (ELISA) was found to be about 100 folds higher than that of human VIG.

Table 27.1 Human and rabbit Vaccinia immune globulins

Antibody		Titer injected to mice	
Product	Against	ELISA	mg (protein)
Rabbit hyper immune IgG	ECTV VACV-WR VACV-Lister	6×10^5	0.3
Human IgG (VIG)	Dryvax	6×10^5	32

The therapeutic potential of IgG is markedly affected by its circulatory residence. The human VIG's half-life in mouse circulation is significantly shorter than that in humans ($t_{1/2}$=7.5 and 21 days respectively), whereas RHIS mouse serum residence was slightly longer than VIG ($t_{1/2}$=10), but still shorter than that of VIG in humans (Hopkins et al., 2004; Lustig et al., 2009). In order to maintain long term high circulatory VIG levels in mice a protocol consisting of an initial injection of 8 mg VIG followed by 9 daily injections of 2 mg each (a total of 26 mg) was applied. This regimen resulted in a maintenance of high circulatory levels for 18 days (2 days following the last injection) (Lustig et al., 2009). This allowed us to overcome the short residence time of the heterologous human antibodies in mice and to reach serum residence time similar to that of VIG in humans.

27.2.2 Protection Efficacy of RHIS Against Respiratory ECTV Infection

The protective efficacy and dose dependence of RHIS preparations (homologous and heterologous) was studied in mice, in comparison to human VIG. In keeping with VIG protective evaluation protocol (Lustig et al., 2009), mice were injected 1 day post ECTV i.n. challenge (10 LD_{50}) with different serum doses. All three RHIS doses efficiently protected mice against an i.n. ECTV challenge, even with doses equivalent to 1/10 of the VIG reference dose (Table 27.2). The results not only substantiate the potential of post exposure antibody prophylaxis against lethal orthopox exposure, but also form the basis for passive cross protection experiments against ECTV, in mice.

Having demonstrated that RHIS can protect mice against an i.n. ECTV challenge when administered 1 day post exposure, we examined whether delayed treatment would still be effective. To this end mice were infected intranasaly with ECTV (5 LD_{50}) and then injected daily with a single injection of 3 mg RHIS (anti VACV-Lister). Full protection was observed up to day 3 and effective protection (60–80% survival and death delayed by about 4 days compared to unprotected mice) was attained even when immune globulins were injected at day 5 post infection (Fig. 27.1). Treatment postponed to day 6 post-infection could neither prevent death nor delay time to death. The protection afforded by 3 mg RHIS was better than of

Table 27.2 Protection against intranasal ECTV challenge in mice using homologous or heterologous sera

	% Survival			
		Antibody titer ELISA units		
Antibody	unprotected	6×10^5 (0.3 mg)	2×10^6 (1 mg)	6×10^6 (3 mg)
α ECTV (Homologous)	0	80	100	100
α WR (Heterologous)	0	80	100	100
α Lister (Heterologous)	0	100	100	100

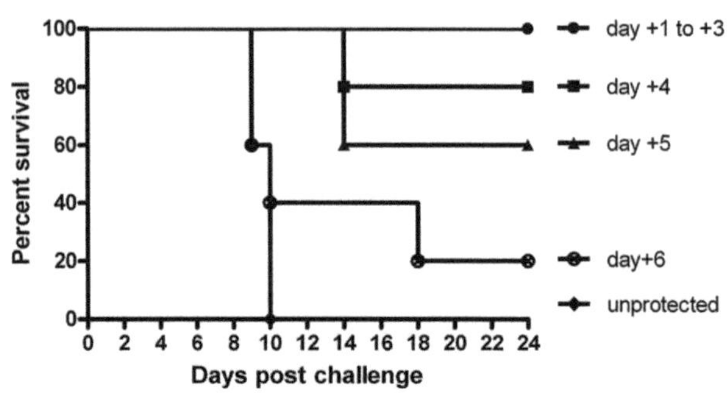

Fig. 27.1 Mice can be rescued from lethal ECTV infection by rabbit anti Lister treatment administered up to 5 days post challenge BALB/c mice were treated with a single IP injection of 3 mg rabbit anti VACV-Lister antiserum which was administered up 6 days post infection. Challenge consisted of intranasal instillation of 5 LD_{50} ECTV (n=5 per each group). Infected, untreated mice served as controls. The protective efficacy of rabbit hyper immune sera was evaluated by monitoring animal mortality

8 mg VIG, where the latter could efficiently protected only up to day 2 (80%) and partially protected (60%) when given 3 days post exposure (Lustig et al., 2009). The enhanced protective efficiency of the RHIS, as compared to VIG, may be attributed to the higher specific activity of the rabbit sera (about 100 times than VIG – see Table 27.1).

27.2.3 Passive Protection Against Systemic and Dermal Manifestations of ECTV Infection in Hairless Mice (SKH-1)

One of the major characteristics of smallpox is the typical disseminated and synchronized skin lesions. In order to evaluate the therapeutic potential of VIG in treatment of poxvirus skin lesions, SKH-1 immune competent hairless mice were

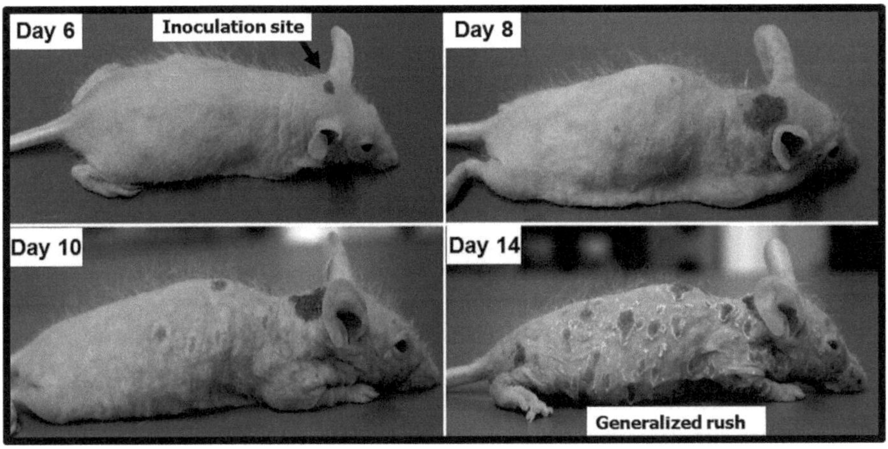

Fig. 27.2 Skin infection with ECTV in hairless mice. Hairless mice were inoculated subcutaneously with 100 LD_{50} ECTV, 6 days post inoculation, a lesion starts to develop at the inoculation site, 2 days later, a rash begins and disseminates to the whole body. In surviving mice (20% day 16), the scabs peels off and the skin heals

Table 27.3 Protection subcutaneous inoculation with ECTV in hairless mice

α Lister Administration	Antibody titer (ELISA)	mg protein	Lesion at inoculation site	Generalized rush	% mortality
Unprotected	–	–	100	100	80
Day −1	3×10^6	3	20	0	0
	1×10^6	1	80	0	0
	3×10^5	0.3	100	20	40
Day +1	3×10^6	3	20	0	0
	1×10^6	1	100	40	0
	3×10^5	0.3	100	100	60

infected subcutaneously with ECTV. Lesion starts to develop at the inoculation site 6 days following post infection with 100 LD_{50} ECTV. Two days later, a rash appears disseminating throughout the whole body (Fig. 27.2).

Under these conditions most of the animals die and in survivors the scabs peels off and the skin heals. Protective efficacy estimates were based on the following criteria: 1. Presence and size of lesions at the inoculation site. 2. Appearance and extent of a generalized rush. 3. Mortality. In unprotected animals, 8 days post infection, a large lesion (about 1 cm^2) was observed at the inoculation site followed 14 days later by a generalized rash. In the above group, 80% of the mice died (Table 27.3). Administration of high titer RHIS at a dose shown to confer protection in the lethal respiratory model, prevented mortality and lesion dissemination and efficiently reduced lesion size at the inoculation site. This antibody dose was essentially equally protective when injected either pre or post exposure. Reducing the protective dose was associated with increased morbidity and mortality (Table 27.3).

27.3 Discussion

Vaccinia immune globulin is the main product currently used for prophylaxis and treatment of severe complications resulting from immunization with anti-smallpox vaccinia virus-based vaccines (Wittek, 2006).

In this study we present findings that underline the protective potential of RHIS in treatment of dermal and lethal respiratory diseases caused by ECTV infection in mice. Smallpox disease involves synchronized disseminated pox skin lesions. Mouse models for smallpox usually utilize hairy mice which does not mimic in full the skin manifestations of the human disease. To better simulate the human disease we employed immune competent hairless mice (SKH-1 strain) which upon subcutaneous infection with ECTV, develop local lesion, followed by disseminated pox-like skin lesions which, depending on the dose, lead in most cases to death. Using this model we were able to demonstrate that passive administration of RHIS allows for post exposure protection. A high RHIS dose limited the lesions to the exposure site, prevented enlargement and the prototypical disseminated rash. RHIS was more potent in prevention of rash dissemination and death than in lesion size control. The exact protective mechanism is currently unknown.

We show that a single dose of RHIS can efficiently protect (80–100%) mice against lethal respiratory disease even if administrated 3–4 days post exposure and that partial protection (60%) is still achieved when treatment is delayed to day 5 post exposure (1 day prior to the appearance of disease symptoms such as weight loss and when substantial levels of viral load are detected in target organs (Paran et al., 2009)). In a previous study we demonstrated that VIG can efficiently protect mice against respiratory ECTV infection, yet efficient protection was achieved only if VIG was injected within 2–3 days post infection (Lustig et al., 2009). A single dose of 3 mg/mouse RHIS (Table 27.2) or 2–8 mg/mouse VIG (Lustig et al., 2009), administrated 1 day post exposure allowed for complete protection against respiratory ECTV infection. These amounts are equivalent to the VIG doses recommended for human use, namely – 100–800 mg/kg (Centers et al., 2003; Hopkins et al., 2004). The comparable mice serum residence of VIG and RHIS ($t_{1/2}$=7.5 and 10 days respectively) could not account for the improved potency of RHIS, suggesting that IgG origin is not the major determinant for the improved results with RHIS. We believe that the 100 fold higher specific activity of RHIS relative to VIG (Table 27.1), may explain the improved protection of the former. In support of this assumption RHIS exhibited higher specific activity in in-vitro neutralization tests, which were shown to correlate with in-vivo protection (Orr et al., 2004).

Recently we have shown that in the ECTV-mouse infection model, human equivalent doses of VIG, even if administrated as a single injection 1 or 2 days post lethal respiratory exposure, are protective, however, our findings indicate that protection by VIG is attained only in immune competent mice (Lustig et al., 2009).

Detailed studies using VIG to treat SCID mice infected with the Dryvax vaccine strain demonstrated that VIG can protect the mice even when administered at the day of challenge, but not post challenge, (Goldsmith et al., 2004; Shearer et al., 2005). Even repeated VIG injections were unable to cure the disease (Lustig et al., 2009) raising the need for a product with higher specific activity. Indeed extremely potent therapeutic IgG preparation protected VACV-WR infected SCID mice (Lustig et al., 2005). It remains to be determined whether RHIS, studied herein, is able to protect immune deficient mice against lethal poxvirus infections.

Acknowledgements The authors thank Dr. Naomi Ariel for her helpful comments, and Ms. Paula Schneider for her excellent technical assistance.

References

Baker, R. O., Bray, M. & Huggins, J. W. (2003) Potential antiviral therapeutics for smallpox, monkeypox and other orthopoxvirus infections. *Antiviral Res,* 57, 13–23.

Centers For Disease Control And & Prevention (2003) Medical management of Smallpox (Vaccinia) vaccine adverse reactions: Vaccinia immune globulin and Cidofovir. In Services, D. O. H. A. H. (ed), U.S. Government Printing Office Washington.

De Clercq, E. (2002) Cidofovir in the therapy and short-term prophylaxis of poxvirus infections. *Trends Pharmacol Sci,* 23, 456–8.

Esteban, D. J. & Buller, R. M. (2005) Ectromelia virus: the causative agent of mousepox. *J Gen Virol,* 86, 2645–59.

Fulginiti, V. A., Papier, A., Lane, J. M., Neff, J. M. & Henderson, D. A. (2003) Smallpox vaccination: A review, part I. Background, vaccination technique, normal vaccination and revaccination, and expected normal reactions. *Clin Infect Dis,* 37, 241–50.

Goldsmith, J. C., Eller, N., Mikolajczyk, M., Manischewitz, J., Golding, H. & Scott, D. E. (2004) Intravenous immunoglobulin products contain neutralizing antibodies to vaccinia. *Vox Sang,* 86, 125–9.

Hopkins, R. J., Kramer, W. G., Blackwelder, W. C., Ashtekar, M., Hague, L., Winker-La Roche, S. D., Berezuk, G., Smith, D. & Leese, P. T. (2004) Safety and pharmacokinetic evaluation of intravenous vaccinia immune globulin in healthy volunteers. *Clin Infect Dis,* 39, 759–66.

Jordan, R. & Hruby, D. (2006) Smallpox antiviral drug development: satisfying the animal efficacy rule. *Expert Rev Anti Infect Ther,* 4, 277–89.

Law, M., Putz, M. M. & Smith, G. L. (2005) An investigation of the therapeutic value of vaccinia-immune IgG in a mouse pneumonia model. *J Gen Virol,* 86, 991–1000.

Lustig, S., Fogg, C., Whitbeck, J. C., Eisenberg, R. J., Cohen, G. H. & Moss, B. (2005) Combinations of polyclonal or monoclonal antibodies to proteins of the outer membranes of the two infectious forms of vaccinia virus protect mice against a lethal respiratory challenge. *J Virol,* 79, 13454–62.

Lustig, S., Maik-Rachline, G., Paran, N., Melamed, S., Israely, T., Erez, N., Orr, N., Reuveny, S., Ordentlich, A., Laub, O., Shafferman, A. & Velan, B. (2009) Effective post-exposure protection against lethal orthopoxviruses infection by vaccinia immune globulin involves induction of adaptive immune response. *Vaccine,* 27, 1691–99.

Orr, N., Forman, M., Marcus, H., Lustig, S., Paran, N., Grotto, I., Klement, E., Yehezkelli, Y., Robin, G., Reuveny, S., Shafferman, A. & Cohen, D. (2004) Clinical and immune responses after revaccination of israeli adults with the Lister strain of vaccinia virus. *J Infect Dis,* 190, 1295–302.

Panchanathan, V., Chaudhri, G. & Karupiah, G. (2008) Correlates of protective immunity in poxvirus infection: where does antibody stand? *Immunol Cell Biol,* 86, 80–6.

Paran, N., Suezer, Y., Lustig, S., Israely, T., Schwantes, A., Melamed, S., Katz, L., Preub, T., Hanschmann, K. M., Kalinke, U., Erez, N., Levin, R., Velan, B., Lower, J., Shafferman, A. & Sutter, G. (2009) Post-exposure immunization with vaccinia virus MVA or conventional Lister vaccine provides solid protection in murine model for human smallpox. *J Infect Dis,* 199, 39–48.

Quenelle, D. C., Buller, R. M., Parker, S., Keith, K. A., Hruby, D. E., Jordan, R. & Kern, E. R. (2007) Efficacy of delayed treatment with ST-246 given orally against systemic orthopoxvirus infections in mice. *Antimicrob Agents Chemother,* 51, 689–95.

Quenelle, D. C., Collins, D. J. & Kern, E. R. (2003) Efficacy of multiple- or single-dose cidofovir against vaccinia and cowpox virus infections in mice. *Antimicrob Agents Chemother,* 47, 3275–80.

Shearer, J. D., Siemann, L., Gerkovich, M. & House, R. V. (2005) Biological activity of an intravenous preparation of human vaccinia immune globulin in mouse models of vaccinia virus infection. *Antimicrob Agents Chemother,* 49, 2634–41.

Smee, D. F. (2008) Progress in the discovery of compounds inhibiting orthopoxviruses in animal models. *Antivir Chem Chemother,* 19, 115–24.

Smee, D. F. & Sidwell, R. W. (2003) A review of compounds exhibiting anti-orthopoxvirus activity in animal models. *Antiviral Res,* 57, 41–52.

Wittek, R. (2006) Vaccinia immune globulin: current policies, preparedness, and product safety and efficacy. *Int J Infect Dis,* 10, 193–201.

World, Health & Organization (1980) The global eradication of smallpox: final report of the Global Commission for the Certification of Smallpox Eradication, Geneva, December 1979. *World Health Organization,* 1–122.

Yang, G., Pevear, D. C., Davies, M. H., Collett, M. S., Bailey, T., Rippen, S., Barone, L., Burns, C., Rhodes, G., Tohan, S., Huggins, J. W., Baker, R. O., Buller, R. L., Touchette, E., Waller, K., Schriewer, J., Neyts, J., Declercq, E., Jones, K., Hruby, D. & Jordan, R. (2005) An orally bioavailable antipoxvirus compound (ST-246) inhibits extracellular virus formation and protects mice from lethal orthopoxvirus Challenge. *J Virol,* 79, 13139–49.

Chapter 28
A Live Plaque Vaccine Candidate with Improved Safety and Immunogenecity

V.A. Feodorova, L.N. Pan'kina, E.P. Savostina, O.S. Kuznetsov, N.P. Konnov, L.V. Sayapina, S.V. Dentovskaya, R.Z. Shaikhutdinova, S.A. Ageev, B. Lindner, A.N. Kondakova, O. Holst, G.B. Pier, Y.A. Knirel, A.P. Anisimov, and V.L. Motin

Abstract The live plague vaccine strain *Yersinia pestis* EV line NIIEG, a well-known derivative of an attenuated pigmentation-negative strain *Y. pestis* EV76, has been widely used for human and animal immunization in Russia and the FSU countries. The main advantage of the EV NIIEG is its ability to induce rapidly a high level of specific immunity against both bubonic and pneumonic plague following a single injection. However, the vaccine possesses a "residual virulence", which causes in vaccinees a number of adverse reactions likely mediated by a highly toxic *Y. pestis* lipopolysaccharide (LPS) with hexa-acylated lipid A. Recently, we generated a lauroyltransferase gene (*lpxM*) mutant of strain EV NIIEG, which produced a less toxic, penta-acylated lipid A of the LPS. The mutant was found to be harmless, safe, and low reactogenic, and elicited an enhanced immunity against experimental plague, surpassing in these characteristics the parental EV NIIEG strain. A single injection of the mutant protected outbred (57.1–85.7%) and BALB/c (42.8%) mice from subcutaneous challenge with 2,000 median lethal doses (MLD), as well as 25–50% of guinea pigs from 1,200 MLD of wild-type strain *Y. pestis* 231. The parental strain was not protective in this range of the challenging doses in both mice and guinea pigs. The *lpxM* mutation caused a marked alteration in expression, immunoreactivity and epitope specificity of major surface proteins and carbohydrate antigens (F1, LcrV, Ymt, Pla, ECA, LPS), and conferred resistance to the plague diagnostic bacteriophage L-413C. The penta-acylated LPS of the *lpxM* mutant showed a reduced capacity in stimulating the macrophage-like cell line J774.A1 for production of TNF-α. Therefore, the pleiotropic effects of the *lpxM* mutation enabled the construction of a new live plague vaccine candidate with improved characteristics.

Keywords *Yersinia pestis* EV NIIEG · *lpxM (msbB)* mutant · Live plague vaccine · Plague immunity · Pleiotropic effect

---

V.A. Feodorova (✉)
Scientific and Research Institute for Medical and Veterinary Biotechnology, Branch in Saratov, 9, Proviantskaya Street, Box 1580, Saratov, 410028, Russia; Russian Research Anti-Plague Institute "Microbe", Saratov, Russia
e-mail: feodorovav@mail.ru

A. Shafferman et al. (eds.), *The Challenge of Highly Pathogenic Microorganisms*,
DOI 10.1007/978-90-481-9054-6_28, © Springer Science+Business Media B.V. 2010

28.1 Introduction

The live plague vaccine used in Russia and elsewhere is based on the *Y. pestis* strain EV line NIIEG, which is a derivative of an attenuated pigmentation-negative strain, *Y. pestis* EV76. This strain has been widely used for protection of plague researchers, people who may have close and prolonged contact with *Y. pestis* in endemic regions in Russia and other countries of the Former Soviet Union (FSU), and also for protection of camels in enzootic areas (Feodorova and Corbel, 2009). The main feature of this vaccine is its ability to induce rather rapidly (on day seven post immunization) a high level of specific immunity against both bubonic and pneumonic plague following a single injection. No reversion of the *Y. pestis* EV76 vaccine to a fully virulent form has been observed (Feodorova et al., 2007, 2009; Feodorova and Corbel, 2009) as the attenuation was performed by a spontaneous deletion of a ~102-kb "pigmentation" (Pgm) region encoding the hemin-storage (*hms*) locus and high pathogenicity island (HPI) (Fetherston et al., 1992). The live EV NIIEG plague vaccine showed superior protective properties over other live and killed plague vaccines developed thus far (Feodorova and Corbel, 2009). The main factor that limits its worldwide licensing is the fact that the vaccine possesses a "residual virulence" (Feodorova et al., 2007, 2009; Feodorova and Corbel, 2009) and can occasionally trigger undesirable local and systemic reactions in immunized individuals, likely mediated by the *Y. pestis* endotoxin (LPS) (Anisimov et al., 2007; Feodorova et al., 2007, 2009; Feodorova and Corbel, 2009).

Recently, we have reported that inactivation of the lauroyltransferase gene (*lpxM*) by a mutation in the *Y. pestis* EV line NIIEG genome results in synthesis of a less toxic penta-acylated lipid A variant in the lipopolysaccharide (LPS), whereas a more toxic hexa-acylated lipid A is produced when the parental strain is grown at 25°C (Anisimov et al., 2007). It has been found that the *lpxM* mutant exhibits improved characteristics as a vaccine, such as decreased endotoxic activity and overall reactogenicity, and an enhanced protective immunity in three animal models as compared with the parental vaccine strain (Anisimov et al., 2007; Feodorova et al., 2007).

The aim of this study was to investigate in detail the synthesis and immunoreactivity of major antigens of the mutant *Y. pestis* EVΔ*lpxM* strain known to be involved in virulence of *Y. pestis* and development of immunity against plague.

28.2 Results

28.2.1 Construction of the lpxM Mutant of the Vaccine Strain Y. pestis EV NIIEG (Pgm⁻)

The mutant was generated by allelic exchange procedure based on a suicide plasmid system. The original copy of the *lpxM* gene was deleted and replaced with the marker of resistance to kanamycin. The success of the *lpxM* inactivation was evaluated by PCR using primers flanking the region of deletion as described by us previously (Anisimov et al., 2007; Feodorova et al., 2007).

28.2.2 LPS Structural Analysis

The LPSs from the *Y. pestis* EVΔ*lpxM* and the parental *Y. pestis* EV NIIEG plasmid-less strain (KM218) were degraded under mild acid conditions to cleave the linkage between the core and lipid A, and lipid A samples were studied by electrospray ionization Fourier transform ion-cyclotron resonance (ESI FT-ICR) mass spectrometry to determine their chemical composition.

The ESI FT-ICR mass spectra revealed different acyl variants of lipid A, from triacyl to hexaacyl (Fig. 28.1). Hexacyl lipid A (LA_{hexa}) includes four primary 3-hydroxymyristoyl groups (3HO14:0) and two secondary acyl groups: one dodecanoyl (12:0) and one palmitoleoyl (16:1). Compared to this, pentaacyl variants ($LA1_{penta}$ or $LA2_{penta}$) lack either of the secondary acyl groups, and tetraacyl lipid

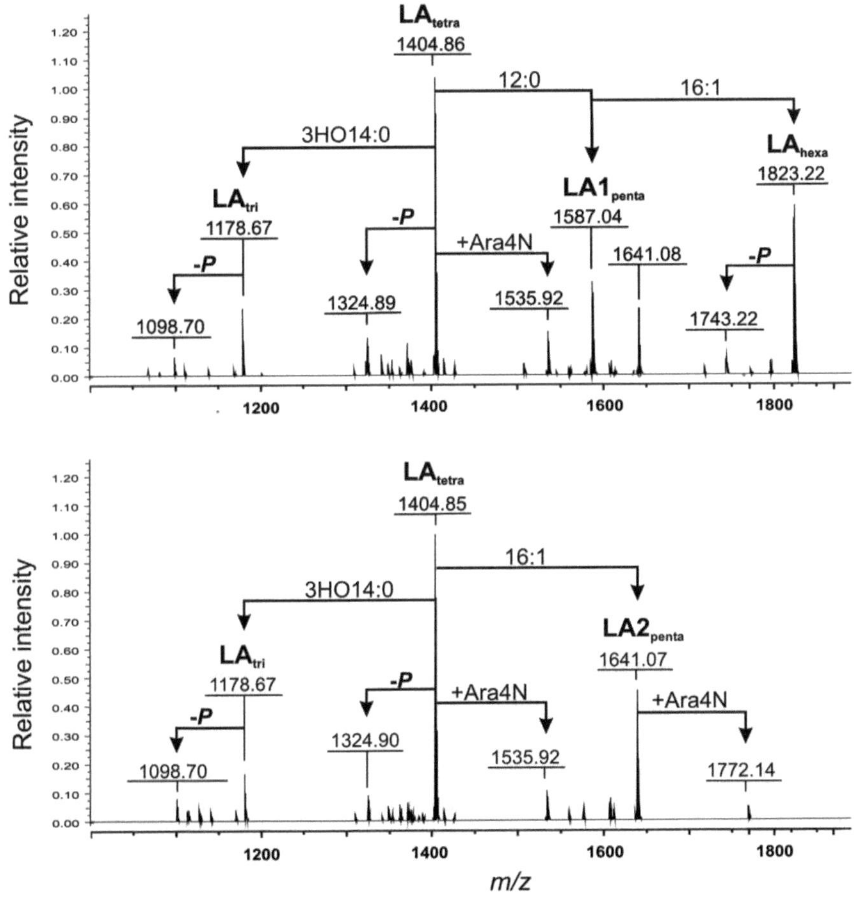

Fig. 28.1 Charge deconvoluted negative ion ESI FT-ICR mass spectra of lipid A samples from *Y. pestis* KM218 (*top*) and *Y. pestis* EVΔ*lpxM* (*bottom*)

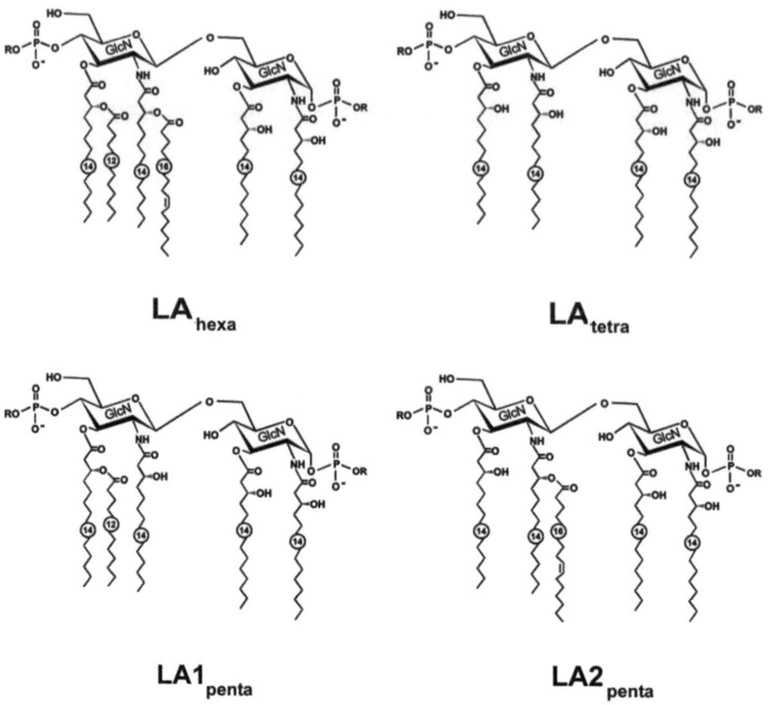

Fig. 28.2 Structures of hexacyl (LA_{hexa}), tetraacyl (LA_{tetra}) and pentaacyl ($LA1_{penta}$ and $LA2_{penta}$) lipid A variants of *Y. pestis*. Numbers inside circles refer to numbers of carbons in fatty acids

A (LA_{tetra}) is devoid of both secondary fatty acids (Fig. 28.2). A minor triacyl variant (LA_{tri}) lacks also one of the 3HO14:0 groups. In the parent strain, LA_{tetra} and LA_{hexa} were predominant (Fig. 28.1a). The EVΔ*lpxM* mutant was distinguished by the absence of LA_{hexa} and $LA1_{penta}$, i.e. the structural variants that include the 12:0 group (Fig. 28.1b). Instead, a significant increase in the content of $LA2_{penta}$ with the single 16:1 secondary group was observed.

28.2.3 "Residual Virulence" and Protective Efficacy of the Y. pestis EVΔlpxM

The EVΔ*lpxM* was compared to the parental *Y. pestis* EV NIIEG and it was found that the mutant is (*a*) completely avirulent and safe in innocuity tests and (*b*) possesses a high protective capacity against experimental bubonic plague (Table 28.1). The parental EV NIIEG strain typically provides protection in the range of 50% of animals against 200 MLD but failed to protect against higher challenge doses (1,200–2,000 MLD) (Feodorova et al., 2007; Feodorova and Corbel, 2009). There was no systemic spread or deaths among outbred and BALB/c mice and guinea

Table 28.1 Residual virulence and protective potency of the EVΔ*lpxM* in different animal models

Animal model	Residual virulence[a], CFU	Protective potency[b], % of survivors – MLD of virulent *Y. pestis* strain inoculated
Outbred mice	1×10^5–1×10^9	85.71–2,000
BALB/c mice	2×10^2–$2{,}5\times10^4$	42.8–1,200
Guinea pigs	1×10^7, 2×10^9 or 1.5×10^{10}	100–1,200

[a]Residual virulence was estimated by the minimal concentration of the bacteria inoculated which can provide the *Y. pestis* bacteria spreading to host organs and tissues as described previously (Feodorova et al., 2007);
[b]The efficacy of vaccination was estimated by the number of surviving animals during 21 days after acute challenge with the fully-virulent *Y. pestis* strain 231

pigs inoculated with the *Y. pestis* EVΔ*lpxM*. At the same time, among the BALB/c mice challenged with the parental *Y. pestis* EV NIIEG strain, there was at least one vaccine-related death in each of the groups which received different challenge doses and overall 8 of 28 tested mice died from infection with this strain.

28.2.4 A Pleiotropic Effect of the Y. pestis EVΔlpxM

The phenomenon was observed in the *Y. pestis* EVΔ*lpxM* when the synthesis and immunoreactivity of major antigens known to be involved in virulence of *Y. pestis* and development of immunity against plague were tested as detailed below.

(i) An alteration in the immunoreactivity and changes in the epitope specificity of major surface proteins were revealed using a panel of relevant monoclonal antibodies (MAbs). Among them there were the antigens that contribute to the resistance to phagocytosis, namely the capsule-like envelope antigen Fraction 1 (F1), Pla protease with versatile virulence-associated functions and V antigen, a multifunctional virulence factor and protective antigen of *Y. pestis*.
(ii) A modification in the biosynthesis or/and epitope specificity of either LPS, or enterobacterial common antigen (ECA), or galactolipid was detected with a panel of relevant MAbs. Only four of 14 tested MAbs showed a similar reactivity with both mutant and parental EV NIIEG strains (Feodorova et al., 2009).
(iii) A marked reduction in production and expression of F1, Pla protease, V antigen, murine toxin Ymt as well as some thermoregulated proteins with molecular masses 24, 31, 34 and 55 kDa was revealed by SDS-PAGE in the mutant as compared with the parental strain cultured at either 37°C or 28°C (Fig. 28.3).

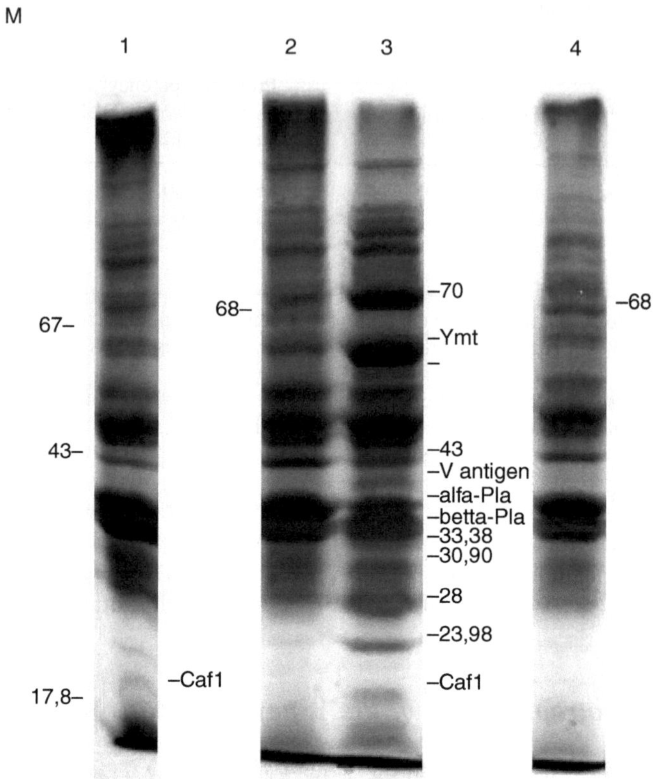

Fig. 28.3 SDS-PAGE of whole-cell lysates of the *Y. pestis* EV NIIEG (lanes 1, 3) and *Y. pestis* EV Δ*lpx*M (lanes 2, 4) grown either at 28°C (lanes 1, 2) or 37°C (lanes 3, 4)

(iv) Alterations in phenotypic activities were observed, namely (*a*) reduced fibrinolytic and auto-processing activities of Pla in the *Y. pestis* EVΔ*lpxM* with no change in coagulase activity and (*b*) a decrease in production of the capsule-like substance by the EVΔ*lpxM* revealed by electron microscopy (Fig. 28.4).

(v) The *Y. pestis* EVΔ*lpxM* had a reduced growth rate and demonstrated an altered resistance to the plague diagnostic bacteriophage L-413C.

(vi) Immunization of mice with the live *Y. pestis* EVΔ*lpxM* resulted in an increased antibody response (at least 3-fold) to the capsular antigen F1.

(vii) The stimulating potency of the *Y. pestis* LPS with hexa-acylated lipid A in the macrophage-like cell line J774.A1 was almost as high as that of the positive control (LPS of *E. coli* O55:B5). The *Y. pestis* LPS with penta-acylated lipid A showed a decreased ability to induce the production of TNF-α, though it was higher than that of the *Francisella tularensis* LPS known for its week endotoxic activity. Similarly, the tetra-acylated LPS demonstrated a reduced stimulating activity as compared with the penta-acylated LPS.

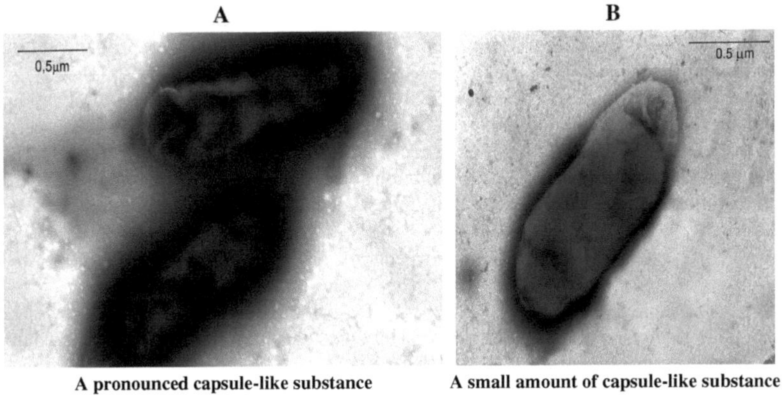

Fig. 28.4 Electron micrographs of *Y. pestis* EV NIIEG (**a**) and its EVΔ*lpxM* derivative (**b**) grown at 37°C

28.3 Conclusions

The *lpxM* mutant of the live plague vaccine strain *Y. pestis* EV NIIEG exhibited improved characteristics, such as decreased endotoxic activity and overall reactogenicity, and an enhanced protective immunity as compared with the parental vaccine strain.

The *lpxM* mutation in *Y. pestis* caused pleiotropic effects resulting in modifications of the biosynthesis and epitope specificity of the major surface proteins and carbohydrate antigens, which were accompanied by alterations in the corresponding phenotypic activities.

The *Y. pestis* EVΔ*lpxM* can be considered as a promising candidate for the development of a new generation of safe, effective and low toxic live plague vaccines.

Acknowledgements This work was supported by the Russian Foundation for Basic Research (grants 06-04-49280 and 03-04-48067), the Western RCE, NIH (award 1-U54-AI-057156), the Council on Grants at the President of the Russian Federation for Support of Young Russian Scientists (grant MK-5304.2007.4), the German Research Foundation (grant LI-448-1), the US National Institutes of Health (grant AI-46706-06) and the International Science and Technology Center (ISTC) (grant 506 3730).

References

Anisimov, A. P., Shaikhutdinova, R. Z., Pan'kina, L. N., Feodorova, V. A., Savostina, E. P., Bystrova, O. V., Lindner, B., Mokrievich, A. N., Bakhteeva, I. V., Titareva, G. M., Dentovskaya, S. V., Kocharova, N. A., Senchenkova, S. N., Holst, O., Devdariani, Z. L., Popov, Y. A., Pier, G. B. & Knirel, Y. A. (2007) Effect of deletion of the *lpxM* gene in *Yersinia pestis* on lipopolysaccharide structure, virulence, and vaccine efficacy of a live attenuated strain. *J Med Microbiol*, 56, 443–53.

Feodorova, V. A., Pan'kina, L. N., Savostina, E. P., Sayapina, L. V., Motin, V. L., Dentovskaya, S. V., Shaikhutdinova, R. Z., Ivanov, S. A., Lindner, B., Kondakova, A. N.,

Bystrova, O. V., Kocharova, N. A., Senchenkova, S. N., Holst, O., Pier, G. B., Knirel, Y. A. & Anisimov, A. P. (2007) A *Yersinia pestis lpxM*-mutant live vaccine induces enhanced immunity against bubonic plague in mice and guinea pigs. *Vaccine,* 25, 7620–28.

Feodorova, V. A., Pan'kina, L. N., Savostina, E. P., Kuznetsov, O. S., Konnov, N. P., Sayapina, L. V., Dentovskaya, S. V., Shaikhutdinova, R. Z., Ageev, S. A., Lindner, B., Kondakova, A. N., Bystrova, O. V., Kocharova, N. A., Senchenkova, S. N., Holst, O., Pier, G. B., Knirel, Y. A., Anisimov, A. P. & Motin, V. L. (2009) Pleiotropic effects of the *lpxM* mutation in *Yersinia pestis* resulting in modification of the biosynthesis of major immunoreactive antigens. *Vaccine*, 27, 2240–50.

Feodorova, V. A. & Corbel, M. J. (2009) Prospects for new plague vaccines. *Exp Rev Vaccine* 8, 1721–38.

Fetherston, J. D., Schuetze, P. & Perry, R. D. (1992) Loss of the pigmentation phenotype in *Yersinia pestis* is due to the spontaneous deletion of 102 kb of chromosomal DNA which is flanked by a repetitive element. *Mol Microbiol*, 6, 2693–2704.

Chapter 29
Protection Against Plague Afforded by Treatment with Polyclonal αLcrV and αF1 Antibodies

Yinon Levy, Yehuda Flashner, Ayelet Zauberman, Avital Tidhar, Moshe Aftalion, Shirley Lazar, David Gur, Sara Cohen, Avigdor Shafferman, and Emanuelle Mamroud

Abstract Plague, initiated by *Yersinia pestis* infection, is a fatal disease that progresses rapidly and leads to high mortality rates if not treated within a short period of time after onset of symptoms. Antibiotics are effectively used for plague therapy yet antibiotic resistance *Y. pestis* strains have been reported and therefore alternative therapies are needed. In this study we demonstrate the ability of rabbit polyclonal antibodies directed against *Y. pestis* F1 and LcrV antigens to confer protection in mouse models of bubonic and pneumonic plague.

Keywords Plague · *Y. pestis* · Passive protection · Anti-F1 · Anti-LcrV

29.1 Introduction

Currently, antibiotics are the only approved treatment against plague. Their effectiveness against *Y. pestis* infection has been demonstrated both in humans and in animal models of bubonic and pneumonic plague (Inglesby et al., 2000). However, the isolation of multi-drug resistant *Y. pestis* strains (Galimand et al., 1997), and the concern that similar strains will be used as agents of bio-terrorism necessitate the examination of alternative therapies. Several studies have demonstrated the therapeutic potential of monoclonal anti-F1 and anti-LcrV antibodies in mouse models of bubonic and pneumonic plague (Hill et al., 2003). Yet, the disadvantage of treatment with mono-specific antibodies is its limitation to pathogenic strains expressing a single serological variant of the protective antigen. Moreover, in cases in which protection requires targeting several epitopes of a given antigen, mono-specific immunotherapy may result in poor protection. Polyclonal antibodies on the

Y. Levy (✉)
Department of Biochemistry and Molecular Genetics, Israel Institute for Biological Research, P.O. Box 19, Ness Ziona 74100, Israel
e-mail: levy-ynon@iibr.gov.il; yinonl@iibr.gov.il

A. Shafferman et al. (eds.), *The Challenge of Highly Pathogenic Microorganisms*,
DOI 10.1007/978-90-481-9054-6_29, © Springer Science+Business Media B.V. 2010

other hand, target several epitopes within a given antigen thereby enabling broad spectrum of interactions with the antigen/pathogen. In this study we evaluated the therapeutic potential of polyclonal anti-F1 and anti-LcrV antibodies administrated following subcutaneous or pulmonary infection with the fully virulent *Y. pestis* strain Kimberley53 (Kim53).

29.2 Results

29.2.1 Generation of Anti-F1 and Anti-LcrV Antibodies, Pharmacokinetics and Protection

Hyper-immune sera was generated in rabbits against the recombinant F1 and LcrV antigens, cloned from the *Y. pestis* Kim53 strain (Holzman et al., 2006; Zauberman et al., 2007). Since anti-F1 ELISA titers usually correlate with protection efficiency (Williamson et al., 1999 and our unpublished data), we selected rabbit anti-F1 hyper-immune sera having high ELISA titer values of 10^6–10^7 and purified their IgG fraction. Anti-LcrV sera having high neutralizing activity (Zauberman et al., 2007) were similarly pooled and their IgG fraction was purified.

The pharmacokinetic profile of rabbit anti-F1 and anti-LcrV antibodies demonstrated half-life of 10–14 days (Fig. 29.1).

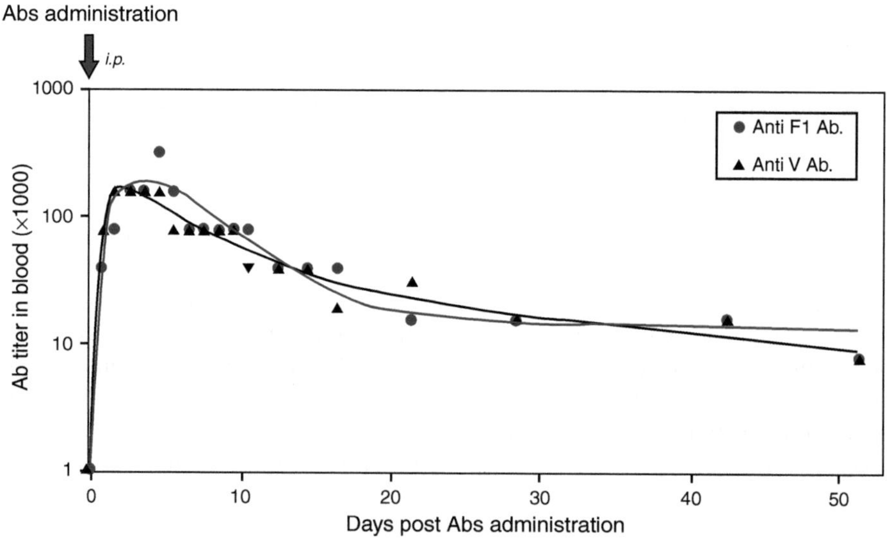

Fig. 29.1 Pharmacokinetic analysis. Anti-F1 (closed circles) and anti-LcrV (closed triangles) titers measured in the blood of naïve OF-1 mice following a single intraperitoneal injection of rabbit anti-F1 or anti-LcrV antibodies (0.5 mg total IgG/mouse). Fit curves and half-life were determined using the PK Solutions software (Summit Research Services)

29.2.1.1 Bubonic Plague Model

Mice infected subcutaneously with 10 or 50 LD_{50} of Kim53, were treated 6 h or 72 h post infection with a single intraperitoneal injection of a mixture of rabbit anti-F1 (1.5×10^5 ELISA units (EU)) and anti-LcrV antibodies (1.5×10^5 EU; 2.2×10^3 neutralizing units). Control, mice were injected with normal rabbit IgG (Fig. 29.2). While all control-treated mice died, 80% of the animals survived after the 6-hour treatment irrespective of the infective dose (Fig. 29.2a, b). Even at the late stages of the disease, a single treatment protected 40 and 30% of the mice infected with 10 LD_{50} or 50 LD_{50}, respectively (Fig. 29.2a, b). Moreover, the time to death of mice treated at this late stage of the infection was significantly delayed (Log-rank, $P<0.01$). It is worth noting that by the time of the late treatment (72 h), all mice displayed significant morbidity signs (hunched posture, ruffled fur, reduced motility) and large numbers of bacteria could be isolated from spleen and blood, reflecting a severe prognosis (Fig. 29.2c).

29.2.1.2 Pneumonic Plague Model

The beneficial effect of the passive treatment in rescuing mice from lethal infection was further evaluated in the more challenging model of pneumonic plague.

In this model, intranasal (i.n) inoculation with 100 LD_{50} of *Y. pestis* Kim53 (55,000 cfu, (Tidhar et al., 2009)) led to development of severe morbidity signs within 48 hours and all mice died within 72 h. Following infection, bacteria rapidly grow in the lungs and disseminate to peripheral organs via the lymphatic system (Fig. 29.3).

The treatment regime of intranasally infected mice (100 LD_{50}, Kim53) included a mixture of 0.5 mg, 1 mg, and 2.5 mg of each of the anti-F1 and anti-LcrV antibodies, 6 h post-infection. Efficient protection of 75% was achieved only when mice were treated with the mixture of 2.5 mg of each antibody type. The lower doses could not rescue the animals but did increase the mean time to death (MTTD; Fig. 29.4a).

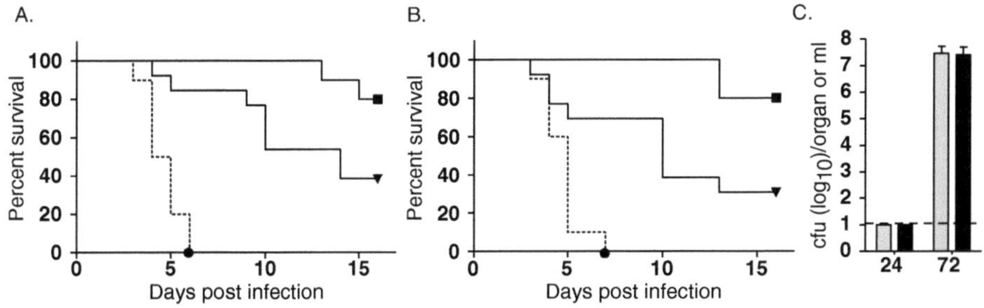

Fig. 29.2 Passive protection in the bubonic plague model. Survival of OF-1 mice infected subcutaneously with (**a**) 10 LD_{50} or (**b**) 50 LD_{50} of *Y. pestis* Kim53 and treated with anti-F1 and anti-LcrV antibody mixture (0.5 mg total IgG of each antibody/mouse), 6 h (squares) or 72 h (triangles) later. Control mice (circles) were injected with NR IgG. (**c**) Bacterial loads in the spleen (gray columns) and blood (black columns) of mice ($n = 5$) infected with 50 LD_{50} of *Y. pestis* Kim53 (1 LD_{50} = 1 cfu). Dashed line indicates the limit of detection

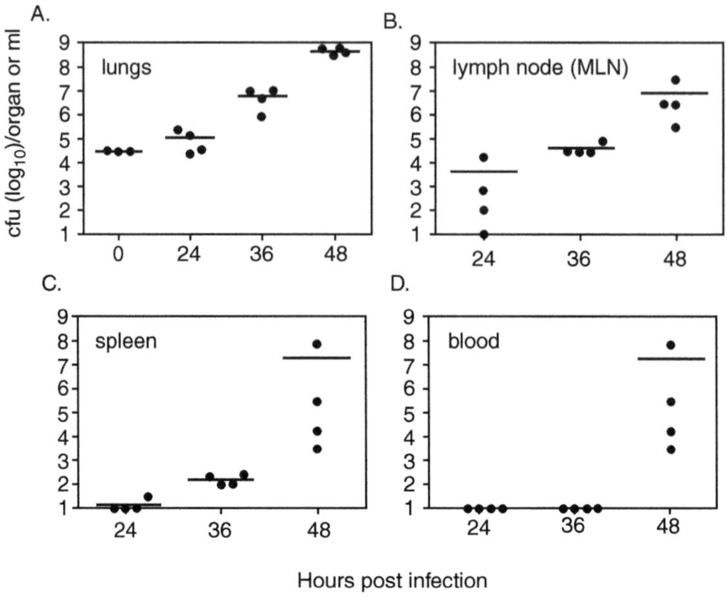

Fig. 29.3 Dissemination of *Y. pestis* Kim53 to internal organs and to the blood in the pneumonic plague model. Mice were infected intranasally with 5.5×10^4 cfu of Kim53. Animals (3–4 per group) were sacrificed at the indicated time points post exposure. Blood was collected and the indicated internal organs were harvested and homogenized in 1 ml PBS. Serial dilutions of organ homogenate or blood were seeded on brain heart infusion agar plates and incubated at 28°C for 48 h. Values represent total bacterial loads in organs (cfu/organ), or bacterial concentration in blood (cfu/ml). Horizontal bars represent the average value of bacterial load in each case. MLN; mediastinal lymph node

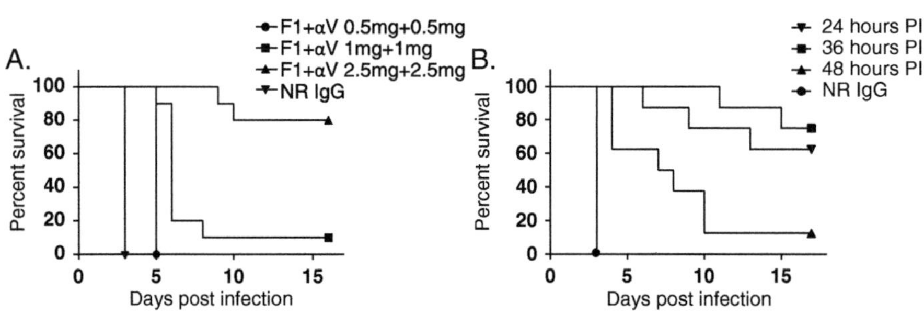

Fig. 29.4 Passive protection in the pneumonic plague model. Survival curves of mice treated with three different doses of rabbit anti-F1 and anti-LcrV antibodies (total IgG of each) or with NR IgG, (**a**) 6 h after intranasal inoculation of mice (n=8) with *Y. pestis* Kim53. (**b**) Survival of mice treated with anti-F1 and anti-LcrV antibodies (2.5 mg + 2.5 mg total IgG), 24, 36 or 48 h following infection with *Y. pestis* Kim53 via the intranasal route (PI; post infection)

To further evaluate the potential of the antibodies in providing post-exposure protection at later stages of pneumonic plague, the protective dose (2.5 mg of each of the anti-F1 and anti-LcrV antibodies) was administered to mice at 24, 36 or 48 h following airways exposure to 100 LD_{50} of *Y. pestis* Kim53. Efficient protection was achieved when mice were treated up to 36 h post infection (Fig. 29.4b). However, if this treatment was postponed by 12 h (i.e. 48 h post infection), 90% of the mice succumbed to infection. Yet, a significant delay in the MTTD was observed (Log-rank; $P<0.01$).

29.3 Conclusions

Our results demonstrate the efficacy of polyclonal anti-F1 and anti-LcrV antibodies as post-exposure therapy against bubonic and pneumonic plague. In both plague models, a single treatment was sufficient to protect mice infected with high-challenge dose of the fully virulent Kim53 strain (Figs. 29.2a–b and 29.4). Moreover, this treatment was able to rescue mice having high bacterial loads within their internal organs (Fig. 29.3a–c). This treatment was less effective when it was given after bacterial dissemination into the blood. Yet, even among bacteremic animals, passive treatment provided a significant delay in the time to death (Figs. 29.2a and 29.4b). This delay provides an opportunity for additional therapeutic intervention, especially in cases involving antibiotic-resistant *Y. pestis* strains.

Acknowledgments We thank Dr. Orit Cohen and Mr. Joseph Shlomovits for their technical assistance with rabbit immunization and harvesting of immune sera.

References

Galimand, M., Guiyoule, A., Gerbaud, G., Rasoamanana, B., Chanteau, S., Carniel, E. & Courvalin, P. (1997) Multidrug resistance in Yersinia pestis mediated by a transferable plasmid. *N Engl J Med,* 337, 677–80.

Hill, J., Copse, C., Leary, S., Stagg, A. J., Williamson, E. D. & Titball, R. W. (2003) Synergistic protection of mice against plague with monoclonal antibodies specific for the F1 and V antigens of Yersinia pestis. *Infect Immun,* 71, 2234–8.

Holzman, T., Levy, Y., Marcus, D., Flashner, Y., Mamroud, E., Cohen, S. & Fass, R. (2006) Production and purification of high molecular weight oligomers of Yersinia pestis F1 capsular antigen released by high cell density culture of recombinant Escherichia coli cells carrying the caf1 operon. *Microb cell fact,* 5, 98.

Inglesby, T. V., Dennis, D. T., Henderson, D. A., Bartlett, J. G., Ascher, M. S., Eitzen, E., Fine, A. D., Friedlander, A. M., Hauer, J., Koerner, J. F., Layton, M., Mcdade, J., Osterholm, M. T., O'toole, T., Parker, G., Perl, T. M., Russell, P. K., Schoch-Spana, M. & Tonat, K. (2000) Plague as a biological weapon: Medical and public health management. Working Group on Civilian Biodefense. *Jama,* 283, 2281–90.

Tidhar, A., Flashner, Y., Cohen, S., Levi, Y., Zauberman, A., Gur, D., Aftalion, M., Elhanany, E., Zvi, A., Shafferman, A. & Mamroud, E. (2009) The NlpD lipoprotein is a novel Yersinia pestis virulence factor essential for the development of plague. *PLoS One,* 4, e7023.

Williamson, E. D., Vesey, P. M., Gillhespy, K. J., Eley, S. M., Green, M. & Titball, R. W. (1999) An IgG1 titre to the F1 and V antigens correlates with protection against plague in the mouse model. *Clin Exp Immunol,* 116, 107–14.
Zauberman, A., Velan, B., Mamroud, E., Flashner, Y., Shafferman, A. & Cohen, S. (2007) Disparity between Yersinia pestis and Yersinia enterocolitica O:8 in YopJ/YopP-dependent functions. *Adv Exp Med Biol,* 603, 312–20.

Chapter 30
Combination of Anthrax-Toxin Neutralizing Antibodies: Analysis of Synergism/Antagonism Effect

Ohad Mazor, Einat Ben-Arie, Adva Mechaly, Ronit Rosenfeld, Hadar Marcus, and Arie Ordentlich

Abstract In order to mimic the neutralizing potency of polyclonal antibodies, combinations of neutralizing monoclonal antibodies (mAbs) are used. The main prerequisite of any preparation which is comprised of several neutralizing mAb is that its antibodies will target distinct epitopes on the antigen molecule. The aim of this study was to elucidate a mixture of antibodies that exhibits a synergistic protective-antigen (PA)-neutralizing activity. We have previously demonstrated that mab29 has a potent PA-neutralizing activity, and that its chimeric form is able to confer full protection against anthrax infection. Here, PA-binding competition assays were applied in order to screen and isolate mAb that recognize different epitopes on PA and can bind the toxin simultaneously with mab29. Two antibodies were selected in this screen, mab33 and mab55, which have demonstrated high neutralizing activity when tested in the anthrax lethal toxin neutralization assay. In order to define the dose-effect relationships of the combination of these antibodies, the multiple-drug effect analysis was applied. It was found that combining mab29 with mab55 results in a slightly synergistic effect (combination index ≤ 0.9). In contrast, combining mab29 with mab33 resulted in an antagonistic effect (combination index >1.2), although when tested alone in vivo, mab33 provides full protection of rats against lethal toxin challenge. The results presented in this study generally support the notion that a cocktail of neutralizing antibodies can result in an improved activity, and suggest that the combination of mab29 with mab55 may expand the breadth of protection against anthrax infection.

Keywords Antagonism · Anthrax · *Bacillus anthracis* · Protective antigen · Synergism

O. Mazor (✉)
Departments of Biochemistry and Molecular Genetics, Israel Institute for Biological Research, Ness-Ziona 74100, Israel
e-mail: ohadm@iibr.gov.il

A. Shafferman et al. (eds.), *The Challenge of Highly Pathogenic Microorganisms*,
DOI 10.1007/978-90-481-9054-6_30, © Springer Science+Business Media B.V. 2010

30.1 Introduction

Bacillus anthracis, the causative agent of anthrax, exerts its toxicity via the dissemination of a tripartite exotoxin comprised of protective antigen (PA), lethal factor (LF) and edema factor (EF). PA plays a critical role in anthrax pathogenesis, where it can associate with either LF to form lethal toxin (LeTx) or with EF to form the edema toxin (EdTx). Currently, a six-dose vaccination schedule for anthrax vaccine adsorbed is indicated, and is now recommended by the FDA as a post-exposure treatment together with a prolonged antibiotic regimen (Friedlander et al., 2002; Grabenstein, 2008). However, antibiotic prophylaxis can be problematic in situations where their use is contraindicated, or in cases involving antibiotic resistant *B. anthracis* strains (Turnbull et al., 2004). Furthermore, in cases where disease has progressed and a substantial amount of anthrax toxins has been delivered to the bloodstream, these treatments are of less value. In these scenarios, the passive transfer of neutralizing antibodies directed against either PA or lethal factor (LF) can provide immediate, specific and low-toxicity protection (Albrecht et al., 2007; Karginov et al., 2004). We have previously isolated a wide panel of PA-neutralizing monoclonal antibodies, from which mab29 demonstrated the most potent activity and was therefore chosen for chimerization. The chimeric antibody conferred full protection to guinea pigs infected with 40 LD_{50} *B.anthracis* spores, and was therefore proposed to serve as an effective immunotherapeutic agent against anthrax (Rosenfeld et al., 2009).

In order to increase the potency of mAbs and to broaden their applicability, it was suggested to use a cocktail of several mAbs which will simultaneously neutralize distinct epitopes, thus mimicking the potency of polyclonal-antibody preparations. Indeed, synergistic neutralization by mAb combination was demonstrated before for several viruses (Li et al., 1998; Meulen et al., 2006; Prabakaran et al., 2009), toxins (Nowakowski et al., 2002) and malignant tumors (Lotenberg, 2007). Moreover, in the case of bio-threat agents there are concerns about the deliberate circumvention of epitope binding sites, which will result in a protein whose activity cannot be inhibited by mAb that neutralize the native form of this agent, enforcing the need for combination of neutralizing antibodies that recognize distinct epitopes.

The aim of this study was to identify PA-neutralizing mAbs that can be combined with mab29 in a therapeutic preparation, and act in a synergistic or additive manner to inhibit the LeTx activity. The mAbs that were tested in this study were chosen based upon distinct PA-epitopes recognition, compared to mab29. In this report, we demonstrate that the combination of mab29 with mab55 results in synergetic LeTx – neutralization, and that its combination with mab33 results in an antagonistic effect.

30.2 Materials and Methods

30.2.1 Antibody Production and Purification

Mouse IgG anti-PA monoclonal antibodies were produced in ascitic fluid (Rosenfeld et al., 2009), purified by affinity chromatography on HiTrap Protein G (GE

Healthcare, Sweden) according to the manufacturer's instruction and dialyzed against PBS pH 7.4.

30.2.2 Antibody Binding Assay

PA labeling by activated Eu-chelate (N-1(ρ-iso-thiocyanatophenyl)diethylene tri-N^1, N^2, N^3,-tetraacetate chelated with europium; Perkin Elmer, USA) followed the procedure described before for EGF (Mazor et al., 2002). Eu^{3+} labeled PA (Eu-PA) was purified on a microcon (millipore, USA).

For binding studies, mab29 or mab55 were immobilized on 96-well microtiter plates (Nunc, Denmark) and incubated (1 h at 37°C) in the presence of constant Eu-PA concentration and increasing concentrations of anti-PA antibodies (10–100,000 ng/ml). Plates were then washed, enhancement solution was added and bound Eu^{3+} was measured using time-resolve fluorescence (Victor3, Perkin Elmer, USA).

30.2.3 In Vitro LeTx Neutralization Assay

Neutralizing antibody activity was determined essentially as described previously (Reuveny et al., 2001; Singh et al., 1989) by assessing their ability to protect murine macrophage J774A.1 cells (ATCC, USA) against PA/LF toxin complex (LeTx) intoxication. Tested samples were serially diluted in TSTA buffer containing PA (5 μg/ml) and LF (2 μg/ml), and after 1 h incubation, 10 μl of each of the reaction mixture dilutions was added to the J774A.1 cells (6–8 $\times 10^5$ cells/ 0.2 ml). Plates were then incubated for 5 h at 37°C in 5% CO_2, and cell viability was monitored by the MTT (3-[4,5-Dimethylthiazol-2-yl]-2,5- Diphenyltetrazolium bromide; Thiazolyl blue) assay (absorbance was measured at 540 nm) (Mosmann, 1983). The percent of cell survival was plotted against antibody concentration, and the effective antibody concentration needed to neutralize 50% of the LeTx (EC_{50}) was calculated from the linear range of the plot.

To quantitatively compare the neutralizing activity of neutralizing antibody combinations, antibodies were mixed in an equimolar ratio and the mixture potency was tested in the LeTx assay as described above. The combination index (CI) and dose reduction index (DRI) were calculated by Calcusyn (Biosoft, USA), in order to quantify synergistic/antagonistic effects, as described previously (Chou and Talalay, 1984).

30.2.4 Animal Studies

All animal experiments were performed in accordance with Israeli law and were approved by the Ethics Committee for animal experiments at the Israel Institute for Biological Research. Animals were maintained at 20–22°C and a relative humidity of 50±10% on a 12-h light/dark cycle, fed with commercial rodent chow (Koffolk Inc.), and provided with tap water ad libitum. Treatment of animals was in accordance with regulations outlined in the USDA Animal Welfare Act and the conditions specified in the Guide for Care and Use of Laboratory Animals (National Institute of Health, 1996).

30.2.5 In Vivo Neutralization Experiments

To evaluate the ability of the mab33 to protect rats against a LeTx challenge, Male Fisher 344 rats (200–250 g, Harlan, Israel) were injected intramuscularly (i.m.) with various doses of the antibody (administration volume of 100 μl in sterile PBS). Seventeen hours later, rats were anesthetized with ketamine:xylazine (50:2 mg/kg), injected via the tail vein (i.v.) with LeTx containing 20 μg PA and 10 μg LF in 500 μl saline, and monitored for survival.

30.3 Results

30.3.1 mab29, mab33 and mab55 Do Not Compete on for Same PA-Epitopes

Previously, we have isolated a panel of 19 PA-neutralizing mAbs that were further subdivided into several groups based upon a distinguishable PA-binding pattern in their direct- and indirect-ELISA formats (Rosenfeld et al., 2009). From this panel, antibody #29 (mab29) has demonstrated the highest neutralizing activity (titer of 2.3×10^6), and its chimeric form (ch29) was used successfully to protect guinea pigs against lethal challenge of 40 LD_{50} *B. anthracis* spores infection. Here, we sought to isolate mAb that recognize different epitopes on PA and can bind the toxin simultaneously with mab29. To this end, PA-binding competition assays were applied in order to test the ability of each mAb from the isolated panel to compete with the binding of eurpoium-labeled PA (Eu-PA) to immobilized mab29. It was found that antibodies 22, 45, 58 and 62 could displace more than 90% of the Eu-PA, suggesting that they share overlapping or closely adjacent epitopes (Table 30.1). Antibody 68 has displaced 30–60% of the labeled PA, and probably binds an adjacent epitope to mab29. Six antibodies (2, 6, 19, 33, 55 and 61) did not compete at all with mab29, implying that they bind distinct epitopes on PA. From these non-competing antibodies, mab55 demonstrated the highest PA-neutralizing activity (Rosenfeld et al., 2009) and was therefore chosen for further analysis. The PA-neutralizing mAbs panel was tested again in a competition assay, where mab55 was immobilized. Under these conditions, only antibodies 6 and 19 could displace the Eu-PA (Table 30.1). By applying a BIAcore-based pair-wise analysis it could be confirmed that mab55 can bind PA simultaneously with mab29 (data not shown), and thus it

Table 30.1 Competition binding assay between PA-neutralizing mAbs[a]

	Group 1[b]			Group 2[b]						Group 3[b]		
Immobilized Ab	6	19	55	22	29	33	45	58	68	2	61	62
29	–	–	–	†	†	–	†	†	†	–	–	‡
55	†	†	†	–	–	–	–	–	–	–	–	–

† >90% Eu-PA displacement; ‡ 30–60% Eu-PA displacement; – No competition
[a]Each mAb was tested for its ability to displace the binding of Eu-PA to the immobilized Ab. Data represent the % of displacement at the highest competitor Ab concentration (100 μg/ml)
[b]Groups were divided based upon their ELISA binding characteristics (Rosenfeld et al., 2009)

was of interest to determine the neutralizing potency of the combination of these two antibodies.

30.3.2 mab29 and mab55 Neutralize LeTx Synergistically

We next examined the ability of each antibody (mab29 and mab55) to neutralize PA-LF complex (LeTx) activity in the J774A macrophage cell line. It has been shown previously that the results of the LeTx assay can serve as a surrogate marker to predict an antibody's neutralization potential against anthrax challenges in vivo (Reuveny et al., 2001). Cells were exposed to a lethal dose of LeTx in the presence of increasing antibody concentration, and cell survival was determined 5 hrs later (Fig. 30.1a). For both antibodies, cell survival positively correlated with antibodies

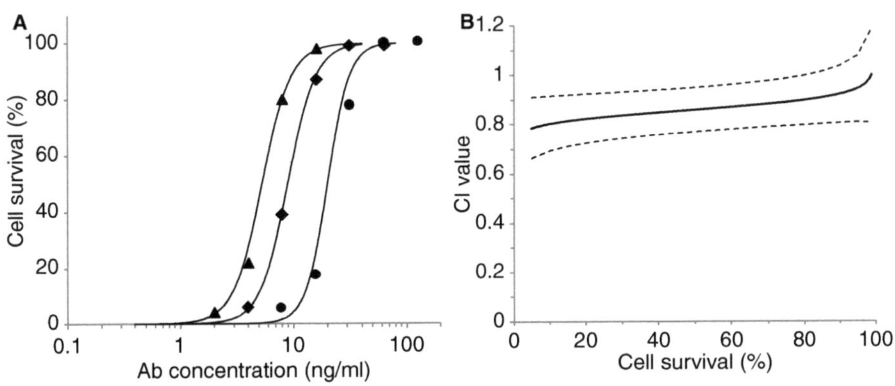

Fig. 30.1 In vitro LeTx neutralization by a combination of mab29 and mab55. (**a**) Toxin complex was pre-incubated with increasing concentrations of either mab29 (*diamonds*), mab55 (*circles*) or their combination in a 1:1 ratio (triangles; plotted against mab29 concentration), and added to J774A.1 mouse macrophage cells. Cell survival is plotted as the percent of untreated control cells. Points represent a mean of triplicate determinations. (**b**) CI value of mab29 and mab55 combination. Broken lines represent upper and lower 2 STD

Table 30.2 ELISA titers and LeTx neutralization activity of selected PA-neutralizing mAbs

	Titer ($\times 10^3$)[a]			
	Direct ELISA	Indirect ELISA	LeTx neutralization	EC_{50} (ng/ml)
mab29	72	820	2300	8
mab33	7	820	820	15
mab55	820	410	290	20

[a]Data reproduced from Rosenfeld et al., 2009

concentrations, and the effective concentrations needed to neutralize 50% (EC_{50}) of the LeTx activity were determined to be 8 ng/ml and 20 ng/ml for mab29 and mab55, respectively (Table 30.2).

In order to evaluate the LeTx neutralizing activity of these two antibodies when combined, the neutralizing potency of a mixture containing mab29 and mab55 in a 1:1 ratio was titrated (Fig. 30.1a). In this kind of assay, the neutralizing response of the mixture is compared to the dose response curve of each antibody when applied individually in the same assay. It was found that in this protocol, the EC_{50} value of mab29 in the mixture is 4 ng/ml (Table 30.3). The data was further analyzed by applying the median effect principle equations as formulated by Chou and Talay (Chou and Talalay, 1984). These equations have provided the basis for the combination index (CI)-isobologram equation that allows quantitative determination of drug interaction. A CI value of <0.9, 0.9–1.1 or >1.1 will indicate synergism, additive effect and antagonism, respectively, for each drug combination. Here, a CI value of 0.86 was observed for the combination of mab29 with mab55, at the 50% neutralization point (Fig. 30.1b), indicating synergistic neutralization by this antibody combination. It should be noted that a CI value of less than 1 was observed over the whole range of antibody doses tested.

Having a synergistic drug combination will allow to reduce the drug doses used while maintaining its efficacy. The dose reduction index (DRI) is a measure of the

Table 30.3 Properties of PA-neutralizing mAbs combination in the LeTx neutralization assay[a]

			DRI at EC_{50}[e]		
	EC_{50} (ng/ml)	CI value[d]	mab29	mab33	mab55
mab29:mab33	5[b]	1.47	0.87	3.2	
mab29:mab55	4[b]	0.86	1.7		3.8
mab33:mab55	10[c]	1.1		1.5	2.2

[a]Antibodies were combined in a 1:1 molar ratio
[b]EC_{50} value of mab29 in the combination
[c]EC_{50} value of mab33 in the combination
[d]Combination index (CI) was calculated from the median effect principle equation.
[e]Dose reduction index (DRI) was measured by comparing the dose required to reach 50% neutralization when the mAb was used alone and in combination

fold reduction of the dose of each drug in a synergistic combination at a given effect level, compared to the dose of each drug alone. The DRI value can be calculated from the same set of equations used earlier, and it was found that combining mab29 and mab55 reduced the required dose significantly for mab55 (DRI value of 3.8), and to a lesser extent for mab29 (DRI value of 1.7; Table 30.3).

30.3.3 Mab33 Antagonizes the Neutralizing Activity of mab29 but Not of mab55

Our next goal was to characterize other PA-neutralizing antibodies that recognize different epitope than those recognized by mab29 and mab55. From the binding assay (Table 30.1), three candidate antibodies could be selected (mAb 2, 33 and 61). When tested in BIAcore-based pair-wise analysis it was found that mab33 can bind PA simultaneously with either mab29 or mab55, while inconclusive results were obtained for the other two antibodies (data not shown). It was shown before (Rosenfeld et al., 2009) that mab33 have ELISA titers of 7×10^3 and 820×10^3 in the direct and indirect formats, respectively (Table 30.2). To further explore the PA-neutralizing activity of mab33, its in vivo protection capacity against a LeTx challenge was tested. In this assay, mab33 was administered (i.m.) at different doses to rats, followed by an i.v. challenge with a lethal dose of LeTx 17 h later. All untreated control animals succumbed to this challenge, with mean time to death (MTTD) of 110 min (Fig. 30.2). Treatment of animals with 100 μg/kg of mab33 could protect 33%, and full protection was achieved by a passive transfer of 150 μg/kg. According to these results, a dose of 105 μg/kg mab33 was needed in order to provide 50% protection (PD_{50}), which is similar to the protective activity of mab29 in the same system (PD_{50} achieved by the administration of 170 μg/kg (Rosenfeld et al., 2009)).

We next examined the LeTx neutralization of a preparation containing combination of mab29 and mab33. To this end, antibodies were tested alone or in a 1:1 mixture, in the in vitro assay. The EC_{50} of mab33 was determined to be 15 ng/ml (Fig. 30.3a), which is in the same neutralizing range as mab29 and mab55 (8 and 20 ng/ml, respectively; Table 30.3). However, when the neutralizing response of the

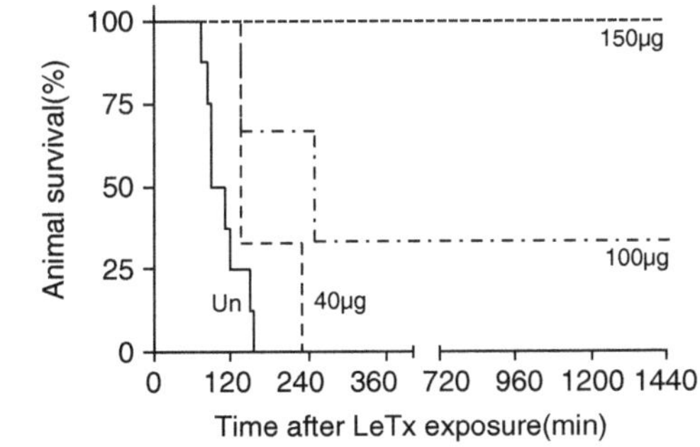

Fig. 30.2 In vivo LeTx neutralization assay with mab33. Rats ($n\leq6$ for each group) were i.m. administered with the indicated doses of mab33, followed by i.v. challenge with LeTx (20 μg PA and 10 μg LF), 17 hours later. Animal survival was monitored for the 24 hours following the LeTx challenge

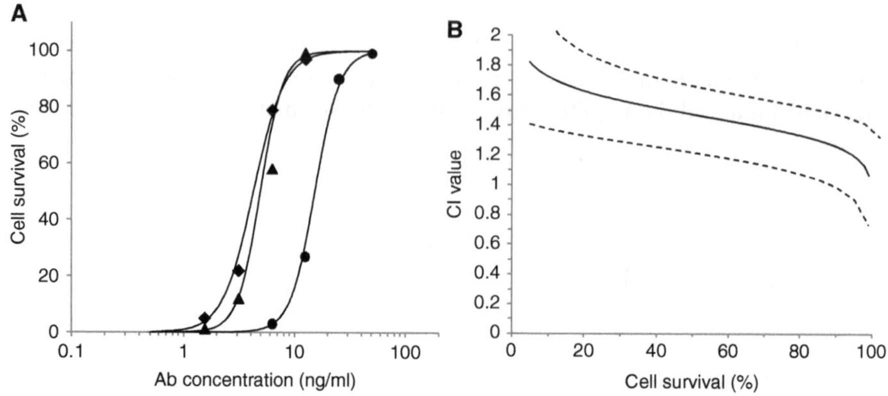

Fig. 30.3 In vitro LeTx neutralization by a combination of mab29 and mab33. (**a**) Toxin complex was pre-incubated with increasing concentrations of either mab29 (*diamonds*), mab33 (*circles*) or their combination in a 1:1 ratio (triangles; plotted against mab29 concentration), and added to J774A.1 mouse macrophage cells. Cell survival is plotted as the percent of untreated control cells. Points represent a mean of triplicate determinations. (**b**) CI value of mab29 and mab33 combination. Broken lines represent upper and lower 2 STD

mixture was compared to the dose response curve of each antibody when applied individually in the same assay, it was found that mab33 had an antagonistic effect on the activity of mab29. The EC_{50} value of mab29 in the mixture is 5 ng/ml, as compared to a value of 4 ng/ml when tested alone in the same assay (Fig. 30.3a). Accordingly, the CI value of 1.47 was observed for the 50% neutralization point, and the CI value was above 1.1 for the whole curve range, indicating a strong antagonism between mab29 and mab33 when mixed (Fig. 30.3b). The DRI values at the

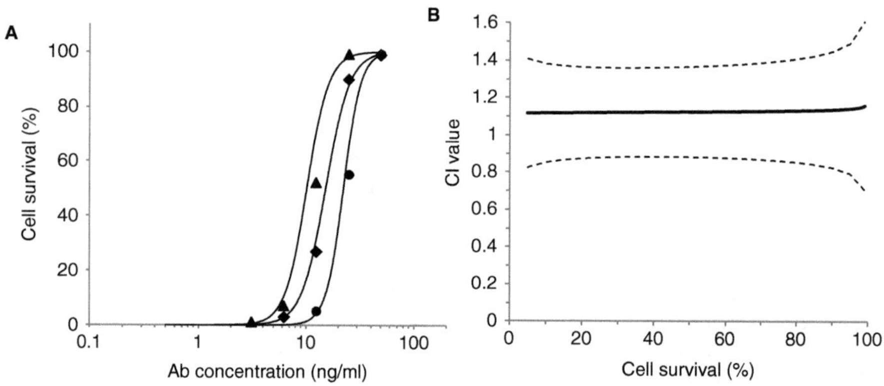

Fig. 30.4 In vitro LeTx neutralization by a combination of mab33 and mab55. (**a**) Toxin complex was pre-incubated with increasing concentrations of either mab33 (*diamonds*), mab55 (*circles*) or their combination in a 1:1 ratio (triangles; plotted against mab33 concentration), and added to J774A.1 mouse macrophage cells. Cell survival is plotted as the percent of untreated control cells. Points represent a mean of triplicate determinations. (**b**) CI value of mab33 and mab55 combination. Broken lines represent upper and lower 2 STD

50% point were 0.87 and 3.2 for mab29 and mab33, respectively, indicating that mab33 indeed interferes with the neutralizing activity of mab29.

As mentioned earlier, mab33 does not compete with mab55 on the binding to the toxin, and these two antibodies can bind PA simultaneously. It was therefore of interest to determine whether mab33 will antagonize the activity of mab55, or is it a unique observation for the combination with mab29. To this end, the neutralizing potency of a mixture containing mab33 and mab55 in a 1:1 ratio was titrated (Fig. 30.4a). The EC_{50} of mab33 in the combination was 10 ng/ml, as compared to a value of 15 ng/ml when tested alone in the same assay. A CI value of 1.1 was calculated for the whole curve range, which according to the median effect principle indicates on an additive effect between the two antibodies. These results suggest that the combination of mab33 with mab55 does not provide any additional benefit over simply increasing the dose of each antibody alone.

30.4 Discussion

In this study we have continued to characterize the PA binding profile of the PA-neutralizing antibodies, and tested the neutralizing activity of several mAbs combination. Previously, the isolated panel of PA-neutralizing mAbs was subdivided into three groups based upon the ELISA-binding characteristics (Rosenfeld et al., 2009). In the same paper, it was also hypothesized that the distinct binding patterns implies that each group of antibodies may recognize different sites of the target molecule, which are unequally exposed in the different ELISA formats, and indeed, our current data supports this notion. Group 1, represented by mab55, recognizes a unique epitope which is not shared by the other groups. In group 2, all except one (mab33) compete with mab29 on the binding to PA, and the situation is reversed in group 3, where one Ab (mab62) did compete with mab29, while the others did not. We have also found that some interference occurs between the binding of mabs 2 and 61 with mab55 (data not shown), which may suggest that they share adjacent or partially over-lapping epitopes. Work is now in progress in order to characterize these clones further and to identify their epitopes.

Synergistic neutralization of LeTx was also demonstrated for the combination of anti-PA mAbs with either anti-LF or anti-EF mAbs (Albrecht et al., 2007; Chen et al., 2009; Staats et al., 2007). Several variables can influence the ability of mAbs combination to act in an additive or synergistic manner, including distinct epitope recognition and antigen binding affinities. Indeed, antibodies #29 and #55 were selected due to the fact that they can recognize different epitopes and bind simultaneously to PA. However, we have also found a pair of antibodies that while recognizing distinct epitopes and possessing high neutralizing LeTx activity, when mixed antagonize each other.

Antagonism between two mAbs was demonstrated in several cases before (Heinz et al., 1983; Massey and Schochetman, 1981; Verrier et al., 2001). Yet, in most cases the mixture contained one mAb with neutralizing activity and one without such activity, in contrast to this study where two neutralizing mAbs are involved. Two main reasons may account for this phenomenon: The first is due to steric hindrance that interferes with the two antibodies binding on the same antigen. However, this is probably not the case in this study, since we have found that mab29 and mab33 can bind simultaneously to PA. The second explanation is that there is a negative cooperativity between the two mAbs, due to conformational changes in the PA structure that is caused by the binding of one of the antibodies. In such case, the subsequent interaction and neutralization activity of the second mAb will be partially inhibited. It is logical to assume that this is the case here, and more study is needed in order to understand how the binding of mab33 to PA influences the consequent ability of mab29 to bind to its epitope. One can also assume that the binding of mab33 to PA does not induce major changes in the toxin conformation, since mixing this antibody with mab55 results in an additive rather than synergistic effect. An epitope mapping study in underway in order to identify the binding domains of the three neutralizing antibodies described in this study.

To summarize, while the results of the present study generally support the notion that a cocktail of neutralizing antibodies results in an improved activity, we have also demonstrated an interesting finding that does not follow this logic. Yet, we suggest that the combination of mab29 with mab55 may expand the breadth of protection against anthrax infection, as will be further examined in animal models.

References

Albrecht, M. T., Li, H., Williamson, E. D., Lebutt, C. S., Flick-Smith, H. C., Quinn, C. P., Westra, H., Galloway, D., Mateczun, A., Goldman, S., Groen, H. & Baillie, L. W. (2007) Human monoclonal antibodies against anthrax lethal factor and protective antigen act independently to protect against *Bacillus anthracis* infection and enhance endogenous immunity to anthrax. *Infect Immun,* 75, 5425–33.

Chen, Z., Moayeri, M., Crown, D., Emerson, S., Gorshkova, I., Schuck, P., Leppla, S. H. & Purcell, R. H. (2009) Novel chimpanzee/human monoclonal antibodies that neutralize anthrax lethal factor: evidence for possible synergy with anti-PA antibody. *Infect Immun,* 77, 3902–08.

Chou, T. C. & Talalay, P. (1984) Quantitative analysis of dose-effect relationships: the combined effects of multiple drugs or enzymes inhibitors. *Adv Enzyme Regul,* 22, 27–55.

Friedlander, A. M., Welkos, S. L. & Ivins, B. E. (2002) Anthrax vaccines. *Curr Top Microbiol Immunol,* 271, 33–60.

Grabenstein, J. D. (2008) Countering anthrax: Vaccines and immunoglobulins. *Clin Infect Dis,* 46, 129–36.

Heinz, F. X., Berger, R., Tuma, W. & Kunz, C. (1983) A topological and functional model of epitopes on the structure glycoprotein of tick-borne encephalitis virus defined by monoclonal antibodies. *Virology,* 126, 525–37.

Karginov, V. A., Robinson, T. M., Riemenschneider, J., Golding, B., Kennedy, M., Shiloach, J. & Alibek, K. (2004) Treatment of anthrax infection with combination of ciprofloxacin and antibodies to protective antigen of *Bacillus anthracis*. *FEMS Immunol Med Microbiol,* 40, 71–4.

Li, A., Katinger, H., Posner, M. R., Cavacini, L., Zolla-Pazner, S., Gorny, M. K., Sodroski, J., Chou, T., Baba, T. W. & Ruprecht, R. M. (1998) Synergitic neutralization of simian-human immunodefiecency virus SHIV-vpu+ by triple and quadruple combinations of human monoclonal antibodies and high-titer anti-human immunodeficiency virus type 1 immunoglobulins. *J Virology,* 72, 3235–40.

Lotenberg, T. (2007) Antibody cocktails: Next-generation biopharmaceuticals with improved potency. *Trends Biotech,* 25, 390–94.

Massey, R. J. & Schochetman, G. (1981) Viral epitopes and monoclonal antibodies: isolation of blocking antibodies that inhibit virus neutralization. *Science,* 213, 447–49.

Mazor, O., Hellairet De Boisferon, M., Lombet, A., Grua-Guyon, A., Gayer, B., Skrzydelsky, D., Kohen, F., Forgez, P., Scherz, A., Rostene, W. & Salomon, Y. (2002) Europium-labeled epidermal growth factor and neurotensin: Novel probes for receptor-binding studies. *Anal Biochem,* 301, 75–81.

Meulen, J. T., Van Den Brink, E. N., Poon, L., Marissen, W. E., Leung, C., Cox, F., Cheung, C. Y., Bakker, A. Q., Bogaards, J. A., Van Deventer, E., Preiser, W., Doerr, H., Chow, V. T., De Kruif, J., Peiris, J. & Goudsmit, J. (2006) Human monoclonal antibody combination against SARS coronavirus: Synergy and coverage of escape mutants. *PLoS Medicine,* 3, e237.

Mosmann, T. (1983) Rapid colorimetric assay for cellular growth and survival: application to proliferation and cytotoxicity assays. *J Immunol Methods,* 65, 55–63.

Nowakowski, A., Wang, C., Powers, D. B., Amersdorfer, P., Smith, T. J., Montgomery, V. A., Sheridan, R., Blake, R., Smith, L. A. & Marks, J. D. (2002) Potent neutralization of botulinum neurotoxin by recombinant oligoclonal antibody. *Proc Nat Acad Sci,* 99, 11346–50.

Prabakaran, M., Prabhu, N., HE, F., Hongliang, Q., Go, H., Qiang, J., Meng, T., Goutma, M. & Kwang, J. (2009) Combination therapy using chimeric monoclonal antibodies protect mice from lethal H5N1 infection and prevents formation of escape mutants. *PLoS ONE,* 4, e5672.

Reuveny, S., White, M. D., Adar, Y. Y., Kafri, Y., Altboum, Z., Gozes, Y., Kobiler, D., Shafferman, A. & Velan, B. (2001) Search for correlates of protective immunity conferred by anthrax vaccine. *Infect Immun,* 69, 2888–93.

Rosenfeld, R., Marcus, H., Ben-Arie, E., Lachmi, B., Mechaly, A., Reuveny, S., Gat, O., Mazor, O. & Ordentlich, A. (2009) Isolation and Chimerization of a Highly Neutralizing Antibody conferring Passive Protection against Lethal *B. anthracis* Infection. *PLoS ONE,* 4(7), e6351.

Singh, Y., Chaudhary, V. K. & Leppla, S. H. (1989) A deleted variant of *Bacillus anthracis* protective antigen is non-toxic and blocks anthrax toxin action in vivo. *J Biol Chem,* 264, 19103–7.

Staats, H. F., Alam, S. M., Scearce, R. M., Kirwan, S. M., Zhang, J. X., Gwinn, W. M. & Haynes, B. F. (2007) In vitro and in vivo characterization of anthrax anti-protective antigen and anti-lethal factor monoclonal antibodies after passive transfer in a mouse lethal toxin challenge model to define correlates of immunity. *Infect Immun,* 75, 5443–52.

Turnbull, P. C., Sirianni, N. M., Lebron, C. I., Samaan, M. N., Sutton, F. N., Reyes, A. E. & Peruski, L. F., Jr. (2004) MICs of selected antibiotics for *Bacillus anthracis*, *Bacillus cereus*, *Bacillus thuringiensis*, and *Bacillus mycoides* from a range of clinical and environmental sources as determined by the Etest. *J Clin Microbiol,* 42, 3626–34.

Verrier, F., Nadas, A., Gorny, M. K. & Zolla-Pazner, S. (2001) Additive effects characterize the infection of antibodies involved in neutralization of the primary dualtropic human immunodeficiency virus type 1 isolate 89.6. *J Virology,* 75, 9177–86.

Chapter 31
The Efficacy of Antibiotic Treatment to Cure Guinea Pigs Intranasally Infected with *Bacillus anthracis* Spores

Zeev Altboum, Haim Levy, Shay Weiss, Avi Pass, Nili Rothschild, Arnon Tal, Ada Barnea, Josef Schlomovitz, and David Kobiler

Abstract Respiratory anthrax is a fatal disease in the absence of early therapy with antibiotics. Guinea pigs are highly sensitive to infection with *Bacillus anthracis* spores by intranasal spore instillation succumbing within 2 to 4 days post infection. Post exposure prophylaxis that initiated 24 h post infection with tetracycline, doxycycline, ciprofloxacin, ofloxacin, erythromycin, gentamicin or imipenem protected the animals, but after termination of treatment the animals developed a fatal anthrax disease. Combined treatment with antibiotics and active immunization with PA based vaccine conferred an efficient immune response that prevented the development of a fatal anthrax disease after termination of antibiotic administration. The CDC recommended in case of a major bioterrorist attack a mass distribution of antibiotics within 48 h. As the onset of symptoms in humans were reported to start on day 1–7 post exposure, this delay in initiation of treatment may result in treatment of sick populations. We evaluated the efficacy of late antibiotic administration starting from 30 h post infections and every 6 h thereafter, to cure respiratory anthrax sick animals. Administration of ciprofloxacin provided efficient protection up to 44 h post infection, at which time 60% of the sick animals survived. Administration of doxycycline cured up to 48 h post infection, when up to 75% of the sick animals survived. The antibiotic administration protected most animals with bacteremia level up to 10^5 CFU/ml, but failed to protect animals exhibiting higher levels of bacteremia. In conclusion these results demonstrate the efficacy of antibiotic treatment to cure animals even after the onset of the disease.

Keywords *Bacillus anthracis* · Antibiotics · Post exposure prophylaxis · Curing sick animals · Respiratory anthrax

---

Z. Altboum (✉)
Departments of Infectious Diseases, Israel Institute for Biological Research,
Ness-Ziona 74100, Israel
e-mail: zeeva@iibr.gov.il

A. Shafferman et al. (eds.), *The Challenge of Highly Pathogenic Microorganisms*,
DOI 10.1007/978-90-481-9054-6_31, © Springer Science+Business Media B.V. 2010

31.1 Introduction

Infection with *Bacillus anthracis* spores via the respiratory route causes a severe and fatal disease. Following pulmonary deposition, the spores are phagocytized by the alveolar macrophages, transferred through the lymphatic channels to the tracheo – bronchial lymph nodes, where the spores germinate; grow as vegetative cells that enter the blood stream causing a systemic disease. Already within the alveolar macrophages, the germinating spores synthesize an antiphagocytic capsule and secrete lethal (LT) and edema (ET) toxins. Following the 2001 bioterrorist attack in USA 22 people were infected with *B. anthracis* Ames spores, 12 developed cutaneous anthrax and 10 inhalation anthrax. The severity of inhalation anthrax was described in detailed in these patients. Following an incubation time of 4–6 days post infection the patients exhibit non specific initial signs of the disease, resembling flu like disease. This stage rapidly develops into specific symptoms that include abnormal chest X ray, pleural infusion and mediastimal widening, a result of bacterial proliferation and activity of the lethal and edema toxins. The patient's condition continues to deteriorate as the disease enters the fulminant phase which is characterized by the abrupt onset of fever, dyspnea, respiratory distress, shock and death. In spite of massive antibiotic administrations 5 patients died, possibly because the treatment started too late into the disease (Jernigan et al., 2001).

Post-exposure protection may be achieved by antibiotic treatment either alone or in combination with active immunization with PA based vaccines. Experiments to cure *B. anthracis* infected animals were performed with guinea-pigs (Jones et al., 1996; Altboum et al., 2002) and rhesus monkeys (Henderson et al., 1956;, Friedlander et al., 1993). The results indicated that early antibiotic treatment by doxycycline, ciprofloxacin and penicillin may prevent the development of a fatal disease. These antibiotics are recommended for treatment of Anthrax patients (Inglesby et al., 1999, 2002).

In this work we monitored the progress of the respiratory anthrax disease in guinea pigs following infection by intranasal inoculation. We determined the development of bacteremia that characterized the systemic phase of the disease, and tested the ability of antibiotic administration to cure infected animals (non bacteremic) and sick animals (bacteremic). This work characterized a therapeutic window time of about 44–48 h post infection till initiation of treatment that allows the cure of most sick animals.

31.2 Results

31.2.1 Progress of Respiratory Anthrax Disease in Guinea Pigs

Following infection with 3×10^6 Vollum spores (75 LD_{50}) via intranasal spore instillation, bacteremia started to develop from 30 h post infection. At each time point afterwards the percentage of the bacteremic animals increases. Death of infected animals started from 48 h post infection and the animals died with a mean time to death of 54 h (Altboum et al., 2002). The progression of the disease could be divided into three stages: incubation phase of 30 h from infection till initiation of bacteremia; systemic phase from 30 h post infection till 48 h post infection during which all the infected animals became bacteremic, and the level of the bacterial concentration in each animal increase; death phase that started from 48 h post infection till approximately 60 h post infection. Simultaneously with the development of bacteremia, PA (the major protein component of the bacterial lethal and edema toxins) is detected in the animal's sera. The concentration of PA is proportional to the bacterial concentration in the blood, making the PA a specific marker to the level of bacteremia and to the severity of the disease (Kobiler et al., 2006). To study the efficiency of curing the infected animals we addressed the following questions:

a. Will initiation of antibiotic administration at the pre bacteremic stage (24 h post infection) prevent the development of a fatal respiratory anthrax disease?
b. Will late initiation of antibiotic administration, (> 30 h post infection), of bacteremic animals cure the sick animals?
c. What is the highest level of bacterial concentration of the sick animals that could be cured by the antibiotic treatment?
d. What is the therapeutic window time from infection till initiation of treatment that will provide an efficient curing?

In order to answer the above questions we performed two types of experiments:

a. Post exposure prophylaxis experiments with various antibiotics,
b. Treatment of sick animals.

31.2.2 Post Exposure Prophylaxis Experiments

Female Hartley guinea pigs (250–300 gr) were inoculated intranasally by unilateral instillation (100 μl) of 3×10^6 Vollum spores. Various antibiotics (described in Table 31.1) were administered s.c. starting 24 h post infection, several times per day, for 30 days. Following cessation of antibiotic administration the animals were observed for at least 30 days, and the surviving animals were challenged intramuscularly with 1.5×10^3 (30 LD_{50}) of Vollum spores.

The efficiency of treatment with tetracycline, ciprofloxacin and erythromycin was described previously (Altboum et al., 2002). As indicated in Table 31.1 the administration of all the antibiotics prevented the death of all the infected animals. However, after cessation of treatment animals developed a fatal respiratory anthrax disease. This phenomenon indicates that 30 days of antibiotic administration did not eradicate all the inhaled spores from the animal's lungs (Jones et al., 1996; Altboum et al., 2002). Upon termination of antibiotic administration the remaining spores germinated, proliferated and caused the respiratory anthrax. The animals

Table 31.1 Efficiency of post exposure prophylaxis of intranasally infected guinea pigs with *B. anthracis* spores, by various antibiotics

Antibiotic group	Antibiotic[a]	Survivors/infected		
		During antibiotic administration	Post cessation of treatment	Post challenge
Tetracycline	Tetracycline	9/9	5/9	0/5
	Doxycycline	10/10	5/10	0/4
Fluoroquinolones	Ciprofloxacin	8/9	4/8	4/4
	Ofloxacin	12/12	11/12	7/10
Carbapenem	Imipenem	8/8	5/8	5/5
Macrolides	Erythromycin	17/18	0/17	–
Aminoglycosides	Gentamicin	9/10	1/9	–

[a]Antibiotic administration started 24 h post infection

that survived after termination of the antibiotic administration are either those with a residual spore concentration lower than the lethal dose, or those that acquired protective immunity during the antibiotic administration. Following cessation of treatment with erythromycin and gentamicin all the animals died a fact which indicated failure of the antibiotics to eradicate the spores and/or lack of acquirement an efficient immune response. Treatments with tetracycline antibiotics eradicated the spore concentration to a level near the lethal dose concentration. In 50% of the animals the remaining spore concentrations were above the lethal dose that killed the animals, whereas in the other animals the remaining spores were at a lower concentration that did not elicit protective immunity, and the animals succumbed to rechallenge with Vollum spores. Treatment with fluroquinolones antibiotics and with imipenem seemed to allow a low level of growth that elicited an efficient protective immune response. After cessation of treatment 50–92% of the animals survived, most of which acquired resistance to rechallenge with Vollum spores.

In order to improve the efficiency of treatment to ensure the survival of all animals upon cessation of antibiotic administration, we tested the therapeutic approach

Table 31.2 Efficiency of post exposure prophylaxis of intranasally infected guinea pigs with *B. anthracis* spores, by combined treatment with antibiotics and active immunization with PA based vaccine

Treatment[a]	Survivors/infected		
	During antibiotic administration	Post cessation of treatment	Post challenge
Tetracycline + PA vaccine	9/9	8/8	8/8
Doxycycline + PA vaccine	9/9	9/9	7/7
Ciprofloxacin + PA vaccine	8/9	7/8	6/7
Ofloxacin + PA vaccine	12/12	11/12	10/10

[a]Treatment with antibiotics started 24 h post infection, on days 8 and 21 post infection the animals were immunized with PA based vaccine

of active immunization with PA based vaccine (Reuveny et al., 2001) that was given twice on days 8 and 21 after initiation of the antibiotic administration. The results, shown in Table 31.2, indicate that such treatment provided protection during antibiotic administration, after cessation of treatment and induced acquired protective immunity to rechallenge with Vollum spores.

These results indicate that antibiotic administration can prevent the development of a fatal respiratory anthrax disease in infected animals, but only active immunization with PA based vaccine during the antibiotic administration can provide protection for a long period of time after cessation of antibiotic treatment.

31.2.3 Curing Bacteremic Sick Guinea Pigs

We tested the efficiency of treatment with doxycycline and ciprofloxacin to cure sick animals already carrying a load of *B. anthracis* bacterial cells in the circulation. Guinea pigs were intranasally infected with 40 LD_{50} of Vollum spores and from 24 h post infection and every 4–6 h afterwards, groups of 4–8 animals were sampled by withdrawal of blood for determination of the bacterial concentration. Immediately after collection of the blood sample the animals were injected with either doxycycline or ciprofloxacin twice daily for 21 days. Following cessation of treatment the animals were observed for additional 30–60 days and survivors were tested for acquiring protective immunity. The efficiency of the treatment to cure the sick bacteremic animals is presented in Table 31.3. The table describes the levels of bacteremia of the guinea pigs immediately before the administration of the antibiotics. Doxycycline cured 43/47 (91.5%) animals exhibiting bacteremia levels of 2 CFU/ml to 6×10^4 CFU/ml and failed to cure animals with bacteremia level above 10^5 CFU/ml. The administration of ciprofloxacin cured 26/30 (86.6%) animals with bacteremia levels of 10 CFU/ml to 10^5 CFU/ml, and failed to rescue animals with higher levels of bacteremia. All the protected animals that survived

Table 31.3 The efficiency of curing systemic sick animals with *B. anthracis* bacteria by treatment with doxycycline or ciprofloxacin

CFU/ml at the beginning of treatment	Doxycycline administration		Ciprofloxacin administration	
	Survivors/treated	% Protected	Survivors/treated	% protected
2–10	5/5	100	–	–
$10–10^2$	6/6		7/8	88.2
$10^2–10^3$	10/10		4/4	
$10^3–10^4$	8/8		4/5	
$10^4–10^5$	14/18	77.8	11/13	84.6
$10^5–10^6$	2/8	25	4/9	44.4
$10^6–10^7$	1/4		2/8	25
$10^7–10^8$	0/6	0	0/1	0
$>10^8$	0/5		0/11	

Table 31.4 Determination of a therapeutic time window from infection till initiation of antibiotic administration

	Doxycycline administration		Ciprofloxacin administration	
Initiation of treatment post infection (hrs)	Survivors/ Treated[a]	% protected	Survivors/ treated*	% protected
24	16/16	100	4/4	100
30	20/20	100	3/4	75
36	20/20	100	5/8	62.5
42	18/20	90	2/8	25
46	–	–	19/28	67.8
48	15/20	75	2/9	22.2
54	3/20	15	–	–

[a]The numbers of animals include also non-bacteremic animals that were treated from various hrs post infection

during the antibiotic administration, survived upon cessation of treatment, and all these animals developed protective immunity and resisted subcutaneous challenge of 100 LD_{50} of Vollum spores.

The therapeutic time window from infection till initiation of treatment is described in Table 31.4. Initiation of administration of doxycycline from 24 h post infection till 48 h post infection cured 89/96 (92.7%) animals, initiation of treatment at 48 h post infection protected 75% of the sick animals, however a later initiation of treatment did not cure the animals.. Administration of ciprofloxacin from 24 h post infection till 46 h post infection cured 33/52 animals, initiation of treatment at 46 h post infection rescued 67.8% of the sick animals, and a later initiation of treatment did not cure the animals. These results indicate that in the guinea pig model there are 2 days from infection till initiation of antibiotic administration that provide an efficient curing.

Monitoring the time of death of the unprotected animals that were treated with doxycycline and ciprofloxacin is summarized in Table 31.5. Animals with bacteremia level of $\geq 10^7$ CFU/ml died a few hours after initiation of antibiotic

Table 31.5 Determination of mean time to death of antibiotic treated systemic sick animals with *B. anthracis* bacteria

	MTTD (h)	
Bacteremia level (CFU/ml) at the initiation of treatment	Doxycycline treated animals	Ciprofloxacin treated animals
10^4–10^5	42.0±19.3	
10^5–10^6	65.6±28.5	75.6±72.9
10^6–10^7	38.0±12.0	43.6±34.8
10^7–10^8	14.7±7.2	<24
>10^8	7.2±4.9	8.0±8.4

administration; while animals with a bacteremia level of <10^7 CFU/ml died 2–3 days post initiation of treatment. This result might indicate that a more intensive treatment that combined different antibiotics, antibodies against toxin components and/or agents that neutralize the edema and lethal toxin activities, might improve the efficiency of treatment and cure animals with up to bacteremia level of 10^7 CFU/ml.

To conclude, this work proved that airway infected guinea pigs could be protected from development of a fatal respiratory anthrax disease by administration of different antibiotics belonging to the Tetracycline, Fluoroquinolones, Macrolides, Carbapenem and Aminoglycosides. However, only combined treatment that involved active immunization with PA based vaccine during the administrations of the antibiotics provided full protection after cessation of antibiotics treatment.

Furthermore, either doxycycline or ciprofloxacin can cure animals with systemic disease, carrying bacteremia levels of up to 10^5 CFU/ml. A therapeutic time window of up to 44 to 48 h post infection to initiate an efficient antibiotic treatment was identified.

References

Altboum, Z., Gozes, Y., Barnea, A., Pass, A., White, M. & Kobiler, D. (2002) Post exposure prophylaxis against anthrax: evaluation of various treatment regimens in intranasally infected guinea pigs. *Infect Immun* 70, 6231–41.

Friedlander, A. M., Welkos, S. L., Pitt, M. L. M., Ezzell, J. W., Worsham, P. L., Rose, K. J., Ivins, B. E., Lowe, J. R., How, G. B., Mikesell, P. & Lawrence, W. B. (1993) Postexposure prophylaxis against experimental inhalation anthrax. *J Infect Dis* 167, 1239–42.

Henderson, D. W., Peacock, S. & Belton, F. C. (1956) Observations on the prophylaxis of experimental pulmonary anthrax in the monkey. *J Hyg* 54, 28–36.

Inglesby, T. V., Henderson, D. A., Bartlett, J. G., Ascher, M. S., Eitzen, E., Friedlander, A. M., Hauer, J., Mcdade, J., Osterholm, M. T., O'toole, T., Parker, G., Perl, T. M., Russell, P. K. & Tonat, K. (1999) Anthrax as biological weapon, medical and public health management. *JAMA* 281, 1735–45.

Inglesby, T. V., O'toole, T., Henderson, D. A., Bartlett, J. G., Ascher, M. S., Eitzen, E., Friedlander, A. M., Gerberding, J., Hauer, J., Hughes, J., Mcdade, J., Osterholm, M. T., Parker, G., Perl, T. M., Russell, P. K., & Tonat, K . (2002) Anthrax as a biological weapon, 2002, updated recommendations for management. *JAMA* 287, 2236–52.

Jernigan, J. A., Stephens, D. S., Ashford, D. A., Omenaca, C., Topiel, M. S., Galbraith, M., et al. (2001) Bioterrorism-related inhalational anthrax: The first 10 cases reported in the United States. *Emerg Infect Diseases* 7, 933–44.

Jones, M. N., Beedman, R. J., Turnbul, P. C. B. & Manchee, R. J. (1996) Antibiotic prophylaxis for inhalational anthrax. *Salisbury Med Bull* (*Special Supplement*) 87, 127–8.

Kobiler, D., Weiss, S., Levy, H., Fisher, M., Mechaly, A., Pass, A. & Altboum, Z. (2006) Protective antigen as a correlative marker for Anthrax in animal models. *Infect Immun* 74, 5871–76.

Reuveny, S., White, M. D., Adar, Y. Y., Kafri, Y., Altboum, Z., Gozes, Y., Kobiler, D., Shafferman, A. & Velan, B. (2001) Search for correlates of protective immunity conferred by anthrax vaccine. *Infect Immun* 69, 2888–93.

Chapter 32
Reverse Vaccinology in *Bacillus anthracis*

Avigdor Shafferman, Orit Gat, Naomi Ariel, Theodor Chitlaru, Haim Grosfeld, Anat Zvi, Izhak Inbar, Galia Zaide, Ronit Aloni-Grinstein, and Sara Cohen

Abstract In search for antigens which may form the basis for improved subunit or live attenuated *B. anthracis* vaccines, extensive genomic, proteomic and serologic analyses coupled with functional screens for surface exposed and/or secreted proteins, were carried out. The screens resulted in selection of over 50 promising novel in-vivo expressed immunogens, classified as S-Layer Homology (SLH) proteins, repeat proteins, hydrolytic enzymes and ABC transporters. DNA vaccination experiments established that most of these novel antigens are indeed able to elicit a strong humoral response. Yet, unlike the major *B. anthracis* immunogen Protective Antigen (PA), none of the selected immunogens could provide protection against a subsequent virulent *B. anthracis* strain challenge. When over-expressed in an attenuated non-toxinogenic and non-encapsulated *B. anthracis* platform strain, at least three of the novel antigens did confer partial protection against a lethal *B. anthracis* challenge.

Keywords *B. anthracis* · Bioinformatics · Functional genomics · Proteomics · Vaccine candidates

32.1 Introduction

The gram-positive spore-forming bacterium *Bacillus anthracis* is the causative agent of anthrax, initiated in its most severe form by spore inhalation. Upon germination, vegetative bacilli invade the blood stream where they multiply massively, secreting toxins and virulence factors. *B. anthracis* is considered one of the most likely biological bioterrorism agents due to the high mortality associated with inhalational anthrax, the ease of respiratory spore transmission and spore stability.

A. Shafferman (✉)
Department of Biochemistry and Molecular Genetics, Israel Institute for Biological Research, P.O. Box 19, Ness-Ziona 74100, Israel
e-mail: avigdors@iibr.gov.il

A. Shafferman et al. (eds.), *The Challenge of Highly Pathogenic Microorganisms*,
DOI 10.1007/978-90-481-9054-6_32, © Springer Science+Business Media B.V. 2010

Most of the documented virulence factors identified so far are coded by genes derived from the two virulence plasmids pXO1 and pXO2. The tripartite toxin is encoded by the pXO1 genes *pagA*, *lef* and *cya* which encode the protective antigen (PA, which mediates toxin binding to and internalization, Bradley et al., 2001); the lethal factor (LF, a zinc protease targeting several MAP kinases; Klimpel et al., 1994) and the edema factor (EF, a calmodulin dependent adenylate cyclase; Leppla, 1982). Genes encoding for the anti-phagocytic capsule are located on the second virulence plasmid, pXO2 (Leppla, 1995). The bacterial binary exotoxin is necessary for the manifestation of fulminant disease, yet other factors may be required for efficient colonization and expansion of the bacteria in the host (Fellows et al., 2001; Brossier et al., 2002; Cendrowski et al., 2004; Gat et al, 2005; Fisher et al., 2006).

Licensed human vaccines are constituted of the anthrax toxin component PA, as the principal protective immunogen. The extent of PA-induced protection correlates with the elicited humoral immune response (Reuveny et al., 2002; Marcus et al., 2004). However, PA based vaccines (acellular or recombinant) require an extensive dosing regimen (Friedlander et al., 2002). Vaccines based on live attenuated bacteria (toxin producing, non encapsulated; $pXO1^+$ $pXO2^-$), have been extensively used for human immunization in the former Soviet Union. Although documented as more efficacious than PA based acellular vaccines, such vaccines are considered to represent high risk to humans and in the Western world they serve for veterinary purposes only. Studies of immune response directed against PA presented in the context of spore antigens either as live attenuated PA-expressing *B. anthracis* vaccines (Cohen et al., 2000; Mendelson et al., 2005; Aloni-Grinstein et al., 2005) or as formalin inactivated spores supplementing PA-based acellular vaccines (Brossier et al., 2002), suggest that spore antigens or additional vegetative antigens indeed may augment PA-mediated protective immunity. A genome-scale multi-disciplinary search for such proteins is the main objective of the study presented in this report.

The recent advent of rapid whole- genome sequencing concomitant with the development of in-silico genome-mining procedures and high throughput global evaluation technologies, provide innovative strategies for a faster and more comprehensive identification of vaccine candidates, disease markers and virulence factors (for a recent review, see Bambini and Rappouli, 2009 and references therein). The first pioneering approach, generally referred to as "reverse vaccinology" (first implemented for *Neisseria meningitides* B by Pizza et al., 2000), exploited the complete genome protein repertoire of a microorganism as the starting material for bioinformatic candidate antigen identification, characterization and selection; followed by experimental evaluation. In contrast to "classic vaccinology" where following extensive biochemical characterization only individual potential antigens are studied, identification of vaccine candidates by the "reverse vaccinology" approach addressing the complete protein repertoire of a pathogen, allows for rapid simultaneous selection of numerous candidates. The in-silico search is directed to selection of host immune system accessible proteins and putative virulence factors. Combined with a rational reductive strategy, this approach allows for down-selection of candidates from essentially the entire genomic protein repertoire (several thousand), to tens-hundred ORFs, manageable for experimental evaluation. Combined with the

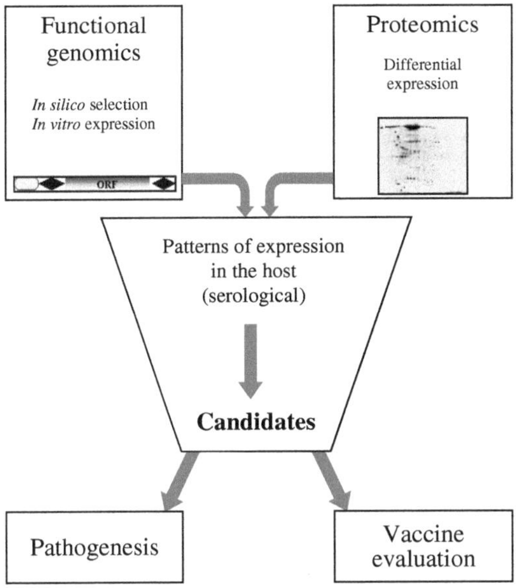

Fig. 32.1 General scheme for genomics and proteomics-based vaccine candidate identification. The search for novel vaccine candidate proteins in the *B. anthracis* complete genome involved two approaches: in-silico screen, coupled with functional genomics and proteomics/ SERPA. Both approaches address surface exposed proteins and involve serological examination of infection related expression and antigenicity

observed microorganism's proteome map, this strategy may provide a panoramic view of the genetic make-up and protein content of a bacterium (Chitlaru and Shafferman, 2009).

In this report, we briefly review results of a a global search for *B. anthracis,* vaccine candidates which involved bioinformatic analysis of the bacterial genome, functional genomic/serological examination of selected antigens and a parallel proteomic/serological inspection of bacterial surface exposed/secreted in-vivo expressed immunogens (Fig. 32.1). Preliminary validation experiments addressing potential protective immunity elicited by administration of novel antigens identified are also presented.

32.2 Results

32.2.1 Bioinformatic Analysis

B. anthracis chromosome and virulence plasmids open reading frames (ORFs) products (Read et al., 2003; Okinaka et al., 1999) predicted to code for surface exposed (membrane-anchored and/or secreted) or putative virulence related proteins, were characterized and selected as vaccine candidates by a multi-step

A

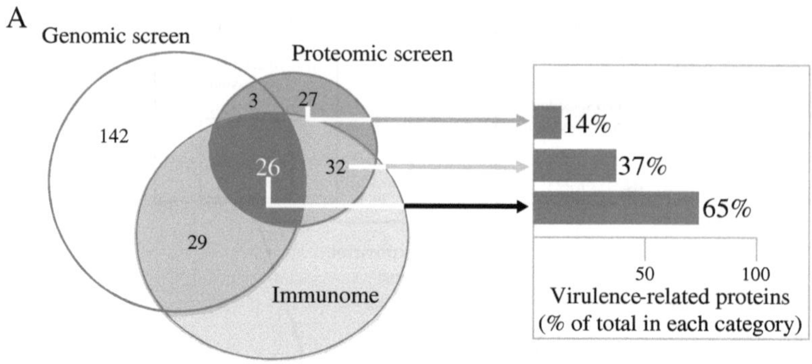

B

Gene name	Function	Locus tag	Exposed	Dual localization (Secreted/membranal)	CO_2 modulated	Proteases	SBP of ABC transporters	Examined by targeted mutagenesis
	Unknown	pXO1-90	•		•			•
LF	***Lethal factor***	***pXO1-107***	•		•	•		
PA	***Protective Antigen***	***pXO1-110***	•		•			
EF	***Edema Factor***	***pXO1-122***	•		•			
AdcA	SBP-ABC [Zn]	pXO1-130	•		•		•	•
	Prolyloligopeptidase	BA0165	•			•		
GroEL	Heat Shock protein	BA0267		•	•			
RocA	d-1-Pyrroline-5-carboxylase	BA0309		•				
AhpC	Hydroperoxide reductase	BA0345		•				
	SBP-ABC [oligopeptide]	BA0656	•		•		•	
	Unknown	BA0796	•		•			
YckK	SBP-ABC [amino acid]	BA0855	•		•		•	
Sap	S-layer protein	BA0885	•	•	•			
EA1	S-layer protein	BA0887	•	•	•			
CwlB	Alanine amidase III	BA0898	•	•	•			
	SBP-ABC [oligopeptide]	BA1191	•		•		•	
InhA1	Metalloprotease	BA1295	•		•	•		
	Alanine amidase IV	BA1818	•	•				
	NlpC/P60 endopeptidase	BA1952	•		•	•		
YjeA	Polysaccharide deacetylase	BA2944	•		•			
MntA	SBP-ABC [Mn]	BA3189	•	•	•		•	•
GamR	ϒ-Phage receptor	BA3367	•		•			•
HtrA	Serine protease	BA3660	•	•	•	•		•
CwlA	Alanine amidase II	BA3737	•	•	•			
YfkN	nucleotide diesterase	BA4346	•	•	•			
IsdK	NEAT domain	BA4787	•		•			•
IsdC	NEAT domain	BA4789	•		•			•
LytE	Endopeptidase	BA5427	•		•	•		
	Collagen adhesion protein	BAS5205	•					

Fig. 32.2 Genomic and proteomic screens identify an overlapping pool of immunogenic proteins with functions potentially involved in bacterial virulence. (**A**) Venn diagram enumerating screen specific as well as common proteins. The number of proteins in each group is indicated in the appropriate segment. The highest percentage of virulence related functions (65%, right panel) were noted in the "common" group. (**B**) The immunogenic proteins identified by the two screens are tabulated in ascending order according to their gene locus tags. Proteins encoded by pXO1 are listed on the top of the table, followed by chromosome-located genes. Gene names, when available, are denoted in the left column. The entries referring to classical anthrax toxin components are

computational screen and rational reduction strategy, combined with comparative genomics (detailed in Ariel et al., 2002, 2003). Out of ~5500 ORF products (>5000 chromosomal and >200 plasmidial), 240 candidate ORF products with assigned function, as well as 280 hypothetical proteins, were selected. The resulting ~500 ORFs were subjected to further manual curation which trimmed the list down to ~200 ORFs, a number amenable for subsequent experimental evaluation.

32.2.2 Functional Genomic/Serologic Analysis

A promising vaccine candidate should be expressed in the course of infection and be able to elicit an immune response. Therefore, the 200 in-silico selected ORF sequences were amplified by PCR using primers suitable for in-vitro synthesis of small amounts of radioactively labeled protein in a coupled transcription/translation system (T&T). The resulting polypeptides were then reacted with a variety of sera collected from *B. anthracis* infected animals, in order to identify in-vivo expressed immunogens. Quantitative immunoprecipitation assays permitted attribution of relative immunogenicity scores for prioritization in subsequent analyses (Gat et al., 2006, 2007). The screen identified 52 novel immunogens which included surface-anchored proteins, adhesins or transporters, cell wall hydrolases, proteins involved in iron acquisition, amino acid and oligopeptide transporters (see Fig. 32.2b). The immunogenic potential of the ORF products was further confirmed by DNA vaccination experiments in mice.

32.2.3 Proteomic/Serologic Analysis

The genomic study was complemented by proteomic analyses of *B. anthracis* surface exposed proteins (Fig. 32.1). To discern in-vivo expressed immunogens from the complex proteome of the pathogen, 2-DE separated proteins identified by mass-spectrometry, were probed with infected animals immune sera (Serological Proteome Analyis, SERPA). A preliminary study examining membranal proteins prepared from a nonvirulent *B. anthracis* strain (Ariel et al., 2002; Chitlaru et al., 2004), revealed several candidate immunodominant proteins. This study was further extended to investigation of the secretomes of virulent and non-virulent (devoid

Fig. 32.2 (continued) denoted in bold and italics. Proteins exhibiting functions related to virulence are framed. Various characteristics of the proteins (indicated in the bottom of the table) are marked by gray dots. Note that seven proteins were investigated by target mutagenesis of their respective loci. MntA, shown to be essential for virulence, is shadowed. SBP of ABC represents the solute binding components of ABC transporters; their putative ligands are indicated by brackets

of the two virulence plasmids) *B. anthracis* strains. Bacteria were cultured under various conditions, including host-mimicking conditions (high bicarbonate/CO_2 concentration). The immunoreactivity of the secreted protein repertoire was probed using immune sera. Fifty eight proteins, exhibiting different levels of immunogenicity (Chitlaru et al., 2006, 2007, see Fig. 32.2b), were identified.

Of the 58 in-vivo expressed immunogens discerned by the proteomic study, 26 were independently identified by the genomic/serologic analysis (Fig. 32.2a). This high percentage of common candidates is quite remarkable, since the genomic screen relied mostly on the prediction of surface localization, unlike the case for SERPA, where only the abundant proteins are analyzed. Thus the two complementary serological screens should probably be both pursued in genome-scale vaccine candidate definitions (Fig. 32.2a).

Interestingly, about 65% of the proteins identified by both global screens have assigned functions invoked as virulence-related in other pathogens (Fig. 32.2b). Selected putative *B. anthracis* virulence factors were directly tested for their possible role in anthrax pathogenesis by examining the phenotype associated with targeted disruption of the respective genes in comparison to the isogenic virulent parental strain (Gat et al., 2005, 2008). Surprisingly, only one of the proteins, the manganese transporter MntA, proved to be essential for virulence manifestation. Disruption of the *mntA* gene resulted in a dramatic attenuation of the mutated strain in a guinea pig model of infection, despite its unaltered ability to express the bacterial classical toxin components and capsule (Gat et al., 2005).

32.2.4 Vaccine Candidate Validation – DNA Immunization

Twelve proteins selected from the common candidate list, administered as DNA vaccines, were evaluated for ability to confer protection in a guinea pig model (Chitlaru et al., 2007, see Fig. 32.3). ORFs were cloned in a suitable eukaryotic expression vector and administered using 4 gene-gun immunizations. All antigens tested induced a strong humoral immune response with levels similar to that elicited by the positive control PA. Yet, with the exception of PA, none of the ORFs conferred significant protection to a subsequent challenge with a virulent *B. anthracis* strain (Fig. 32.3).

32.2.5 Evaluation of Protection Conferred by Selected Immunogens Using a Live-Vaccine Attenuated Platform

In the past, we have developed an attenuated *B. anthracis* strain, devoid of the virulence plasmids, based on the ATCC14185 non-encapsulated and non-proteolytic strain. This platform strain designated Δ14185, has been successfully used for expression of recombinant versions of PA, providing long-lasting immunity in guinea pigs following its administration as spores (Cohen et al., 2000; Gat et al., 2003; Mendelson et al., 2005; Aloni-Grinstein et al., 2005).

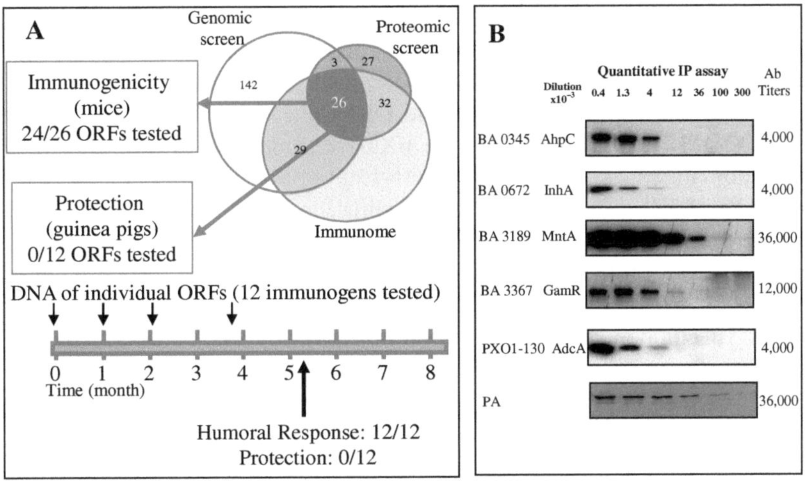

C

Antigen	Ab. Titer	Survival	MTD
PA	>36,000	11/12	>12.2
pCI (expression vector)	<10	0/12	2.7
BA0345/AhpC	4,000	0/12	2.7
BA0672/InhA2	4000	0/12	3.2
BA0796	>500	0/12	2.8
BA3367/GamR	12,000	0/12	2.8
BA4787	12,000	0/12	3.2
BA1295	>500	0/12	3.3
BA2805	36,000	0/12	2.8
BA3189/MntA	36,000	0/12	2.8
BA3660/HtrA	>500	0/12	2.7
POX1-130/AdcA	4,000	0/12	3.3
POX1-54	12,000	0/12	2.9
BA2849/NlpC/P60	>500	0/12	2.9

Fig. 32.3 Evaluation of the protective potential of selected immunogens by DNA vaccination. (**A**) Twenty four of the twenty six "common" proteins were studied as DNA vaccines in mice and twelve in guinea pigs. The time scale of the 4 administrations gene-gun delivery protocol is schematically depicted. Of the twenty four ORFs, twenty two exhibited a strong humoral response. All vaccinated guinea pigs developed significant titers of specific antibodies (see table C). These animals were challenged sub- cutaneously with 30 LD_{50} of virulent Vollum strain *B. anthracis* spores. (**B**) Representative immunoprecipitation titers for T&T radioactive products of selected ORF products, for quantitation of relative immunogenicity. (**C**) Antibody titers, survival and mean time to death (MTD) of the various experimental groups. Note the almost full protection conferred by administration of the PA-encoding DNA vector positive control

In the preliminary study presented in this report, nine of the immunogens selected by the genomic as well as the proteomic screens (Fig. 32.4) were evaluated for their protective ability using the Δ14185 spore immunization platform as an antigen delivery vehicle. For optimal expression, strains expressing the ORFs under investigation were engineered using heterologous promoters of various potencies

(Gat et al., 2003). The vaccination protocol is depicted in Fig. 32.4. Protective ability was probed by sub-cutaneous administration of 100 Lethal Dose 50% (LD_{50}). The data depicted in Fig. 32.4 shows that under conditions where the platform strain alone did not confer protection, 3 of the ORFs marked in black arrows, elicited partial protection. These ORFs are the pXO1 encoded genes pXO1-90 (a putative SLH-domain adhesion protein), pXO1-130 (a putative secreted Zn transporter AcdA-like) and the chromosome gene encoding for GamR, a membrane-associated protein shown to represent the *B. anthracis* γ–phage receptor (Davison et al., 2005). Notably, the novel protective immunogens elicited a protective response similar to that promoted by administration of platform-strain spores engineered to express the bacterial toxin component LF. This observation may imply that the partial protection exhibited by the novel antigens-platform construct, represents the upper protective limit which one may expect from a *B. anthracis* protective protein other than PA. The contribution of these novel vaccine candidates to improved efficacy of a PA-based vaccine is currently under investigation.

32.3 Conclusions

As demonstrated in this study, combining bioinformatic whole-genome screening with serological proteomic analysis of the *B. anthracis* secreteome, allows for judicious selection of in-vivo expressed immunogens. The list of immunogenic proteins identified in this study constitutes a novel pool of antigens, amenable for further evaluation of their protective potential as part of an effort to improve existing PA-based anthrax vaccines. Preliminary vaccination experiments in a guinea pig model, using the live attenuated spore platform, strongly suggest that at least three novel immunogens possess promising protective abilities. Reverse vaccinology screens coupled with different postgenomic approaches, have been successfully applied for

Table 32.1 Examples of reverse vaccinology screens carried out in various pathogens

Pathogen	ORFs in genome	Predicted targets	Expressed (analysed)	Immuno-reactivity	Immuno-protection	Reference
Neisseria meningitides serogroup B	2063	600	300	28	5	Serruto et al. (2009)
Streptococcus pneumoniae	2687	130	97	6	6	Weizmann et al. (2001)
Porphyromonas gingivalis	1909	120	109	40		Ross et al. (2001)
Streptococcus suis serotype 2	2914	153	10	4	4	Liu et al. (2009)
Bacillus anthracis	5900	500	280 200 (T&T) 80 (proteomics)	87	3	This study

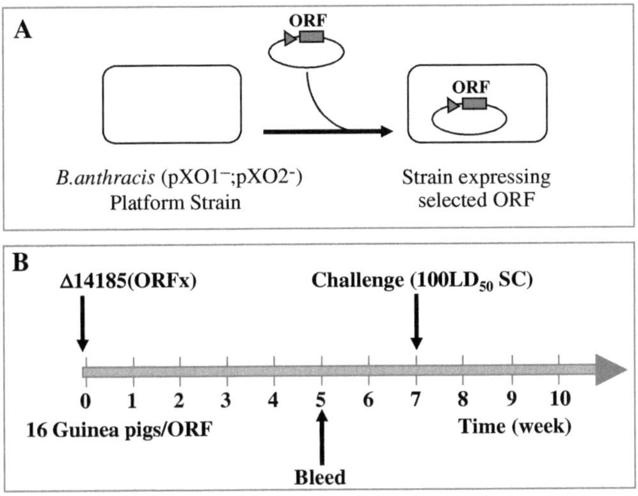

C

	Antigen	anti-ORF Ab.	Survival (live/total)	MTTD (day)	
Controls	Naïve		0/32	3	
Controls	Platform strain		0/32	4	
Controls	PA (*Ppag* promoter)	+	12/32	6	
Controls	PA (*Pamy* promoter)	+++	32/32		
	LF	++	4/32	5	
	pXO1-54	+++	0/16	4	
	pXO1-90	+++	3/16	5	←
	pXO1-130	+	4/16	6	←
	IsdX2	+++	0/16	3	
	AhpC	+++	0/16	4	
	InhA	–	0/16	4	
	MntA	++	0/16	4	
	GamR	+++	5/16	5	←
	Endolysin	+++	0/16	4	

Fig. 32.4 Evaluation of the protective potential of selected immunogens expressed in spores of a live-attenuated platform strain. (**A**) The attenuated platform strain used for expression of selected ORFs was generated by curing of the pXO1 virulence plasmid from the pXO2⁻ ATCC14185 – train. (**B**) Time-scale representation of the protection experiments. (**C**) Number of surviving animals and mean time to death (MTID) for each experimental group. The control groups are shadowed. The anti-ORF antibody response was determined by a quantitative immuno-precipition assays of T&T products (Gat et al., 2007). The specific antibody titers are denoted as strong (+++,>10,000), medium (++, 1,000–10,000) or low (+, 100–1,000)

target identification in other bacterial pathogens (Table 32.1). The proportion of immunogens identified in the course of the present study appears to be similar to that reported for other pathogens. The number of potentially protective antigens identified (three immunogens) appears to be low, yet these three immunogens belong to a group of only nine antigens selected for evaluation via the live vaccine spore platform delivery mode, therefore their actual proportion is remarkably high. Notably, *Bacillus anthracis* differs from all of the pathogens listed in Table 32.1, by the facts that (i) anthrax is a toxinogenic disease in the sense that the bacterial toxin has a pivotal role in its pathogenic manifestation, and (ii) the toxin component PA is a immune-dominant *B.anthracis* protein. These two aspects may strongly mask experimentally the possible contribution to protection/virulence of other antigens.

In addition to identification of putative vaccine targets, the proteomic survey has allowed for identification of three novel diagnostic markers (Sela-Abramovich et al., 2009). Gene disruption experiments carried out with selected immunogens identified by the genomic screens (Gat et al., 2005; 2008), as well as the proteomic determination of various regulons (Chitlaru et al., 2007), clearly expanded our understanding of *B. anthracis* pathogenicity.

At least two lessons can be learned from the present study: (i) The genomic and proteomic screens are complementary and exploit independent criteria for reductive analyses. Most importantly, immunogens discerned by both approaches have a high probability of being valid targets. Notably, the three potential protective antigens, the three novel disease biomarkers and the novel virulence factor (MntA), all belong to the group of immunogens identified by both screens (Fig. 32.2). (ii) In assessing the protective ability of novel antigens, one should take into consideration that certain rapid modes of antigen delivery for immunization purposes, e.g. DNA vaccines, although useful as a screening tool may not necessarily be suitable for discerning the true protective potential of a protein.

In conclusion, the functional genomic and the proteomic/ serological studies of *B. anthracis* emerged as valuable approaches for identification of infection-related immunogens which may serve as the basis for development of next generation anthrax vaccines or diagnostic measures as well as to promote our understanding of *B. anthracis* pathogenicity

References

Aloni-Grinstein, R., Gat, O., Altboum, Z., Velan, B., Cohen, S. & Shafferman, A. (2005) Oral spore vaccine basd on live attenuated nontoxinogenic *Bacillus anthracis* expressing recombinant mutant protective antigen. *Infect Immun,* 73(7), 4043–53.

Ariel, N., Zvi, A., Grosfeld, H., Gat, O., Inbar, Y., Velan, B., Cohen, S. & Shafferman, A. (2002) Search for potential vaccine candidate ORFs in *the B. anthracis* virulence plasmid pXO1- in silico and in vitro screening. *Infect Immun,* 70(12), 6817–27.

Ariel, N., Zvi., A, Makarova, K., Chitlaru, T., Elhanany, E., Velan, B., Cohen, S., Friedlander, A. & Shafferman, A. (2003) Genome-based bioinformatic selection of chromosomal *B. anthracis* putative vaccine candidates coupled with proteomic identification of surface-associated antigens. *Infect Immun,* 71(8), 4563–79.

Bambini, S. & Rappuoli, R. (2009) The use of genomics in microbial vaccine development. *Drug discov today,* 14(5/6), 253–60.

Bradley, K., Mogridge, J., Mourez, M., Collier, B. & Young, J. (2001) Identification of the cellular receptor for anthrax toxin. *Nature*, 414(6860), 225–9.

Brossier, F., Levy, M. & Mock, M. (2002) Anthrax spores make an essential contribution to vaccine efficacy. *Infect Immun,* 70(2), 661–4.

Cendrowski, S., Macarthur, W. & Hanna, P. (2004) *Bacillus anthracis* requires siderophore biosynthesis for growth in macrophages and mouse virulence. *Mol Microbiol*, 51(2), 407–17.

Chitlaru, T., Ariel, N., Zvi, A., Lion, M., Velan, B., Shafferman, A. & Elhanany, E. (2004) Identification of chromosomally encoded membranal polypeptides of *Bacillus anthracis* by a proteomic analysis: Prevalence of proteins containing S-layer homology domains. *Proteomics*, 4(3), 677–91.

Chitlaru, T., Gat, O., Gozlan, Y., Ariel, N. & Shafferman, A. (2006) Differential proteomic analysis of the *Bacillus anthracis* secretome: Distinct plasmid and chromosome CO_2-dependent cross-talk mechanisms modulate extracellular proteolytic activities. *J Bact*, 188(10), 3551–71.

Chitlaru, T., Gat, O., Gozlan, Y., Grosfeld, H., Inbar, I. & Shafferman, A. (2007) Identification of in vivo expressed immunogenic proteins by serological proteome analysis of *Bacillus anthracis* secretome. *Infect Immun,* 75(6), 2841–52.

Chitlaru, T. & Shafferman, A. (2009) Proteomic) Proteomic studies of *Bacillus anthracis*. *Future Microbiol*, 4(8), 983–98.

Cohen, S., Mendelson, I., Altboum, Z., Kobiler, D., Elhanany, E., Bino, T., Leitner, M., Inbar, Y., Rosenberg, H., Gozes, Y., Barak, R., Fisher, M., Kronman, C., Velan, B. & Shafferman, A. (2000) Attenuated nontoxinogenic and nonencapsulated recombinant *Bacillus anthracis* spore vaccines protect against anthrax. *Infect Immun,* 68(8), 4549–58.

Davison, S., Couture-Tosi, E., Candela, T., Mock, M. & Fouet A. (2005) Identification of the *Bacillus anthracis* γ Phage receptor. *J Bact*, 187(19), 6742–9.

Fellows, P. F., Linscott, M. K., Ivins, B. E., Pitt, M. L. M., Rossi, Gibbs, P. H. & Friedlander, A. M. (2001) Efficacy of a human anthrax vaccine in guinea pigs, rabbits, and rhesus macaques against challenge by *Bacillus anthracis* isolates of diverse geographical origin. *Vaccine,* 19(23–24), 3241–7.

Fisher, N., Shetron-Rama, L., Herring-Palmer, A., Heffernan, B., Bergman, N. & Hanna, P. (2006) The dltABCD operon of *Bacillus anthracis* Sterne is required for virulence and resistance to peptide, enzymatic, and cellular mediators of innate immunity. *J Bact,* 188(4), 1301–9.

Friedlander, A., Welkos, S. & Ivins, B. (2002) Anthrax Vaccines. *Curr Top Microbiol Immun,* 271, 34–60.

Gat, O., Grosfeld, H., Ariel, N., Inbar, I., Zaide, G., Broder, Y., Zvi, A., Chitlaru, T., Altboum, Z., Stein, D., Cohen, S. & Shafferman, A. (2006) Search for *Bacillus anthracis* potential vaccine candidates by a functional genomic-serologic screen. *Infect Immun,* 74(7), 3987–4001.

Gat, O., Grosfeld, H. & Shafferman, A. (2007) In vitro screen of bioinformatically selected *Bacillus anthracis* vaccine candidates by coupled transcription, translation and immunoprecipitation analysis. In: Grandi, G (ed), In vitro transcription and translation protocols. Methods in molecular biology, Humana Press, Totowa, NJ, 375, pp. 211–33.

Gat, O., Inbar, I., Aloni-Grinstein, R., Zahavy, E., Kronman, C., Mendelson, I., Cohen, S., Velan, B. & Shafferman, A. (2003) Use of a promoter trap system in *B. anthracis* and *B. subtilis* for the development of recombinant protective antigen-based vaccines. *Infect Immun,* 71(2), 801–13.

Gat, O., Mendelson, I., Chitlaru, T., Ariel, N., Altboum, Z., Levy, H., Weiss, S., Grosfeld, H., Cohen, S. & Shafferman, A. (2005) The solute-binding component of a putative Mn(II) ABC transporter (MntA) is a novel *Bacillus anthracis* virulence determinant. *Mol Microbiol,* 58(2), 533–51.

Gat, O., Zaide, G., Inbar, I., Grosfeld, H., Chitlaru, T., Levy, H. & Shafferman, A. (2008) Characterization of *Bacillus anthracis* iron-regulated surface determinant (Isd) proteins containing NEAT domains. *Mol Microbiol,* 70(4), 983–99.

Klimpel, K. R., Arora, N. & Leppla, S. H. (1994) Anthrax toxin lethal factor contains a zinc metalloprotease consensus sequence which is required for lethal toxin activity. *Mol Microbiol*, 13(6), 1093–100.

Leppla, S. H. (1982) Anthrax toxin edema factor: A bacterial adenylate cyclase that increases cyclic AMP concentrations of eukaryotic cells. *Proc Natl Acad Sci U S A*, 79(10), 3162–6.

Leppla, S. H. (1995) Anthrax toxins. In: Moss, I., Iglewski, J., Vaughan, M. & Tu, A. T. (ed), Bacterial toxins and virulence factors in disease. Marcel Dekker, New York, NY, pp. 543–72.

Liu, L., Cheng, G., Wang, C., Pan, X., Cong, Y., Pan, Q., Wang, J., Zheng, F., Hu, F., Tang, J. (2009) Identification and experimental verification of protective antigens against *Streptococcus suis* serotype 2 based on genome sequence analysis. *Current Microbiol,* 58, 11–7.

Marcus, H., Danieli, R., Epstein, E., Velan, B., Shafferman, A. & Reuveny, S. (2004) Contribution of immunological memory to protective immunity conferred by a *Bacillus anthracis* protective antigen-based vaccine. *Infect Immun,* 72(6), 3471–7.

Mendelson, I., Gat, O., Aloni-Grinstein, R., Altboum, Z., Inbar, I., Kronman, C., Bar-Haim, E., Cohen, S., Velan, B. & Shafferman, A. (2005) Efficacious, nontoxigenic *Bacillus anthracis* spore vaccines based on strains expressing mutant variants of lethal toxin components. *Vaccine,* 23(48–49), 5688–97.

Okinaka, R., Cloud, K., Hampton, O., Hoffmaster, A., Hill, K., Keim, P., Koehle, T., Lamke, G., Kumano, S., Manter, D., Martinez, Y., Ricke, D., Svensson, R. & Jackson, P. (1999) Sequence, assembly and analysis of pXO1 and pXO2. *J Applied Microbiol,* 87(2), 261–2.

Pizza, M., Scarlato, V., Masignani, V., Giuliani, M. M., Arico, B., Comanducci, M., Jennings, G. T., Baldi, L., Bartolini, E., Capecchi, B., Galeotti, C. L., Luzzi, E., Manetti, R., Marchetti, E., Mora, M., Nuti, S., Ratti, G., Santini, L., Savino, S., Scarselli, M., Storni, E., Zuo, P., Broeker, M., Hundt, E., Knapp, B., Blair, E., Mason, T., Tettelin, H., Hood, D. W., Jeffries, A. C., Saunders, N. J., Granoff, D. M., Venter, J. C., Moxon, E. R., Grandi, G. & Rappuoli, R. (2000) Identification of vaccine candidates against serogroup B meningococcus by whole- genome sequencing. *Science*, 287, 1816–20.

Read, T. D., Peterson, S. N., Tourasse, L. & 49 additional authors (2003) The genome sequence of *Bacillus anthracis* Ames and comparison to closely related bacteria. *Nature,* 423(6935), 81–6.

Reuveny, S., White, M., Adar, Y. Kafri, Y., Altboum, Z., Gozes, Y., Kobiler, D., Shafferman, A. & Velan, B. (2002) Search for correlates of protective immunity conferred by anthrax vaccine. *Infect Immunol,* 69(5), 2888–93.

Sela-Abramovich, S., Chitlaru, T., Gat, O., Grosfeld, H., Cohen, O. & Shafferman A. (2009) Novel and unique diagnostic biomarkers for *Bacillus anthracis* infection. *Applied Environ Microbiol,* 75(19), 6157–67.

Serruto, D., Serino, L., Masigniani, V., Pizza, M. (2009) Genome-based approaches to develop vaccines against bacterial pathogens. *Vaccine*, 27, 3245–50.

Weizmann, T. M., Heinrichs, J. H., Adamou, J. E., Erwine, A. L., Kunsch, C., Choi, G. H., Barash, S. C., Rosen, C. A., Masure, R. H., Tuomanen, E., Gayle, A., Brewah, Y. A., Walsh, W., Barren, P., Lathigra, R., Hanson, M., Langermann, S., Johnson, S., Koenig, S. (2001) Use of a whole genome approach to identify vaccine molecules affording protection against *Streptococcus pneumonia* infection. *Infect Immun,* 69(3), 1593–8.

Index

A. Shafferman et al. (eds.), *The Challenge of Highly Pathogenic Microorganisms*, Mechanisms of Virulence and Novel Medical Countermeasures,
DOI 10.1007/978-90-481-9054-6, © Springer Science+Business Media B.V. 2010

MIX
Papier aus verantwortungsvollen Quellen
Paper from responsible sources
FSC® C105338

If you have any concerns about our products, you can contact us on
ProductSafety@springernature.com

In case Publisher is established outside the EU, the EU authorized representative is:
Springer Nature Customer Service Center GmbH
Europaplatz 3, 69115 Heidelberg, Germany

Printed by Libri Plureos GmbH
in Hamburg, Germany